Sustainable Water Resources Planning and Management Under Climate Change

Elpida Kolokytha • Satoru Oishi
Ramesh S.V. Teegavarapu
Editors

Sustainable Water Resources Planning and Management Under Climate Change

Springer

Editors
Elpida Kolokytha
School of Civil Engineering
Division of Hydraulics and Environmental Engineering
Aristotle University of Thessaloniki
Thessaloniki, Greece

Satoru Oishi
The Research Center for Urban Safety and Security
Kobe University
Kobe, Hyogo, Japan

Ramesh S.V. Teegavarapu
Department of Civil, Environmental and Geomatics Engineering
Florida Atlantic University
Boca Raton, FL, USA

ISBN 978-981-10-2049-0 ISBN 978-981-10-2051-3 (eBook)
DOI 10.1007/978-981-10-2051-3

Library of Congress Control Number: 2016955299

Printed on acid-free paper

This Springer imprint is published by Springer Nature
The registered company is Springer Science+Business Media Singapore Pte Ltd.

Preface

This book, entitled *Sustainable Water Resources Planning and Management Under Climate Change*, deals with a fundamental component of engineering design and practice under changing climate. Due to climate change and other sources of non-stationarity, the traditional approaches that are currently available for understanding, modeling and evaluating hydrological processes are not adequate. Complexity creates non-linear behavior in the entire system, which needs new knowledge to be addressed. Mitigation and adaptation policies are important to tackle climate change, extreme events, and water scarcity. This book contains chapters by distinguished scientists from different regions of the world covering topics that range from hydrologic modeling to management and policy responses to climate variability and change with primary emphasis on water resources management. The authors also provide insight into regional problems supported by results from several real-life case studies. The book also touches upon the issue of climate variability, applications and limitations of climate change models, and scenarios related to precipitation projections that ultimately relate to the future availability of water resources. It also includes discussions on policy options to deal with climate change through interesting study results and examples of applications of theory from almost all over the world.

This work is respectfully dedicated to the late Professor Toshiharu Kojiri by the authors who were his friends, research collaborators, or students. Professor Kojiri was born in Kyoto, Japan. He graduated from the Graduate School of Engineering, Kyoto University, and he received the Doctor of Engineering degree from Kyoto University based on his thesis work entitled "Systems approach on quantity and quality control of water by using multi-dams". Dr. Kojiri started his academic career as a research associate in the Department of Civil Engineering, Kyoto University. He later became an associate professor in the Disaster Prevention Research Institute (DPRI) of Kyoto University. After serving some time in Kyoto University he moved to Gifu University as associate professor and soon became full professor there. He returned to the DPRI, Kyoto University, as a full professor in 1997. Professor Kojiri took the role of chair of the Committee on Water Resources Management in the International Association for Hydro-Environment Engineering and Research

(IAHR). He and his co-authors published several original and seminal articles on water resources management and hydrological modeling that are referenced in a number of research publications and studies throughout the world. His outstanding contributions in advanced research focus on issues related to impacts of climate change on the hydrologic cycle and water resources management. Unfortunately, the water resources community lost a well-accomplished researcher and leader in November of 2011, when Professor Kojiri passed away after a long battle with cancer.

Professor Kojiri had always advocated progress in water sciences and development of new methods and their practical applications for the benefit of society. We believe that this book, as a small token of respect towards his outstanding work and contributions to water resources engineering, will serve to honor his research, and promote and motivate many readers around the world to strive for a better understanding of water resources management under changing climate. We hope the book will provide significant help to practicing engineers, academic researchers, engineering professionals, and students from different regions of the world. We expect to receive objective feedback from readers of the book. We also strongly believe that the book will in its own way promote and fulfill the late Professor Kojiri's wishes for the water resources community to continue the efforts of developing new approaches, handling and solving challenging multidisciplinary water resources problems, and lay a path for water resources management under a future uncertain climate.

Thessaloniki, Greece — Elpida Kolokytha
Kobe, Japan — Satoru Oishi
Boca Raton, FL, USA — Ramesh S.V. Teegavarapu

Contents

Part I
Hydrological Aspects of Climate Change

Chapter 1
Climate Variability and Changes in Precipitation Extremes and Characteristics

Ramesh S. V. Teegavarapu

Abstract Climate variability and change are expected to bring several changes to hydrologic cycles and regimes in different parts of the world. Natural climate variability based on large-scale, global inter-year, quasi-decadal and decadal, and multidecadal-coupled oceanic–atmospheric oscillations (e.g., El Niño Southern Oscillation (ENSO), Atlantic Multidecadal Oscillation (AMO) and Pacific Decadal Oscillation (PDO), Madden–Julian Oscillation (MJO), Indian Ocean Dipole (IOD)) contribute to regional variations in extremes and characteristics of essential climatic variables (e.g., temperature, precipitation, etc.) in different parts of the globe. These oscillations defined based on climate anomalies that are related to each other at large distances (referred to as *teleconnections*) are known to impact regional and global climate. Linkages of these teleconnections to the variability in regional precipitation patterns have been well documented in several research studies. This chapter focuses on evaluation of climate variability influences on precipitation extremes and characteristics. Several indices and metrics are discussed for such evaluation, and a few results from case studies are presented.

Keywords Climate variability • Precipitation extremes and characteristics • Coupled oceanic and atmospheric oscillations • Hydrologic design

1.1 Climate Variability: Introduction and Background

Our terrestrial environment continues to transform under the natural cyclic variations of climate and evolve due to changing climate mainly influenced by human activities. Climate variability generally denotes deviations in climatological statistics over a given period, and these deviations are usually referred to as anomalies. Variability can be associated with natural internal processes within a climate system or anthropogenic influences referred to as external forcings. Climate change on the other hand refers to a significant variation in state of the climate over an extended

R.S.V. Teegavarapu (✉)
Department of Civil, Environmental and Geomatics Engineering,
Florida Atlantic University, Boca Raton, FL, USA
e-mail: rteegava@fau.edu

E. Kolokytha et al. (eds.), *Sustainable Water Resources Planning and Management Under Climate Change*, DOI 10.1007/978-981-10-2051-3_1

period of time again linked to both internal and external anthropogenic influences. Essential climatic variables (ECVs) that have been observed, reconstructed, and projected in future by climate change models tell different stories of our changing planet's climate. Understanding these changes from the past based on limited observations and adapting to future changes based on uncertain projections of future climate derived from climate models (Teegavarapu 2010) are two main challenges faced by water management agencies. Separating clean signal of natural cyclical changes of climate from noise of human-induced changes is a difficult task we need to understand and undertake.

1.2 Coupled Oceanic–Atmospheric Oscillations

Natural climate variability on multiple timescales (ranging from inter-annual, multidecadal, and longer geologic timescales) is a major obstacle to the reliable characterization of global climate changes resulting from human activities (Ghil 2002; Gurdak, et al. 2009). Quantifying the human fingerprint on climate change and predicting future changes are two of the greatest challenges facing all scientists who are involved in understanding variability of hydrologic cycles in different regions around the world. Detection and attribution which deal with identification of trends in essential climatic variables (ECVs) and address the causes, respectively, generally lead the current climate change and impact studies. Climate change is expected to bring several changes to hydrologic cycles and regimes in different parts of the world. In addition to uncertain sea level rise rates and future changes in temperature and precipitation patterns, human-induced climate change masks the natural climate variability. This is primarily because they are dependent on large-scale, global decadal oscillation weather systems. Some of these are limited to large multidecadal sea surface temperature (SST) anomalies which have significant impacts on regional and global climate.

1.3 Inter-year, Decadal, and Multidecadal Oscillations

The following sections provide brief descriptions of inter-year, decadal, and multidecadal oscillations.

1.3.1 El Niño Southern Oscillation (ENSO)

The ENSO is a slow oscillation in which the atmosphere and ocean in the tropical Pacific region interact to produce a slow, irregular variation between two phases: the warm and cool phase of ocean temperatures. ENSO is a major source of inter-annual climatic variability in many regions of the world. One of the indices used for

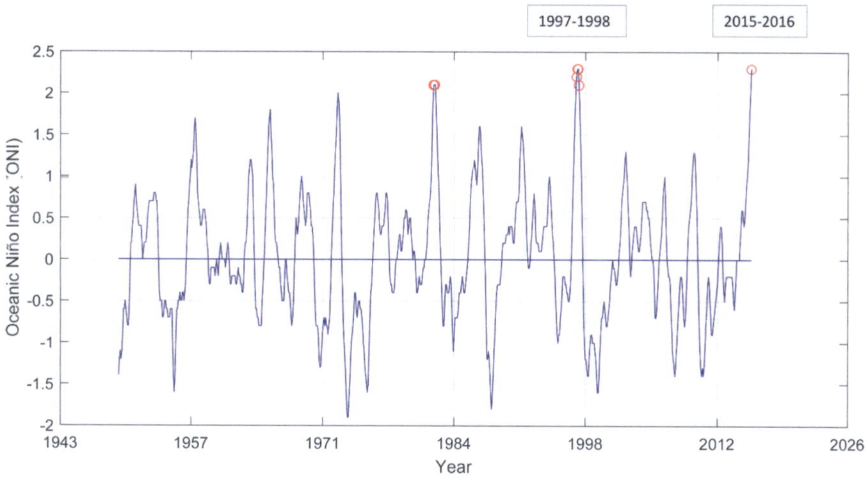

Fig. 1.1 Variation of ONI used for the determination of ENSO phases

defining two phases of ENSO is Oceanic Niño Index (ONI). The index is calculated based on observed sea surface temperature in the region that spans a swath from 5°N to 5°S latitude and 120°W to 170°W longitude. Figure 1.1 shows the ONI values and especially values above 2 indicating strong El Niño years. These variations are more commonly known as El Niño (the warm phase) and La Niña (the cool phase). Even though ENSO is centered in the tropics, the changes associated with El Niño and La Niña events affect climate around the world. ENSO events tend to form between April and June and typically reach full strength in December. ENSO is by far the most studied teleconnection and probably most publicized due to strong El Niño events of 1982–1983 and 1997–1998 which lead to the highest damages to agricultural and other sectors. ENSO is considered as the single largest cause of extremes in precipitation (as well as the cause of inter-annual variability) accounting for 15–20 % of the global variance of precipitation (Dai et al. 1997). The contribution is higher than 20 % in ENSO regions. These conclusions were based on monthly gridded datasets from 1900 to 1988. Studies performed by Goly and Teegavarapu (2012) indicated that El Niño is responsible for higher precipitation totals compared to La Niña in the months of December to February in Florida, USA. The conclusions were based on gridded precipitation datasets and also historical precipitation data from several rain gauges.

1.3.2 Atlantic Multidecadal Oscillation (AMO)

The Atlantic Multidecadal Oscillation (AMO) is a naturally occurring oceanic–atmospheric phenomenon on the North Atlantic Ocean that manifests in variability of SST. The AMO index is calculated using SST anomalies calculated in the region between the latitude of 75.0°N and 0.0°S and between the longitudes of 10.0°E and 75.0°W. Temperature variations have been instrumentally observed for over 150 years, but considering paleoclimatic evidence, such as ice cores and tree rings, it can be concluded that AMO has been present for the last millennium. To better understand this naturally occurring oscillation, and its effect on extreme events, it is helpful to study past phases. SST is an indicator of a specific phase. During the warm phase, elevated SST is generally observed, while the cold phase experiences lower SST. Since instrumental temperature records are available only from 1880, the existence of two phases of AMO, dating back to the nineteenth century or earlier, can be confirmed only possible using sediment core or tree ring data. United States Geological Survey (USGS) 2011 used data from sediment core from the Gulf of Mexico and coral core from Puerto Rico and found that these two sources showed a similar variability and correlated with the instrumental temperature data of the twentieth century, concluding that SST oscillations have existed at least since the 1800s. AMO has two phases, warm and cool, with each phase lasting about 20–40 years, yielding in an approximate of 70 years long cycle. Between the extreme values of the cool and warm phases, an SST difference of 1 °F can be observed. The effects of AMO can either obscure or exaggerate the global warming due to anthropogenic sources, depending on the current phase. AMO is known to have impacts on temperature, rainfall, hurricanes, drought, or floods. There are several different studies and agencies that determine warm and cool phases for the AMO, some of which determine the intervals with gaps between them, such as Enfield et al. (2001), which lists the cool periods to be 1905–1925 and 1970–1990 and the warm periods to be 1930–1960 and 1995–2010. Some studies (e.g., Koch-Rose et al. 2011; Teegaravapu et al. 2013) have used different intervals for AMO, without leaving any gaps between them. The intervals are consecutive and include every year from 1985 to 2010. Since the phases of AMO can last up to 20 to 30 years, it is possible to make a clear distinction about the effects when studying historical data of precipitation extremes. A study published by USGS (2011) looked at the frequency of hurricane occurrences, under the different phases of AMO. Using historical data provided by the National Oceanic and Atmospheric Administration (NOAA), the number of category 4 and 5 hurricanes was counted. It was concluded that during the 26 years long negative phase, eight hurricanes made landfall, while the positive phase lasting only 13 years experienced 14 hurricane landfalls.

1.3.3 Pacific Decadal Oscillation (PDO)

The Pacific Decadal Oscillation (PDO) is an inter-decadal climate variability phenomenon characterized by changes in sea surface temperature, sea level pressure, and wind patterns. The warm and cold phases are defined by positive and negative values of PDO index, respectively. The oscillation was discovered in 1997 by Steven Hare who has conducted a study on salmon fisheries in the Pacific Northwest. PDO is characterized by the sea surface temperature and sea level pressure change that occurs over the North Pacific Ocean (Mantua et al. 1997; Mantua and Hare 2002). Tree ring analysis suggests that this phenomenon has existed and reoccurring with 50–70 years intervals (Mantua et al. 1997). To determine the driving forces of the PDO, statistical analysis was applied on the SST, the sea level pressure (SLP), and the wind stress across the North Pacific Ocean (Schneider and Cornuelle 2005). This was done to study the temporal and spatial effects of this climate variability. The climate variations related to PDO have significant effects across the North Pacific, as well as the Americas with influences on the water resources, fisheries, and other natural habitats.

1.3.4 North Atlantic Oscillation (NAO)

The North Atlantic Oscillation (NAO) (Wallace and Gutzler 1980) is characterized by low pressure occurring over Iceland and high pressure over the Azores, which is centrally located between Portugal and North America. NAO is in the positive phase when the Icelandic low pressure and the Azores subtropical high pressure are strongly dominant. During this phase the Atlantic experiences stronger westerly winds, which bring storms in higher frequency to Europe. When NAO is in the positive (warm) phase, the east coast of North America has a milder winter with above-average temperatures and more precipitation. During the positive phase, the crossing storms are stronger and more frequent, in a northern direction. Winter conditions are warm and wet across Europe, while Canada and Greenland experience cold and dry conditions. Negative phase will occur when the pressure areas, the Icelandic low and the Azores high, are not as dominant. In this phase, the westerly winds are weaker, allowing the cold Arctic air to enter the USA and reach southern areas. There are fewer storms over the Atlantic, and the east coast of North America has a colder winter and precipitation in the form of snow. Snowstorms with subfreezing conditions occur in higher frequency over the USA. Characteristics of the negative NOA index include colder temperatures over northern Europe, while the Mediterranean experiences more moisture and milder winters. Fewer and weaker storms are crossing over the Atlantic in the west to east direction. The winter along the East Coast of the USA has more cold air outbreaks, as well as snowy weather conditions.

1.3.5 Other Major Oscillations

A number of other high- and low-frequency oscillations influence hydroclimatic variables in different regions of the world. These oscillations include Arctic Oscillation (AO), Madden–Julian Oscillation (MJO), Pacific North American (PNA) pattern, and Indian Ocean Dipole (IOD). The IOD, referred to as the Indian Niño, is an irregular oscillation of sea surface temperatures with cool and warm phases. Exhaustive discussion about these oscillations is available elsewhere (Cronin 2009; Teegavarapu 2013; Rosenzweig and Hillel 2008).

1.4 Regional and Global Influences of Oscillations

Regional and global influences of oscillations on ECVs are documented by several studies (Cronin 2009; Teegavarapu 2013). AMO, ENSO, PDO, and NAO have influences on precipitation and temperature characteristics of the USA. The effects of PDO can be felt during winter and spring, between November and March across the USA. When PDO is in a warm phase, higher temperatures are observed across the Northwestern USA, while southeast America experiences cool temperatures. PDO is dependent upon the ENSO, because it showed more decadal variability in response to ENSO (Newman et al. 2003). The results of this study also showed that the oscillation has a strongest effect on SST across the northern part of the pacific during winter and spring. Previously it was believed that the PDO has greater effect on the same geographical area during summer (Zhang et al. 1996). Zhang studied the inter-annual variability of SST and SLP and the Southern Oscillation Index time series. Based upon historical data, a change in the temporal variability was observed and analyzed with several techniques. This variation showed a trend that was inter-annual and consistent with an ENSO-like oscillation. In addition, a linearly independent decadal variability was also observed. This inter-decadal variability has similar properties like the ENSO, except it shows effects over the North Pacific and not confined to the equatorial area. A shift was observed in the data, around 1977, which is consistent with the phase change of PDO from cool to warm. Hurrell and Van Loon in their research paper published in 1995 studied the changes in distribution of precipitation and surface temperature over the Northern Hemisphere, more specifically the North Atlantic. These changes can be correlated with the current phase of NAO. Data about the oscillation is available for the past 150 years. When analyzing the data and the changes occurring, it was established that NAO has been in the positive phase since the 1980s. It was concluded that the precipitation anomalies of the same period can be correlated with the warm NAO phase. Anomalies include changes in temperature and precipitation. Wintertime warming over Europe and wintertime cooling of the northwest Atlantic have been recorded. Northern Europe has experienced winters wetter than usual, while winters are dryer in southern Europe. It was also concluded that as a result of the current positive phase of NAO, the storm tracks over the Atlantic have experienced a northward shift (Hurrell and Van Loon 1995). Teleconnections can be analyzed using several different parameters. Wallace and Gutzler (1980) used sea level pressure and geopotential height to find evidence of oscillation patterns of at least a month, on the Northern Hemisphere. Correlation statistics were used to find the strongest teleconnections. After the analysis, it was concluded that the NAO and the Pacific/North Oscillation have strong presence. Also, there is a correlation between the Atlantic Jet Stream and the NAO (Wallace and Gutzler 1980). The IOD is known to have an opposing effect or neutralizing effect on the influences of El Niño on the Indian subcontinent that reduces monsoon precipitation amounts. El Niño is linked with wetter conditions in the southeastern USA, a few regions in South America and Northern Africa during the months of December–February. Also, dry conditions are known to exist in several regions of Asia including India and wet conditions in the Northwestern USA and southwestern part of South America during the months of June–August. La Niña is associated with wet conditions in few regions of the USA and drier conditions in the southeastern USA during the months of December–February. Parts of India and Asia experience wet conditions during June–August months associated with La Niña. AMO influences on precipitation with increases in extremes during warm phase in several regions of the southeastern USA are documented by Teegaravapu et al. (2013) and Goly and Teegavarapu (2014). Increased hurricane landfalls were also noted during warm phase of AMO. The temporal windows associated with cool and warm phases of ENSO and AMO are provided in Table 1.1. Similarly, the temporal windows associated with cool and warm phases of PDO and NAO are provided in Table 1.2. In many regions limited information about temporal variations in influences of oscillations are available, and spatial extent is not clearly defined.

Table 1.1 Years identified as cool and warm episodes for AMO and ENSO (1950–2010)

	AMO	ENSO[a]
Cool (La Niña)	1970–1994	1950, 1954, 1955, 1964, 1967, 1970, 1971, 1973, 1974, 1975, 1983, 1984, 1988, 1995, 1998, 1999, 2000, 2005, 2007, 2008
Warm (El Niño)	1950–1969, 1995–2010	1951, 1953, 1957, 1958, 1963, 1965, 1968, 1969, 1972, 1976, 1977, 1982, 1986, 1987, 1991, 1994, 1997, 2002, 2004, 2006, 2009
Neutral	–	1952, 1956, 1959, 1960, 1961, 1962, 1966, 1978, 1979, 1980, 1981, 1985, 1989, 1990, 1992, 1993, 1996, 2001, 2003

Source: Goly and Teegavarapu (2014), adopted with permission
[a]Denotes for the current year–subsequent year, for example, 1950–1951 is represented by 1950.

Table 1.2 Years identified as cool and warm episodes for AMO and ENSO (1900–2010)

	PDO	NAO
Cool	1900–1925	1952–1972, 1977–1980
	1946–1976	
	2000–2010	
Warm	1926–1945	1950–1951, 1973–1976, 1981–2001
	1977–1999	

Source: Pierce (2013)

1.5 Evaluation of Changes in Precipitation Extremes and Characteristics

Evaluation of precipitation characteristics and extremes will involve a number of steps ranging from data collection to development of inferences about the influences of climate variability using statistical tests. The following steps are recommended for analysis of precipitation data:

- Collect and evaluate the precipitation data for different temporal and spatial scales.
- Assess missing data lengths, nonhomogeneity issues, and erroneous data.
- Fill missing data using appropriate spatial or temporal interpolation methods using available data from single- or multisensor precipitation estimates.
- Check the homogeneity of data after any infilling and note change points (if any) in the time series.
- Apply corrections to estimates (i.e., infilled data) and reevaluate the homogeneity of the time series.
- Identify temporal windows for climate variability analysis.
- Identify a list of indices that can characterize the changes in the variables (or time series of variables).
- Identify and select statistical methods for analysis of these indices: examples of these include statistical inference tests (parametric and nonparametric).
- Identify, select, and execute parametric and nonparametric trend tests if trend evaluation is required.
- Report statistically significant results from inference and trend analysis tests.
- Assess the spatial and temporal influences of climate variability on precipitation.
- Understand and document potential implications associated with the influences (noted in the previous step) of climate variability on hydrologic design, water resources management.

Spatial and temporal changes and trends in precipitation extremes and characteristics due to climate variability can be evaluated using a number of indices. These indices reflect different characteristics of precipitation time series, and they include (1) inter- and intra-annual variations, (2) seasonality, (3) spatial and temporal variability of extremes, (4) nature of extremes (based on events related to the type of storm: convective, frontal), (5) transition states as defined by rain or no-rain dichotomous events, (6) temporal persistence as defined by serial autocorrelation, (7) intra-event temporal distribution of precipitation, (8) antecedent moisture conditions (AMC) preceding extreme events, (9) temporal occurrences of extremes, (10) number of extremes over a specific threshold, (11) inter-event time definition (IETD) (based events), and (12) individual and coupled influences of internal modes of climate variability. One major question that needs to be answered related to precipitation extremes and characteristics is: How does the inter-annual, decadal, and multidecadal climate variability affect the occurrence of precipitation extremes relating to magnitude and frequency?

1.5.1 Extreme Precipitation Indices

Indices for precipitation extremes defined by the Expert Team on Climate Change Detection and Indices (ETCCDI) (WMO 2009) can be computed at each site to gain a clear understanding of changes in precipitation extremes during different phases. A total of 27 indices were developed by ETCCDI (WMO 2009) to describe particular characteristics of extremes, including frequency, amplitude, and persistence for temperature and precipitation. Nine extreme precipitation indices as defined by ETCCDI in Table 1.2 are explained in this section. The indices RX1day and RX5day refer to the maximum one-day and five-day precipitation in a given time period. The indices R10mm, R20mm, and Rnnmm are used to calculate the number of times a given value of threshold (viz., 10 mm, 20 mm, and "nn" mm) (WMO 2009) is exceeded. A threshold value of 25.4 mm is considered for "nn" in this study. Simple daily intensity index (SDII) and total precipitation in wet days (PRCPTOT) refer to average and total precipitation amounts of all wet days in a given time period, respectively. The time period used for the analysis can vary from a season to year. Consecutive dry days (CDD) and consecutive wet days (CWD) indices provide the largest number of consecutive dry and wet days in a given time period, respectively. Precipitation depth greater than or equal (less than) to 1 mm is used to categorize wet (dry) days in the calculation of SDII, PRCPTOT, CDD, and CWD indices. A few indices described in Table 1.3 require serially continuous (i.e., gap-free) precipitation datasets. Some of the indices can also be obtained from data with gaps (i.e., missing records), and they include RX1day, RX5day, CDD, and CWD. However, the indices may be underestimated due to the infilling process. Two recent studies (Goly and Teegavarapu 2014 and Teegaravapu et al. 2013) have documented the changes in several extreme precipitation indices in two phases of AMO and ENSO in the state of Florida, USA.

Table 1.3 Extreme precipitation indices and their explanation (WMO 2009)

Index	Description
RX1day	Maximum 1-day precipitation
RX5day	Maximum 5-day precipitation
SDII	Simple daily intensity index
R10mm	Count of precipitation days with *DR* greater than 10 mm
R20mm	Count of precipitation days with *DR* greater than 20 mm
Rnnmm	Count of days where *DR* greater than a threshold value
CDD	Consecutive dry days ($DR < 1$ mm)
CWD	Consecutive wet days ($DR \geq 1$ mm)
R95pTOT	Total precipitation due to wet days (>95th percentile)
R99pTOT	Total precipitation due to extremely wet days (>99 percentile)
PRCPTOT	Total precipitation in wet days ($DR > 1$ mm)
	DR: Daily Rainfall

1.5.2 Drought Characterization

1.5.2.1 Standard Precipitation Index (SPI)

Standard Precipitation Index (SPI) (WMO 2012), an internationally recognized index, developed by McKee et al. (1993, 1995), is useful to evaluate dry and wet conditions. It is generally used for drought monitoring; however, it is also very effective in analyzing wet periods. The only input for this conceptually simple index is the precipitation, requiring monthly data without gaps, with a minimum length of 20 to 30 years, but optimally a longer period, 50 or 60 years or more is recommended. The confidence level of the analysis and the length of the data are positively correlated. The length of water deficit or abundance due to drought and heavy precipitation can have different effects on soil moisture, streamflow, or groundwater supply on different timescales. SPI can be calculated for different intervals, to capture and analyze the effects, based on the point of interest. Standard Precipitation Index (SPI) calculation involves fitting a probability distribution (typically a gamma distribution) to 1-, 3-, 6-, and 12-month precipitation totals and then use of standard normal distribution to obtain SPI values. Probability density function of gamma distribution in standard form is given in Eq. (1.1). The variable α is the shape parameter. SPI values can be used to define dry and wet conditions as explained in Table 1.4.

$$f(x) = \frac{1}{\Gamma(\alpha)} x^{\alpha-1} \mathrm{e}^{-x},\ x \geq 0 \quad (1.1)$$

An example of SPI calculation based on monthly precipitation observations at a rain gauge site (site name, Wakkanai and location; latitude, 45.4025000; longitude, 141.6686111) in Japan is provided in Fig. 1.2. A total of 78 years of monthly data is used for developing 3-month SPI. Teegavarapu (2016) has evaluated the changes in SPI for 155 sites in Japan and indicted more drought occurrences in ENSO warm phase (El Niño). Goly and Teegavarapu (2014) observed an opposite effect of ENSO in their study of drought occurrences in Florida.

Table 1.4 Identification of dry or wet conditions based on SPI

SPI Value	Dry or Wet Condition
2.0 and greater	Extremely wet
1.99 to 1.50	Very wet
1.49 to 1.00	Moderately wet
0.99 to −0.99	Near normal
−1.0 to −1.49	Moderately dry
−1.50 to −1.99	Severely dry
−2 and less	Extremely dry

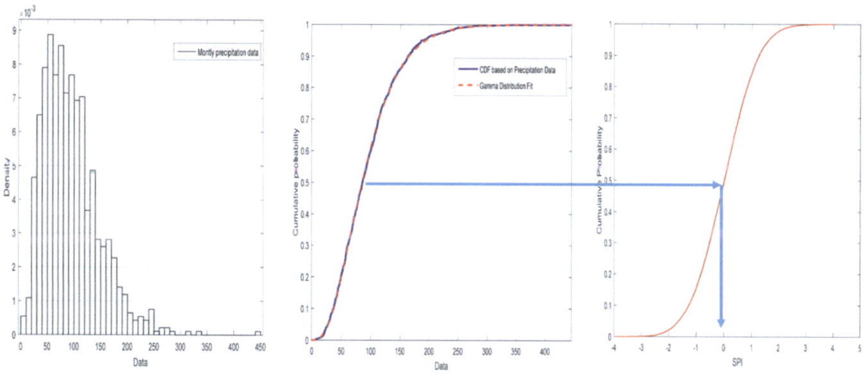

Fig. 1.2 Calculation of a 3-month SPI for a rain gauge site in Japan

1.6 Precipitation Characteristics

Changes in precipitation characteristics based on available historical data can be evaluated for influences of oscillations using a number of indices. These indices are discussed in the following sections.

1.6.1 Inter-year and Intra-year Variations

Precipitation characteristics that vary within a year as well as over several years can be evaluated for influences of climate variability. Within year variations can be assessed at different temporal scales ranging from sub-hourly time intervals to seasons.

1.6.2 Storm Events Based on Inter-event Time Definition (IETD)

Rainfall time series can be considered as a series of rainfall pulses through time. Isolation of individual storm event from such a long record of p requires application of specific criteria to determine when an event begins and ends. One such criterion is inter-event time definition (IETD): Minimum temporal spacing without rainfall required to consider two rainfalls as belonging to different events, used for the statistical analysis of rainfall records. The concepts of IETD are illustrated in Figs. 1.3 and 1.4. If the period between pulses of rainfall is less than or equal to IETD, then the two pulses of rainfall are categorized as belonging to one event.

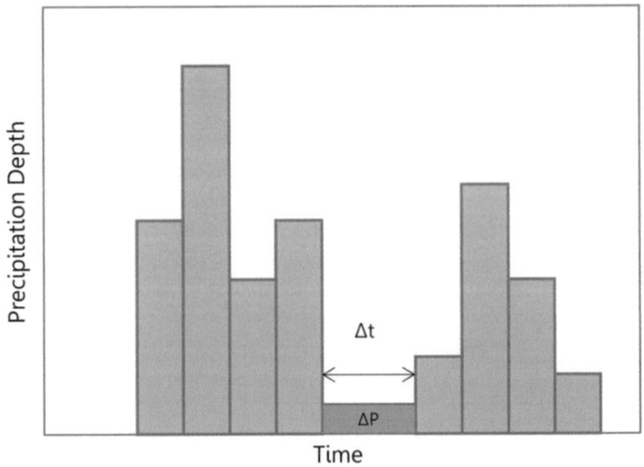

Fig. 1.3 Two storm events separated by a time interval

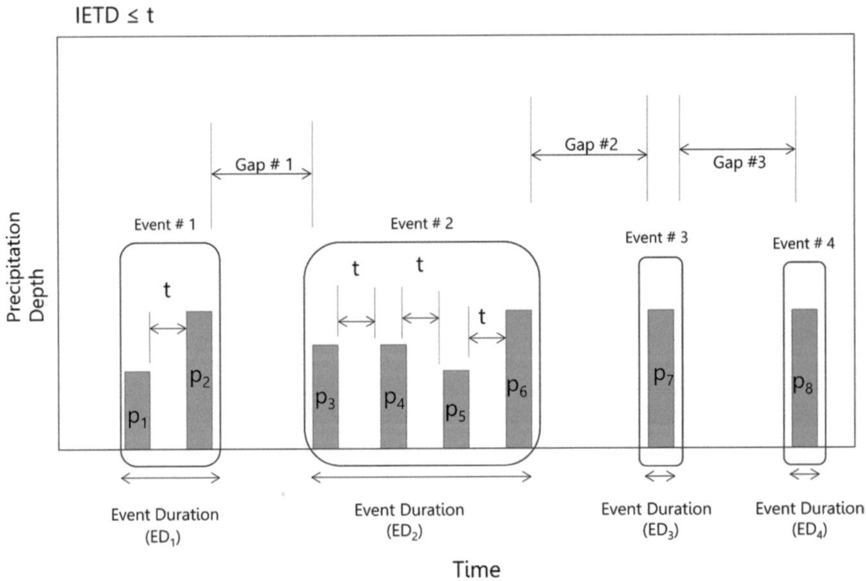

Fig. 1.4 Two storm events separated by a time interval

IETD can be equal to lag –time when autocorrelation is equal to zero. The inter-event time definition (IETD) is defined as the minimum temporal spacing without rainfall required to consider two rainfall events as belonging to different events (Adams and Papa 2000).

Two rainfall events are considered as distinct events if:

1. The precipitation (ΔP) that falls during a time interval between the events is less than a specific threshold value.
2. The time interval (Δt) is greater than a selected time interval (e.g., time of concentration).

The use of design storms based on IDF curves for stormwater management was evaluated by Adams and Howard (1986). The analytical probabilistic models for stormwater management models prescribed by Adams and Papa (2000) describe the need for identification of individual storms using inter-event time definition. Rainfall volumes, durations, intensities, and inter-event times can be characterized using exponential or gamma distributions (Behera et al. 2010) for use in analytical probabilistic models. The statistics of storm event characteristics are influenced by the values of IETD, and these can be analyzed in the context of climate variability and change. Pierce (2013) has documented changes in IETD of storms for AMO, PDO, and NAO in Florida, USA.

1.6.3 Wet and Dry Spells

Determining wet and dry spells can provide further information about the precipitation characteristics. Mean or total monthly rainfall values will give indication about how wet or dry the month was; however, determining the distribution of the rainfall can be essential when managing watersheds and flood or drought conditions. The conditions in the watershed will vary based on the distribution of the total rainfall over any interval. Precipitation threshold values can be established for evaluation of dry and wet thresholds; in general, a zero value of precipitation is ideal for consideration. Once the threshold is established, consecutive wet and dry days can be estimated as wet and dry spells, respectively. The length of each wet and dry spell can be used to calculate the mean length of wet and dry spell individually. The lengths of spells considering a threshold level are representative of regional rainfall patterns and can be evaluated for changes. The number of wet or dry spells that is equal or longer than a prefixed threshold value can be evaluated in different temporal windows that coincide with different phases of oscillations.

1.6.4 Transitions of Wet and Dry States

Transition probabilities associated with dry and wet spells are calculated based on conditions specified in Table 1.5. These probabilities are referred to as two-state first-order Markov chain probabilities.

Table 1.5 Rain or no-rain states in two consecutive time intervals for determination of transition probabilities

		Time interval $(i+1)$	
Time	$R_i > 0$	$R_{i+1} > 0$ [Wet–Wet]	$R_{i+1} = 0$ [Wet–Dry]
Interval (i)	$R_i = 0$	$R_{i+1} > 0$ [Dry–Wet]	$R_{i+1} = 0$ [Dry–Dry]

Two-state first-order Markov chain probabilities are given in Eqs. (1.2), (1.3), (1.4), and (1.5):

$$P_{11} = P_r\left(R_{i+1} > 0 \mid R_i > 0\right) \tag{1.2}$$

The variable P_{11} refers to probability of occurrence of positive precipitation in time interval $i + 1$ given the occurrence of positive precipitation in the previous interval, i:

$$P_{10} = P_r\left(R_{i+1} = 0 \mid R_i > 0\right) \tag{1.3}$$

The variable P_{10} refers to probability of no precipitation in time interval $i + 1$ given the occurrence of positive precipitation in the previous interval, i:

$$P_{01} = P_r\left(R_{i+1} > 0 \mid R_i = 0\right) \tag{1.4}$$

The variable P_{01} refers to probability of occurrence of precipitation in time interval $i + 1$ given no precipitation in the previous interval, i:

$$P_{00} = P_r\left(R_{i+1} = 0 \mid R_i = 0\right) \tag{1.5}$$

The variable P_{00} refers to probability of no precipitation in time interval $i + 1$ given no precipitation in the previous interval, i:

1.6.5 Persistence

Precipitation data can be assessed for serial autocorrelation using the time series at different temporal resolutions. The autocorrelation coefficient is also referred to as serial correlation coefficient. The first-order autocorrelation coefficient can be referred to as correlation coefficient of the first *N-1* observations (observations), $\theta_1 \ldots\ldots \theta_{N-1}$, and the next *N-1* observations, $\theta_2 \ldots\ldots \theta_N$. These two series are used for calculations of average values, and they are referred to as $\overline{\theta}_{(1)}$ and $\overline{\theta}_{(2)}$, respectively. The autocorrelation values can be obtained for different lag (t) values as given in Eq. (1.6):

$$\rho_t = \frac{\sum_{i=1}^{N-t}\left(\theta_i - \overline{\theta}_{(1)}\right)\left(\theta_{i+t} - \overline{\theta}_{(2)}\right)}{\sqrt{\sum_{i=1}^{N-t}\left(\theta_i - \overline{\theta}_{(1)}\right)^2}\sqrt{\sum_{i=2}^{N}\left(\theta_i - \overline{\theta}_{(2)}\right)^2}} \tag{1.6}$$

For sufficiently large N, the autocorrelation at a specific lag can be defined by Eq. (1.7).

$$\rho_t = \frac{\sum_{i=1}^{N-t}\left(\theta_i - \overline{\theta}\right)\left(\theta_{i+t} - \overline{\theta}\right)}{\sum_{i=1}^{N-t}\left(\theta_i - \overline{\theta}\right)^2} \tag{1.7}$$

The variable $\overline{\theta}$ is the mean (average) of the entire available time series data. Autocorrelograms can be evaluated for two-sample datasets from two time periods that coincide with the temporal windows of the oscillation. Spatial variations in lag-1 autocorrelation values in Japan were noted in different phases of ENSO and PDO in a recent study reported by Teegavarapu (2016).

1.6.6 *Seasonality*

The seasonality index (SI) defined by Walsh and Lawler (1981) can be used to determine the intra-annual monthly distribution of precipitation. The SI can also be used for spatial representation of seasonal variability over regions, providing a better understanding of rainfall regimes. The SI, as given in Eq. (1.8), is the sum of the absolute deviations of the monthly rainfall from the mean monthly rainfall, divided by the total annual precipitation of the given year:

$$SI_i = \frac{1}{R_i}\sum_{n=1}^{n=12}\left|x_{i,n} - \frac{R_i}{12}\right| \tag{1.8}$$

where R_i is the total annual precipitation in a particular year and $x_{i,n}$ is the actual monthly rainfall in month n. The SI in Eq. (1.2) yields yearly indices, which can be qualified based on the established index values, to determine the degree of seasonality, shown in Table 1.6. Teegavarapu (2016) and Pierce (2013) evaluated seasonality index values for Japan and Florida, respectively, and concluded that warm and cool phases of PDO have strong influences on the spatial variability of seasonality of precipitation.

Table 1.6 Classification of seasonality index values and links to precipitation regimes (Walsh and Lawler 1981)

Seasonality index	Precipitation regime
<0.19	Precipitation spread throughout the year
0.20–0.39	Precipitation spread throughout the year, but with a definite wetter season
0.40–0.59	Rather seasonal with a short drier season
0.60–0.79	Seasonal
0.80–0.99	Marked seasonal with a long dry season
1.00–1.19	Most precipitation in <3 months
>1.20	Extreme seasonality, with almost all precipitation in 1–2 months

1.6.7 Changes to Extremes of Specific Duration and Frequency

Changes to extreme values of precipitation for different temporal durations in different phases of oscillations can be evaluated by developing depth–duration–frequency (DDF) or intensity-duration-frequency (IDF) curves. Data for two different temporal windows that coincide with the phases of oscillations can be used to fit probability distribution functions (PDFs) to characterize the precipitation extremes of different durations. Svensson and Jones (2010) in a recent survey of evaluation of rainfall frequency distributions have indicated that generalized extreme value (GEV) distribution is most frequently used to characterize rainfall extremes. Besides GEV, lognormal, three-parameter lognormal, and Pearson and log-Pearson distributions should also be evaluated for characterizing extreme precipitation data. Goodness-of-fit (GOF) hypothesis tests and performance measures such as mean absolute deviation (MAD) and mean square deviation (MSD) (Jain and Singh 1987) can be used to measure the relative goodness-of-fits of distributions to the data. GEV with a flexible three-parameter model expressed by a probability density function (PDF) given in Eq. (1.9) (Teegaravapu et al. 2013) is generally used to characterize precipitation extremes.

$$f(x) = \begin{cases} \frac{1}{\sigma}\exp\left(-(1+k^{o}z_{o})^{-\frac{1}{k^{o}}}\right)(1+k^{o}z_{o})^{-1-\frac{1}{k^{o}}} & k^{o} \neq 0 \\ \frac{1}{\sigma}\exp\left(-z_{o}-\exp\left(-z_{o}\right)\right) & k^{o} = 0 \end{cases} \tag{1.9}$$

The variables k^o, σ, and μ refer to the shape, scale, and location parameters, respectively, and the value of $z_o = (x - \mu)/\sigma$. The parameters of the distribution can be estimated using maximum log-likelihood estimation (MLE) method or L-moment method. The maximum precipitation depth for each time interval is related with the corresponding return period from the cumulative distribution function (CDF). The maximum precipitation depth can be determined using a theoretical distribution function that is used to characterize the distribution of precipitation extremes.

1.6.7.1 Changing Intensity–Duration–Frequency Relationships

An example of DDF curves developed for four regions in the state of Florida, USA, in a recent study by Teegaravapu et al. (2013) using GEV is shown in Fig. 1.5. Precipitation extremes obtained from DDF curves developed for a specific return period indicate that the selection of temporal window coinciding with a specific phase of AMO is critical for hydrologic design. Regional differences in extreme precipitation depths based on DDF curves during different AMO phases are evident. Underestimation of design storms is possible when entire available historical data is used compared to the data obtained from one AMO phase alone.

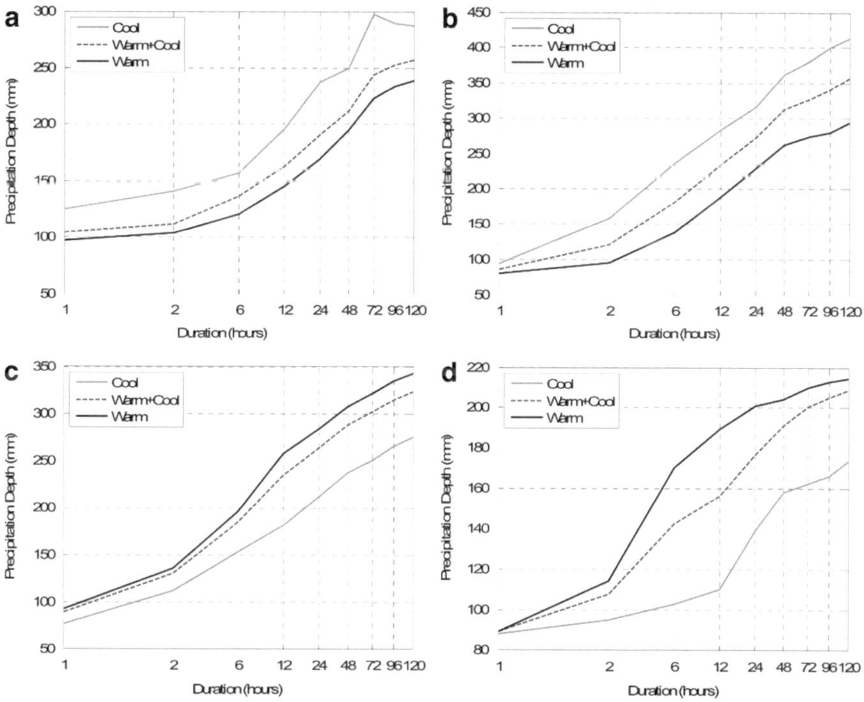

Fig. 1.5 Precipitation depth–duration–frequency curves for a 25-year return period during AMO warm, cool, and combined phases (cool and warm) for different stations: (**a**) North Florida, (**b**) Key West, (**c**) Palm Beach, and (**d**) Lake Okeechobee (Adopted from Teegaravapu et al. 2013)

1.6.8 Variations in Temporal Occurrences of Extremes

Changes in intra-year temporal occurrences of extremes caused due to climate variability of change have wide range of implications on water resources management. An example of such variations in occurrences is shown in Fig. 1.6 based on results from evaluation of precipitation extremes in two phases of AMO. Kernel density estimates (KDEs) using Gaussian kernels showing temporal occurrences of precipitation extremes for different durations are shown. It is evident from the figure that higher densities are seen in cool phase during earlier months of the year compared to those in warm phase. This suggests that flood realization potential is higher in earlier months of the year in the cool phase, and this requires adequate planning and preparation for any flood control management. In general, it has been noticed that precipitation extremes with higher magnitudes are occurring in AMO warm phase than in cool phase especially at durations higher than 24 h.

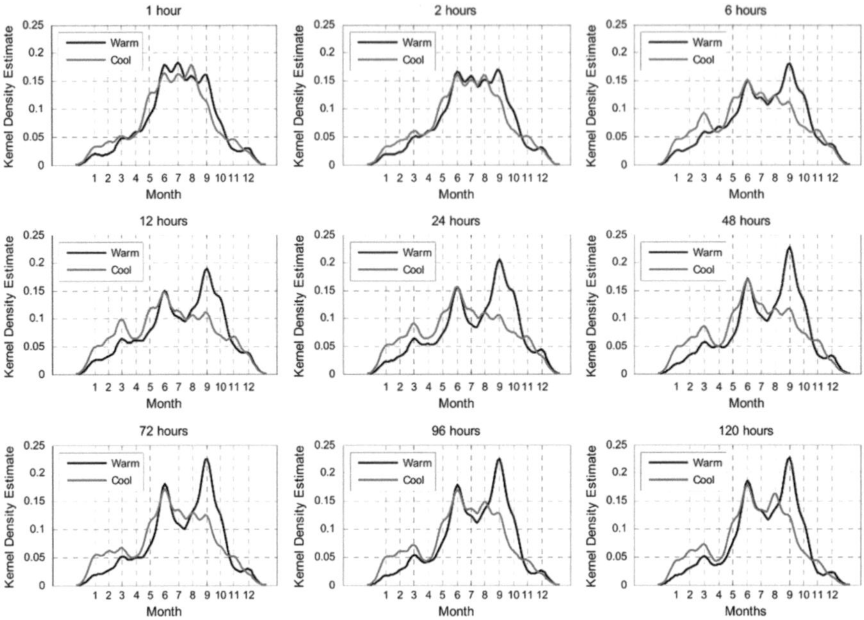

Fig. 1.6 Kernel density estimates of occurrences of precipitation extremes during AMO phases for nine temporal durations for cool (1970–1994) and warm (1942–1969) phases of AMO (Adapted from Teegavarapu et al. 2013)

1.6.9 Temporal Distributions of In-Storm Precipitation

Temporal distribution defines the time distribution of rainfall amounts within a storm event. Synthetic rainfall distributions are commonly used for hydrologic design in many regions of the world. For example, the Soil Conservation Service (SCS) of the USA provides four types of curves referred to as types I, IA, II, and III that are applicable to different regions of the USA. The time distribution will provide information about early, central, and late peaking storms. Changes in temporal distribution of in-storm precipitation totals are noted for different storm events in Florida by Goly and Teegavarapu (2014). Late peaking storms are known to increase flood peaks and are of concern for disaster management agencies.

1.6.10 Antecedent Moisture Conditions (AMC)

Antecedent moisture conditions preceding extreme precipitation events of specific temporal duration can be evaluated for changes in two different phases of oscillations. Higher AMCs will lead to larger peak runoff volumes and discharges based on

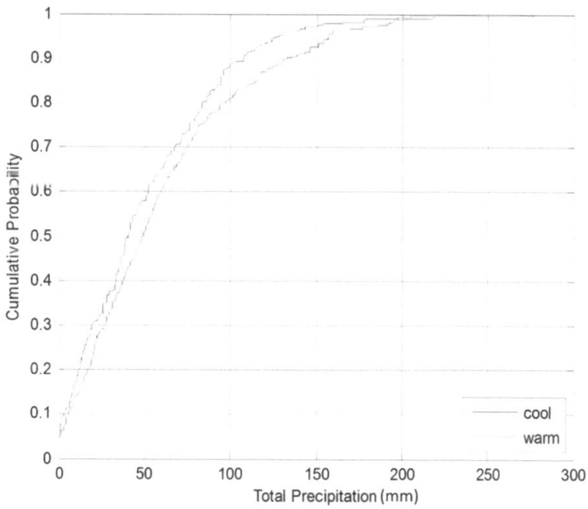

Fig. 1.7 Non-exceedance probability curves for AMO cool and warm phases 5-day antecedent precipitation amounts (Adapted from Goly and Teegavarapu 2014)

extreme precipitation events, and these runoff discharges may sometimes exceed the design discharges that were used for hydrologic/hydraulic infrastructure. In a recent study, Goly and Teegavarapu (2014) have investigated the variations in AMC for multidecadal (e.g., AMO) and inter-year oscillations (e.g., ENSO). Figure 1.7 shows the non-exceedance plots of a 5-day antecedent precipitation amounts for AMO cool and warm phases. Higher exceedance probabilities can be noted for warm phase. Similar conditions of AMC for ENSO influences on precipitation patterns in Japan were noted in a recent study (Teegavarapu 2016). Hydrologic design discharges need to be reevaluated considering the influences of climate variability on extreme precipitation events.

1.7 Evaluations of Precipitation Variability Influenced by Teleconnections

1.7.1 Precipitation Data

Precipitation data at different temporal and spatial resolutions can be used for evaluation of influences of climate variability on extremes and characteristics. Serially continuous (data without gaps) and error-free and chronologically complete precipitation dataset is needed for evaluation of some of the indices discussed earlier in this chapter. Monthly data can be used for a handful of indices (i.e., seasonality, standard precipitation index). Daily data and data at finer temporal resolutions can be evaluated for short duration precipitation extremes and characteristics (Teegaravapu et al. 2013) including transition probabilities and autocorrelation.

1.7.2 Homogeneity Analysis and Tests

Precipitation data are initially evaluated for serial continuity, outliers, and homogeneity. Exploratory data analysis techniques (mostly graphical) can be used in the first step of preliminary assessment of data. Homogeneity evaluation can be carried out using Buishand's (Buishand 1982) or Alexandersson's standard normal homogeneity test (SNHT) (Alexandersson 1986) or von Neumann ratio test (Von Neumann 1941). Randomness of time series values can be tested with the help of runs test (Wald–Wolfowitz test) for use in statistical hypothesis tests.

1.7.3 Point Precipitation Data

Point-based precipitation data refers to observed data collected using a recording or a non-recording gauge. Gauge measurements are influenced by random and systematic errors and several others including gauge catch. In many instances extreme precipitation events are not recorded by gauges. Radar-based precipitation datasets can be used for assessments, with an acknowledgment of limitation that only comprehensive data from most recent decades are available. Infilling of precipitation datasets is required to obtain serially complete precipitation datasets in many situations. These datasets are critical for calculation and evaluation of a number of indices that are discussed earlier in this chapter. Some of those indices include autocorrelation, transition probabilities, IETD calculations, etc. A number of deterministic and stochastic interpolation methods that are available for estimation of missing precipitation data are documented by Teegavarapu and Chandramouli (2005). New methods based on improvised universal function approximation-based kriging (Teegavarapu 2007), association rule mining (Teegavarapu 2009), mathematical programming (Teegavarapu 2012), nearest neighbor, and clustering (Teegavarapu 2013) approaches are available for infilling missing data. Corrections to spatially interpolated data are recommended. These corrections may involve the use of single-best estimator methods (Teegavarapu 2009) or quantile-based methods (Teegavarapu 2014).

1.7.4 Gridded Precipitation Data

Gridded precipitation data based on spatially interpolated estimations from point observations from single-sensor (i.e., rain gauge) or multisensor estimates are widely available for spatial evaluation of precipitation. It is important to note that using such data may result in underestimation of higher-end extremes and overestimation of lower-end extremes. Goly and Teegavarapu (2014) indicate that both point and spatially complete gridded precipitation datasets are valuable for analysis.

Justifiable inferences about spatial variability in precipitation characteristics and extremes can be drawn when the two datasets are used independently. Goly and Teegavarapu (2014) indicated that variance deflation; over- and underestimation of lower- and higher-end extremes, respectively; and alteration of statistical distributions are inevitable consequences of spatial interpolation methods often used for generation of gridded datasets. In conclusion, inferences based on gridded precipitation data should be interpreted carefully.

1.7.5 Trend Evaluation

Trend analysis will help to (1) understand and confirm the existence (or nonexistence) of statistically significant changes in precipitation extremes or characteristics; (2) develop defensible (statistically) parametric and nonparametric methods for evaluation of changes in space and time; (3) develop and evaluate new methods or variants of existing trend evaluation methods considering issues related to homogeneity, data length, and others; and (4) ascertain and confirm any attributable reasons (natural climate variability influenced or anthropogenic changes) to the changes or trends. The number of sites selected for any study will depend on a number of factors: (a) long historical record length, (b) availability of error- and gap-free data records, and (c) availability of sites in watersheds with exhaustive hydrometeorological data. Smoothing methods and filters (e.g., moving average, LOWESS (derived from "locally weighted scatter plot smooth") and Savitzky–Golay filter, and variants of LOWESS) can be used to assess the nonlinear variation of precipitation extremes with time. The following are the tasks that need to be completed for evaluation of trends:

1. Collect and evaluate the precipitation time series data at several sites in the region of interest for homogeneity issues.
2. Extract extremes and conduct analysis to check for existence of change points in the time series.
3. Develop and test nonparametric and parametric approaches for trend assessment considering issues such as serial autocorrelation, length of the data, temporal windows (moving and constant) and temporal windows linked to different phases of natural climate variability, or anthropogenic activity.
4. Assess natural or anthropogenic-based variations in extremes using detection and attribution methods.

Trend assessment can be carried out using nonparametric tests such as Spearman's rho and Mann–Kendall tests. Modified Mann–Kendall test can be used if ranked data based on the sample indicates several ties. If strong serial autocorrelation exists at several lags that are higher than confidence limits as evaluated using autocorrelogram and by Ljung–Box Q-test (Ljung and Box 1978), trend-free pre-whitening can be employed. Serial autocorrelation affects the null distribution of trend tests, and therefore these tests and corrective procedures are required.

Change points in the precipitation time series can be identified using Pettitt's test (Pettitt 1979). Step change in mean or median values can be evaluated using several tests (e.g., distribution-free cumulative sum control chart (CUSUM) (nonparametric test), cumulative deviation (parametric), Worsley's likelihood ratio (parametric)). Parametric (e.g., two-sample *t*-test) and nonparametric (e.g., Rank Sum or Mann–Whitney) tests will also be used to evaluate statistically significant changes in mean and median values in two different data periods (two different temporal slices). A robust estimator of linear trend using Thiel-Sen trend line can be developed for each flow extreme time series. In case of parametric modeling, linear regression models can be developed and hypothesis tests can be used to evaluate the statistically significant slope parameter. Exhaustive evaluation of residuals needs to be carried out using Durbin–Watson test to check for autocorrelation in residuals, probability plots, and goodness-of-fit (GOF) hypothesis tests (e.g., Kolmogorov–Smirnov (KS) and Lilliefors test) for normality of residuals and autocorrelation function plots for visual assessment of serial autocorrelation at different lags. These tests can help in validating the assumptions of linear regression analysis. Spatial evaluation of trends can be carried out either by grouping the data into pooled datasets considering homogenous subregions defined by hydroclimatology and other physical features of the watersheds or basins in the region of interest or by using site-based trend assessment results. Methods for attribution and detection may be based on fingerprinting technique or its variants and time series analysis-based methods wherein noise is removed from the signal.

1.7.6 Biases Due to Missing and Filled Precipitation Data

Existence of missing data in precipitation time series will influence analysis of short- and long-term variations in precipitation based on different indices discussed in this chapter. Data filling may not always alter site-specific statistics of imputed data when spatial interpolation methods are used for temporal scales larger than a day (e.g., monthly, seasonal, or annual). However, data filling can lead to changes in probability distribution of data and significant biases when event-based analysis (such as daily or hourly extreme precipitation analysis) is performed. In some cases, spatial interpolation is only approach as temporal interpolation fails due to lack of high serial correlation at several lags in daily precipitation data. At temporal resolutions of a day or less, spatial interpolation alters probability distributions of data and changes the autocorrelation structure and dry and wet spell transitions (Teegavarapu 2014). Many research efforts have focused on analyzing the trends in precipitation data, but biases introduced in these trends due to data infilling techniques are rarely investigated. Therefore, evaluation of (1) bias in extreme precipitation data due to infilling of data gaps, (2) changes in long-term trends in extreme precipitation indices due to infilling, and (3) variations in probability distributions of infilled and unfilled datasets are essential. In a recent study, Teegavarapu et al. (2011) indicated that infilling may lead to underestimation of both magnitude and frequency of heavy and very heavy precipitation events. Results show that infilling may also affect the spatial characteristics of extreme precipitation in the region. In general, it was noted that bias introduced by the data infilling increases as gaps (i.e., amounts of missing data) in precipitation data increase. Therefore, care should be taken while analyzing extreme precipitation events from precipitation data wherein gaps have been infilled or data with gaps is analyzed without infilling. Also, results from their study indicate that analysis of precipitation time series with missing data gaps infilled and unfilled data will lead to different conclusions about precipitation extremes and characteristics in different temporal windows of coupled oceanic–atmospheric oscillations.

1.8 Statistical Tests

Parametric and nonparametric statistical inference tests can be used to evaluate statistically significant differences in indices related to precipitation characteristics in two different phases of oscillations or two temporal windows.

1.8.1 Parametric and Nonparametric Tests

Samples for statistical hypothesis tests are generally identified based on datasets corresponding to different phases or temporal windows of oscillations. Parametric and nonparametric tests can be used to determine whether there is any statistically significant difference between the two population means or medians, respectively.

1.8.2 Parametric Test: Two-Sample Unpaired T-Test

A two-sample parametric *t*-test can be used to test null hypothesis that the two-sample datasets come from normal distributions with equal means but unknown variances. Satterthwaite's modified *t*-test (Satterthwaite 1946) is used when variances of two samples are unequal. Descriptions of two-sample parametric *t*-test and its modified version are provided in this section. Prior to the use of two-sample unpaired *t*-test, a two-sample F-test is used to evaluate the sample variances. The F-test evaluates the null hypothesis that the two-sample datasets come from normal distributions with the same variance. If the F-test confirms the null hypothesis, then the *t*-test statistic provided in Eq. (1.10) is calculated. The *t*-test statistic is calculated using Eqs. (1.10) and (1.11) when the sample variances based on two different sampling periods are equal:

$$t = \frac{|\overline{x}_1 - \overline{x}_2|}{\sqrt{\frac{n_1 + n_2}{n_1 n_2}\left(\frac{(n_1 - 1)\,S_1^2 + (n_2 - 1)\,S_2^2}{n_1 + n_2 - 2}\right)}} \tag{1.10}$$

The degrees of freedom (*df*) is defined by Eq. (1.11):

$$df = n_1 + n_2 - 2 \tag{1.11}$$

The variables n_1 and n_2 are the number of samples in dataset 1 and dataset 2, S_1^2 and S_2^2 are sample variances, and $\overline{x}_1$ and $\overline{x}_2$ are mean values of datasets 1 and 2, respectively. The unpaired two-sample *t*-test used for unequal sample variances is defined by Eqs. (1.12) and (1.13). The test is referred to as Satterthwaite's modified *t*-test (Satterthwaite 1946). The Welch–Satterthwaite modification Welch (1947) for degrees of freedom (*df*) in the case of this *t*-test is given in Eq. (1.13):

$$t = \frac{|\overline{x}_1 - \overline{x}_2|}{\sqrt{\frac{S_1^2}{n_1} + \frac{S_2^2}{n_2}}} \tag{1.12}$$

$$df = \frac{\left[\frac{S_1^2}{n_1} + \frac{S_2^2}{n_2}\right]^2}{\frac{S_1^4}{n_1^2\,(n_1 - 1)} + \frac{S_2^4}{n_2^2\,(n_2 - 1)}} \tag{1.13}$$

Application of parametric two-sample *t*-test requires normality of the samples as requisite condition. Sample data individually from two datasets or groups can be tested for normality. Normality can be confirmed using visual checks using normal probability plots (Mage 1982; McBean and Rovers 1998) initially. After initial confirmation of normality, several statistical hypothesis tests such as Kolmogorov–Smirnov (Smirnov 1939; Sheskin 2003), Lilliefors (1967), Jarque–Bera (1987), and chi-Square goodness-of-fit (Corder and Foreman 2009) tests can be used for additional confirmations. A well-known Lilliefors test as a two-sided goodness-of-fit test for normality is applicable for situations where a fully specified null distribution is not known. In case of the Kolmogorov–Smirnov (KS) test, the null distribution of the sample needs to be completely specified. Alternatively, Jarque–Bera test can help check the validity of null hypothesis that the data comes from a normal distribution with an unknown mean and variance, and chi-square test can be used to evaluate the null hypothesis that the data follows a normal distribution using parameters (mean and variance) estimated from the sample. It is important to note that chi-square test is sensitive to the number of bins used for grouping the sample data. If datasets do not conform to normality, logarithmic, square, square root, and several other transformations recommended by Helsel and Hirsch (2002) are initially evaluated. If none of these transformations are helpful to achieve normality,

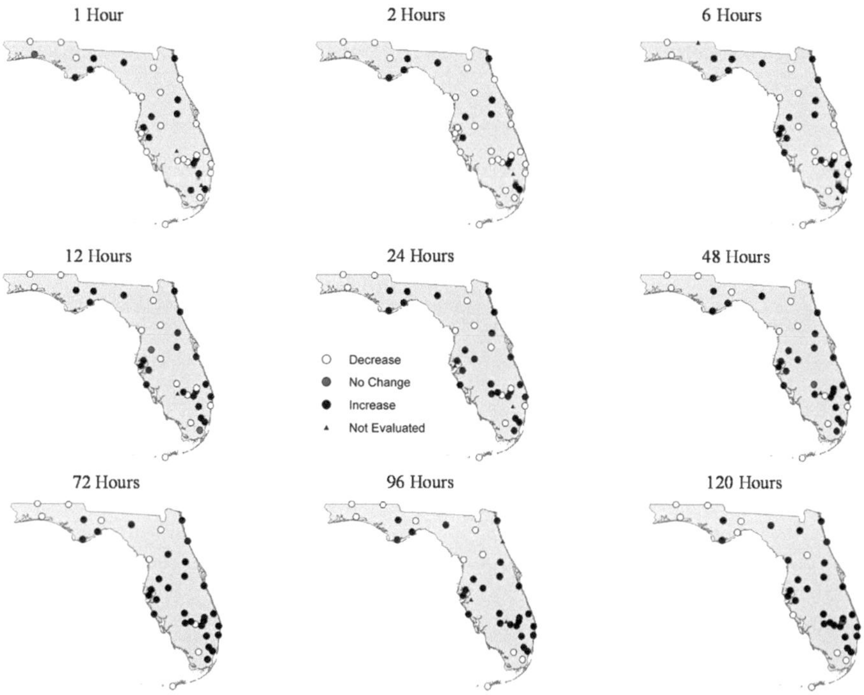

Fig. 1.8 Statistically significant changes in the extreme precipitation depths during two AMO phases for nine different durations (Adapted from Teegavarapu et al. 2013, Journal of Hydrology)

then a power transform such as Box–Cox (Box and Cox 1964) transformation can be used. The parameter of such power transform is obtained by optimization of a log-likelihood function with an objective of maximization of log-likelihood function. If all transformations fail, nonparametric tests can be adopted. Results for statistically significant changes in extreme precipitation depths (Teegaravapu et al. 2013) during two phases of AMO using two-sample unpaired t-test are shown in Fig. 1.8.

1.8.3 Nonparametric Test: Mann–Whitney U-Test

The Mann–Whitney U-test can be used to evaluate the null hypothesis (H_o) that data from two samples are from continuous distributions with equal medians, against the alternative (H_a) that they are not. The test assumes that the two samples are independent. The samples can be of different lengths. In order to apply the Mann–Whitney U-test, the precipitation datasets for two samples (e.g., data from El Niño and La Niña phases or two phases of any oscillation) can be used. The variables n_1 and n_2 are used to refer to the number elements in each sample. The data in each sample l, n_l are then ranked from lowest to highest, including tied rank values where appropriate. The equations related to Mann–Whitney U-test statistic are as follows as defined by Corder and Foreman (2009):

$$U_l = n_1 n_2 + \frac{n_i (n_i + 1)}{2} - \sum R_l \tag{1.14}$$

$$\bar{x}_u = \frac{n_1 n_2}{2} \tag{1.15}$$

$$S_u = n_1 n_2 + \sqrt{\frac{n_1 n_2 (n_1 + n_2 + 1)}{12}} \tag{1.16}$$

$$Z^* = \frac{U_l - \bar{x}_u}{S_u} \tag{1.17}$$

The variable S_u is the standard deviation and R_l is the rank from the sample l of interest and $\bar{x}_u$ is the mean. The variable Z^* is the z-score for a normal approximation of the data. The null hypothesis is rejected if the calculated Z^* statistic is greater than the selected critical value at 5 % significance level (α) obtained from the standard normal distribution table. The notation used in Eqs. (1.14), (1.15), (1.16), and (1.17) in this section is borrowed from Teegavarapu et al. (2013).

1.8.4 Bootstrap Sampling and Confidence Intervals

The use of bootstrap sampling methods (Efron 1979; Efron and Gong 1983) and finally the generation of confidence intervals can help make inferences about the sample statistics when limited numbers of datasets related to precipitation extremes or indices exist due to missing data or other reasons. Bootstrap sampling method (Efron and Tibshirani 1993) can be used to obtain samples from data in each phase of the oscillation or a specific temporal window, and confidence intervals on sample mean statistic can be developed. A general procedure of bootstrap sampling methodology explained by Davison and Hinkley (1997) and Teegavarapu et al. (2013) is adopted in this section. The notation used for explaining the bootstrap confidence interval generation is also based on Davison and Hinkley (1997). The sample values $y_1, y_2, \ldots\ldots, y_n$ are thought of as the outcomes of independent and identically distributed ($\widetilde{iid}$) random variables $Y_1, Y_2, \ldots\ldots, Y_n$ whose cumulative distribution function (CDF) is denoted by F. The estimate of F denoted by $\widehat{F}$ is obtained using data $y_1, y_2, \ldots\ldots, y_n$. The following steps are carried out (Davison and Hinkley 1997) to obtain bootstrap sampling confidence intervals:

- Bootstrap (re) sample $y_1^*, y_2^*, \ldots\ldots, \; y_n^* \; \widetilde{iid} \; \widehat{F}$ are obtained from the original samples allowing repetitions.

- $\widehat{F}$, an estimator of F, is obtained nonparametrically using empirical distribution function (EDF) of the original data, i.e., by placing a probability of "*1/n*" at each data value from sample $y_1, y_2, \ldots\ldots, y_n$.
- Sample mean statistic $\widehat{\theta*}$ is computed from bootstrap sample $y_1^*, y_2^*, \ldots\ldots, y_n^*$.
- The above steps are repeated "*N*" times, to obtain N sample means $\widehat{\theta_1^*}, \widehat{\theta_2^*}, \ldots\ldots, \widehat{\theta_N^*}$. The practical size of "*N*" depends on the tests to be run on the data.

The size of "*N*" recommended by Chernick (2007) is 1,000 and 10,000 for evaluating the sample statistic and confidence intervals, respectively. In the current study, these values are used. After N samples are obtained, normally approximated confidence intervals are computed for the uncertainty assessment. If $\widehat{\theta}$ (estimated mean of original data) is approximately normal, then $\widehat{\theta} \sim N(\theta + \beta, \nu)$. The confidence interval (*CI*) of θ for known bias ($\beta = \beta(F)$) and variance ($\nu = \nu\ (F)$) (Davison and Hinkley 1997) is given in:

$$CI = \widehat{\theta} - \beta \pm Z_\alpha . \nu^{1/2} \tag{1.18}$$

$$\text{where, } \beta(F) \doteq \beta\left(\widehat{F}\right) \doteq b = \overline{\widehat{\theta*}} - \widehat{\theta}, \tag{1.19}$$

$$\nu(F) \doteq \nu\left(\widehat{F}\right) \doteq \upsilon = \frac{1}{N-1}\sum_{i=1}^{N}\left(\widehat{\theta_i^*} - \overline{\widehat{\theta*}}\right)^2 \tag{1.20}$$

$$\text{at 95 \% confidence interval, } \alpha = 0.025,\ Z_\alpha = -1.96$$

The variable $\overline{\widehat{\theta*}}$ is the mean of $\widehat{\theta_1^*}, \widehat{\theta_2^*}, \ldots\ldots, \widehat{\theta_N^*}$ and Z_α is the α quantile of the standard normal distribution.

In a recent study, uncertainty assessment of mean precipitation extremes is performed by Teegaravapu et al. (2013) using bootstrap sampling methodology. The 95 % confidence intervals using bootstrap resampling with normal approximation are computed using a total of 10,000 bootstrap samples that are obtained from data for each phase. The flowchart shown in Fig. 1.9 provides the steps required to estimate confidence intervals. These intervals can be computed at different spatial scales (individual rain gauges, different homogeneous rainfall areas and regions). Teegaravapu et al. (2013) report that in general considering the lower limit, the higher values of extremes are observed in AMO warm phase than in the cool phase for all durations, and for the upper limit, higher precipitation extremes are realized in AMO cool phase compared to those in AMO warm phase up to a 12-h duration. However, for durations equal or above 24 h, precipitation extremes are higher in warm phase compared to those in cool phase.

1.9 Wavelet-Based Methods and Analysis

Wavelet analysis can be used to evaluate the temporal patterns of oscillations and precipitation (Goly and Teegavarapu 2014). Continuous wavelet transform (CWT) allows the study of the temporal structure of precipitation and makes inferences on the influence of oscillation patterns. An example of the continuous wavelet transform power spectrum plots of total precipitation during dry season for continental and peninsular regions of Florida, USA, adopted from a recent study by Goly and Teegavarapu (2014) is shown in Fig. 1.9. The spectrum plots can be compared to those of AMO and ENSO oscillations mean index during dry season. A common time span of 1915–2011 is used for generation of the plots for

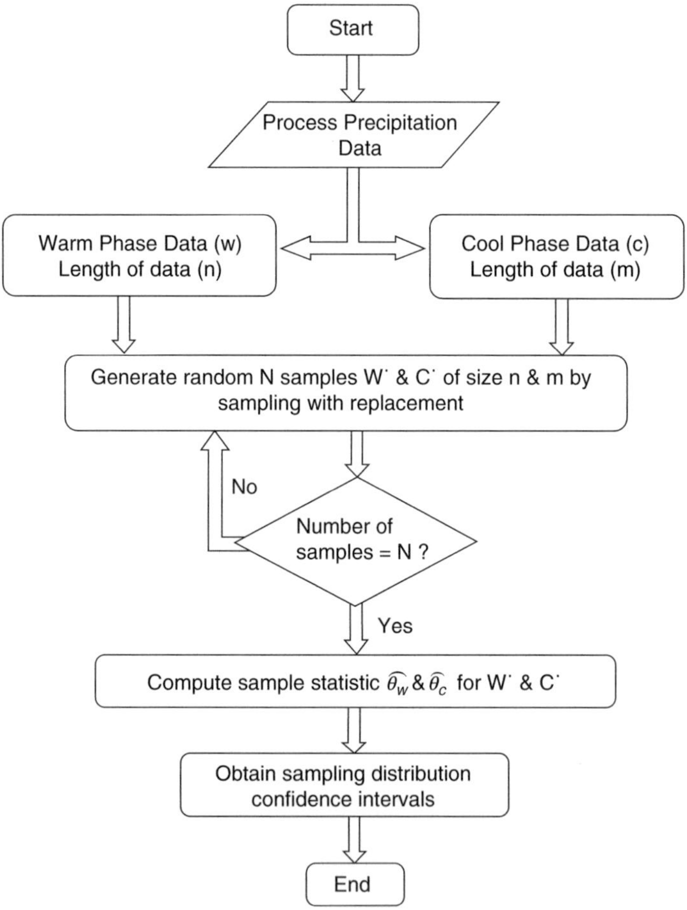

Fig. 1.9 Bootstrap sampling approach for the determination of confidence intervals (Adapted from Teegaravapu et al. 2013)

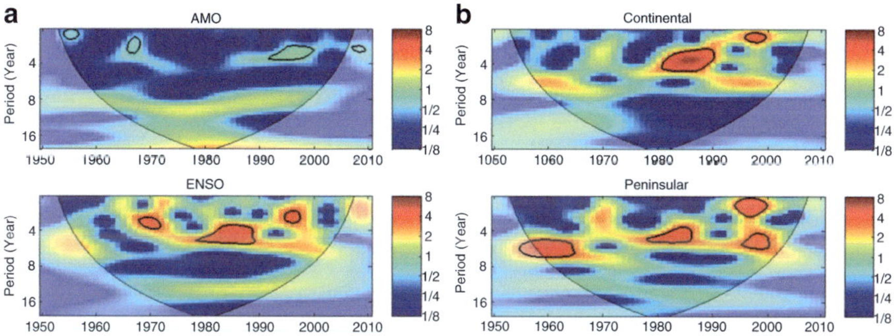

Fig. 1.10 Continuous wavelet power spectrum of (**a**) oceanic–atmospheric oscillation climate indices (dry season) and (**b**) precipitation (dry season) for continental and peninsular regions of Florida. The *thick black* contours designate the 5 % significance level against red noise (Adapted with permission from Goly and Teegavarapu 2014)

both the oscillation indices. The statistical significance of the peaks in the wavelet spectrum was tested using Monte Carlo methods against a lag-1 autoregressive red noise background. The peaks with greater than 95 % confidence interval are shown by thick black contours. AMO with a cycle of approximately 70 years is clearly apparent in this wavelet power spectrum plot shown in Figure 1.10. Similar plot for ENSO shows a much smaller wavelength with a recurrence period of 2–7 years. Significant power in this band was observed during 1940–1960, 1965–1972, 1978–1990, and 1996–1999 years in the wavelet spectrum of ENSO. A look at each power spectrum plot will suggest that in both continental and peninsular regions of state of Florida, significant power at the 5 % significance level is evidenced and coincident with the patterns exhibited by ENSO is noted.

1.10 Regional Hydroclimatology Influences

In many regions around the world, local hydroclimatology may play a major role in restricting spatial and temporal influences of oscillations. Temporal shift in the occurrences of extremes from cool to warm phases or vice versa and distribution of intra-annual extremes will have impact on operation of hydrologic and hydraulic structures. Synthetic design storms generally used for hydrologic design and DDF curves need to be revisited and revised considering the occurrences and magnitudes of precipitation extremes in different phases of oscillations.

1.11 Influences of Individual and Coupled Oscillations

Two or multiple teleconnections influencing regional hydrology simultaneously in specific temporal windows may increase or decrease the frequency and magnitudes of extreme precipitation events and influence intra-annual temporal occurrences of extremes. Teegaravapu et al. (2013) suggest that emphasis should be placed on those temporal windows in which the combined influences of two or more teleconnections lead to rare extremes and data selection for design should be representative of these extremes. Site-specific extremes for hydrologic design confined to one specific region are routinely carried out. However, consideration of region-specific influences of climate variability at different spatial and temporal scales is recommended. Goly and Teegavarapu (2014) in a recent work documented an exhaustive study of combined influences of AMO and ENSO on precipitation extremes and characteristics in the state of Florida, USA. They conclude that precipitation extremes and characteristics are influenced by the two-coupled oceanic–atmospheric phenomena, with seasonally and spatially varying signatures in Florida. Essential climatic variables in many regions across the world are influenced more than one oscillation. A few examples of such multiple influences in different regions of the world include PDO and ENSO in Japan; ENSO, IOD, and MJO in India; and PDO, AMO, ENSO, and NAO in the USA. When multiple oscillations influence the precipitation regimes in a particular region, it is often difficult to evaluate combined influences or associate a particular pattern or change in precipitation to one specific oscillation. This is mainly due to lack of long-term precipitation data and the varying and overlapping temporal windows of intra-year, decadal, and quasi-decadal and multidecadal oscillations.

1.12 Influences of Oscillations on Precipitation: Spatial Extent

Regional and global influences of oscillations on precipitation and temperature extremes are documented in several studies (e.g., Teegavarapu 2013). However, in any given region, the spatial extent of any oscillation is not clearly delineated due to lack of dense observational network of rain gauges or reliable gridded data at a spatial resolution that is adequate to define the spatial extent of the influences. In recent studies availability of gridded precipitation data has helped in specifying spatial extents of influences. Teegaravapu et al. (2013) and Goly and Teegavarapu (2014) documented spatially uniform and nonuniform influences of ENSO and AMO in the state of Florida, USA, respectively, using gridded precipitation data. In the case of AMO, differences in the nature of influences especially in spatial extents in two phases were noted due to thermic and hyperthermic (in continental and peninsular regions) soil regimes in the state of Florida. In few regions of the world (e.g., the southeastern USA, Japan, and India), the paths of hurricane or cyclone landfalls will influence the spatial variability of precipitation extremes. In the southern part of Florida, USA, precipitation extremes of longer durations are limited to regions that frequently experience hurricane landfalls.

1.13 Meteorologically Homogenous Areas

Evaluation of climate variability influences can be evaluated for regions that are classified as meteorologically homogeneous areas. These areas are developed by hydrologists and meteorologists based on information about the regional variations of precipitation, temperature, and other hydroclimatic variables. Teegaravapu et al. (2013) have used meteorologically homogeneous areas developed by a local water management agency in South Florida for evaluation of influences of AMO on precipitation extremes and characteristics. They found that there are differences in how AMO influences precipitation extremes in different areas. In many instances, Köppen–Geiger climate classification (Kottek et al. 2006) for a specific region can be beneficial in delineating the region into several homogenous climate zones.

1.14 Relating Indices and Precipitation Depths

Relationship between monthly oscillation indices of AMO, ENSO, PDO, and others and precipitation depths can be established and evaluated. A simple correlation-based analysis using monthly precipitation totals and lagged values of indices can reveal useful relationships that can help in forecasting seasonal precipitation totals. Spatially and temporally variable SST anomalies and variations in other climatic variables in different regions of the world can be used to establish the links.

1.15 Forecasts Based on Oscillations

Information about the hydroclimatic variables and their links to coupled oceanic–atmospheric oscillations can be beneficial to water management agencies involved in the planning for future water uses. Strong relationships between rainfall occurrences and streamflows have been observed in several parts of the world with variables that relate to the manifestation of teleconnections. Strong correlations between sea surface temperatures (SSTs) and streamflows (Dawod and El-Rafy 2002; Beek 2010) are examples of such relationships. Dawod and El-Rafy also report the links between the annual River Nile flows and SSTs at different locations in Indian Ocean and Pacific Ocean. An example of predictability of the flow is in the next hydrologic year (Beek 2010) using information available at the end of June. A high correlation between predicted set of flows and observed flows suggests the utility of multiple linear regression equation linking flow and SSTs. Seasonal forecasts using climate change information are linked to analogue years through the use of historical climate records (Ludwig 2009). ENSO is strongly correlated with one or more hydroclimatic variables in several regions around the globe. These strong correlations can be used for seasonal rainfall and streamflow forecasts. Souza and Lall (2003) report the utility of using NINO3.4 (an indicator used to define the ENSO state) and North Atlantic Dipole index to forecast streamflows in northeast Brazil. Australian rainfall amounts were linked to El Niño by Chiew et al. (1998) and Southern Oscillation (SO) index (Chiew et al. 2003). Similar links in general will help in futuristic seasonal to yearly forecasts of rainfall helping water resources management professionals. In general teleconnections can be used for seasonal climate forecasts with benefits to water resources management. Water allocation can be improved if seasonal forecasts are available (Stone et al. 1996), and similarly future rainfall forecasts can be extremely beneficial to agrarian communities especially in arid and semiarid regions. Seasonal forecasts may help in water pricing and also developing plans for water use restrictions (Ludwig 2009).

1.16 Conclusions

Evaluation of influences of individual and coupled inter-year, decadal and multi-decadal, coupled oceanic–atmospheric oscillations on precipitation characteristics and extremes is the focus of this chapter. Precipitation regime changes are known to be influenced by several coupled oceanic–atmospheric oscillations around the globe. Exhaustive evaluation of individual and combined influences of these oscillations, descriptive extreme index-based assessment of precipitation extremes and changes in rainfall characteristics, identification of spatially varying influences of oscillations on dry and wet spell transition states, antecedent precipitation prior to extreme events, intra-event temporal distribution of precipitation, and changes in temporal occurrences of extremes is discussed in this chapter. Understanding these oscillations and their influences focusing on different spatial scales and temporal resolutions considering multiple duration-specific precipitation extremes is critical for flood control, water supply management, and hydrologic design. Parametric and nonparametric statistical tests using long-term precipitation data at point and grid scale can be used to assess statistically significant changes in the precipitation characteristics from one phase to another of each oscillation. Understanding of influences of oscillations on precipitation variability with regional hydroclimatology defining the spatial extent of these influences is critical for hydrologic analysis, design, and flood control management. This chapter presented an overview of oscillations and their possible influences on precipitation characteristics and extremes. Wherever possible, results from precipitation data analysis from regions influenced by single and multiple oscillations are used to explain the nature of variability in precipitation extremes and characteristics.

Acknowledgments The author would like to acknowledge the assistance provided by Ms. Milla Pierce in the preparation of this chapter.

References

Adams BJ, Howard CDD (1986) Pathology of design storms. Canad Water Resour J 11(3):49–55

Adams BJ, Papa F (2000) Urban stormwater management planning with analytical probabilistic models. Wiley, New York

Alexandersson HA (1986) Homogeneity test applied to precipitation data. J Climatol 6(6):661–675

Beek EV (2010) Managing water under current climate variability. In: Ludwig F, Kabat P, Schaik H, Valk M (eds) Climate change adaptation in water sector. Earthscan, London

Behera P, Guo Y, Teegavarapu RSV, Branham T (2010) Evaluation of antecedent storm event characteristics for different climatic regions based on inter-event time definition (IETD). In: Proceedings of the world environmental and water resources congress 2010, ASCE Conference Proceedings. doi:10.1061/41114(371)251

Box GEP, Cox DR (1964) An analysis of transformations. J R Stat Soc 26(2):211–252

Buishand TA (1982) Some methods for testing the homogeneity of rainfall records. J Hydrol 58(1–2):11–27

Chernick MR (2007) Bootstrap methods: a guide for practitioners and researchers. Wiley-Interscience, Hoboken

Chiew FHA, Piechota TC, Dracup JA, McMahon TA (1998) El Niño southern oscillation and Australian rainfall, streamflow and drought: links and potential for forecasting. J Hydrol 204:138–149

Chiew FHS, Zhaou SL, McMahon TA (2003) Use of seasonal streamflow forecasts in water resources management. J Hydrol 270:135–144

Corder WG, Foreman DI (2009) Nonparametric statistics for non-statisticians. Wiley, Hoboken

Cronin TM (2009) Paleoclimates: understanding climate change past and present. Columbia University Press, New York

Dai A, Fung J, Del Genio A (1997) Surface observed global land precipitation variations during 1900–1988. J Clim 11:2943–2962

Davison AC, Hinkley DV (1997) Bootstrap methods and their applications

Dawod MAA, El-Rafy MA (2002) Towards long range forecast of Nile Flood, Proceedings of the fourth conference on meteorology and sustainable development. Meteorologist Specialist Association, Cairo

Efron B (1979) Bootstrap methods: another look at the jackknife. Ann Stat 7:1–26

Efron B, Gong G (1983) A leisurely look at the bootstrap, the jackknife, and cross validation. Am Stat 37:36–48

Efron B, Tibshirani R (1993) An introduction to the bootstrap. Chapman and Hall, London

Enfield DB, Mestas-Nunez AM, Trimble PJ (2001) The Atlantic multidecadal oscillation and its relation to rainfall and river flows in the continental U.S. Geophys Res Lett 28(10):2077–2080

Ghil M (2002) Natural climate variability. In: MacCraken MC, Perry JS (eds) Encyclopedia of global environmental change—Volume 1, The earth system—physical and chemical dimensions of global environmental change. Wiley, Chichester, pp 544–549

Goly A, Teegavarapu RSV (2012) Influence of teleconnections on spatial and temporal variability of extreme precipitation events in Florida. World Environ Water Resour Congress 2012:1899–1908. doi:10.1061/9780784412312.190

Goly A, Teegavarapu RSV (2014) Individual and coupled influences of AMO and ENSO on regional precipitation characteristics and extremes. Water Resour Res 50. doi: 10.1002/2013WR014540

Gurdak JJ, Hanson RT, Green TR (2009) Effects of climate variability and change on groundwater resources of the United States, Fact Sheet 2009–3074., United States Geological Survey

Helsel DR, Hirsch RM (2002) Statistical methods in water resources, Book 4, hydrologic analysis and interpretation, United States Geological Survey
Hurrell JW, Van Loon H (1995) Decadal Variations in Climate Associated with the North Atlantic Oscillation. National Center for Atmospheric Research, Boulder
Jain D, Singh VP (1987) Comparison of some flood frequency distributions using empirical data. In: Singh VP (ed) Hydrologic frequency modeling. Reidel, Dordrecht, pp 467–485
Jarque CM, Bera AK (1987) A test for normality of observations and regression residuals. Int Stat Rev 55(2):163–172
Koch-Rose M, Berry L, Bloetscher F, Hernández Hammer N, Mitsova-Boneva D, Restrepo J, Root T, Teegavarapu R (2011) Florida water management and adaptation in the face of climate change. Florida Climate Change Task Force
Kottek M, Grieser J, Beck C, Rudolf B, Rubel F (2006) World map of Köppen-Geiger climate classification updated. Meteorol Z 15(3):259–263
Lilliefors HW (1967) On the Kolmogorov-Smirnov test for normality with mean and variance unknown. J Am Stat Assoc 62(318):399–402
Ljung G, Box G (1978) On a measure of lack of fit in time series models. Biometrika 65:297–303
Ludwig F (2009) Using seasonal climate forecasts for water management. In: Ludwig F, Kabat P, Schaik HV, Van Der Valk M (eds) Climate change adaptation in the water sector. Earthscan Publishers, London
Mage DT (1982) An objective graphical method for testing normal distributional assumptions using probability plots. Am Stat 36(2):116–120
Mantua NJ, Hare SR, Zhang Y, Wallace J, Francis R (1997) A Pacific interdecadal oscillation with impacts on salmon production. Bull Am Meteorol Soc 78:1069–1079
Mantua NJ, Hare SR (2002) The Pacific decadal oscillation. J Oceanogr 58(1):35–44
McBean AE, Rovers FA (1998) Statistical procedures for analysis of environmental monitoring data and risk assessment. Prentice Hall, Upper Saddle River
McKee TB, Doesken NJ, Kleist J (1995) Drought monitoring with multiple timescales. In: Proceedings of the ninth conference on applied climatology, Dallas, TX, 15–20 January 1995. Boston American Meteorological Society, pp 233–236
McKee TB, Doesken NJ, Kleist J (1993) The relationship of drought frequency and duration to time scale. In: Proceedings of the eighth conference on applied climatology, Anaheim, California, 17–22 January 1993. American Meteorological Society, Boston, pp 179–184
Newman M, Compo G, Alexander M (2003) ENSO-forced variability of the Pacific decadal oscillation. NOAA-CIRES Climate Diagnostic Center, Boulder. NOAA. 013
Pettitt AN (1979) A non-parametric approach to the change-point problem. Appl Stat 28(2): 126–135
Pierce M (2013) Influences of decadal and multidecadal oscillations on regional precipitation extremes and characteristics, Thesis, Florida Atlantic University
Rosenzweig C, Hillel D (2008) Climate variability and the global harvest: impacts of El Nino and other oscillations on agroecosystems. Oxford University Press, New York
Satterthwaite FE (1946) An approximate distribution of estimates of variance components. Biom Bull 2:110–114
Schneider N, Cornuelle B (2005) The forcing of the Pacific decadal oscillation. J Clim 18(2005):4255–4273
Sheskin DJ (2003) Handbook of parametric and nonparametric statistical procedures. Chapman and Hall/CRC, Boca Raton, FL
Smirnov NV (1939) On the estimation of the discrepancy between empirical curves of distribution for two independent samples. (Russian). Bull Moscow Univ 2:3–16
Souza FA, Lall U (2003) Seasonal to interannual ensemble streamflow forecasts for Ceara, Brazil: application of a multivariate, semiparametric algorithm. Water Resour Res 39
Stone RC, Hammer GL, Marcussen T (1996) Prediction of global rainfall probabilities using phases of the Southern Oscillation Index. Nature 384:252–255
Svensson C, Jones D (2010) Review of rainfall frequency estimation methods. J Flood Risk Manag 3:296–313

Teegaravapu RSV, Goly A, Obeysekera J (2013) Influences of Atlantic multidecadal oscillation phases on spatial and temporal variability of regional precipitation extremes. J Hydrol 495: 74–93

Teegavarapu RSV (2009) Estimation of missing precipitation records integrating surface interpolation techniques and spatio-temporal association rules. J Hydroinf 11(2):133–146

Teegavarapu RSV (2010) Modeling climate change uncertainties in water resources management models. Environ Model Softw 25(10):1261–1265

Teegavarapu RSV (2012) Spatial interpolation using non-linear mathematical programming models for estimation of missing precipitation records. Hydrol Sci J 57(3):383–406

Teegavarapu RSV (2007) Use of universal function approximation in variance-dependent interpolation technique: an application in Hydrology. J Hydrol 332:16–29

Teegavarapu RSV (2013) Floods in a changing climate: extreme precipitation. Cambridge University Press, Cambridge, UK

Teegavarapu RSV (2014) Statistical corrections of spatially interpolated missing precipitation data estimates. Hydrol Process 28(11):3789–3808

Teegavarapu RSV (2016) Evaluation of trends and variability in monthly sea level, temperature and precipitation of Japan: links to climate variability and change. Technical Report, Research Ceneter for Urban Safety and Security (RCUSS). Kobe University, Kobe

Teegavarapu RSV, Chandramouli V (2005) Improved weighting methods, deterministic and stochastic data-driven models for estimation of missing precipitation records. J Hydrol 312:191–206

Teegavarapu RSV, Nayak A, Pathak C (2011) Assessment of long-term trends in extreme precipitation: implications of in-filled historical data and temporal window-based analysis. In: Proceedings of the 2011 world environmental and water resources congress, doi:10.1061/41173(414)419

USGS (2011) Evidence of multidecadal climate variability in the Gulf of Mexico. USGS

Von Neumann J (1941) Distribution of the ratio of the mean square successive difference to the variance. Ann Math Stat 12:367–395

Wallace JM, Gutzler DS (1980). Teleconnections in the geopotential height field during the northern hemisphere winter. Department of Atmospheric Sciences, University of Washington, Monthly Weather Review

Walsh PD, Lawler DM (1981) Rainfall seasonality: description, spatial patterns and change through time. Weather 36:201–208

Welch BL (1947) The generalization of student's problem when several different population variances are involved. Biometrika 34(1–2):28–35

WMO (2009) Guidelines on analysis of extremes in a changing climate in support of informed decisions for adaptation. WMO, Geneva

WMO (2012) Standard precipitation index user guide, World Meteorological Organization (WMO), WMO-No: 1090. WMO, Geneva

Zhang Y, Wallace J, Battisti D (1996) ENSO-like interdecadal variability: 1900–93. Department of Atmospheric Sciences, University of Washington, Seattle, Washington

Chapter 2
Hydrologic Extremes Under Climate Change: Non-stationarity and Uncertainty

Arpita Mondal and P. P. Mujumdar

Abstract Hydrologic designs, hazard mitigation, and water resources management rely heavily on frequency analysis and risk assessment of extremes such as floods and droughts. Return periods and corresponding return levels of such extremes have been traditionally derived under the assumption of stationarity that has been challenged by recent studies. The presence of non-stationarity, due to various natural or anthropogenic causes, necessitates accurate modeling of the time-varying behavior of extremes, frequency analysis taking time evolution of statistical distributions into consideration, and reformulation of the definition of hydrologic risk under transient conditions. This chapter synthesizes various methodologies for investigating climate change-induced non-stationarity in hydrologic extremes using the statistical extreme value theory. Information on available computational packages to apply such methodologies is provided. Additionally, some fundamental limitations of such methodologies for deterministically modeling real-world observed time series are also discussed. Further, through an illustrative example, modeling approaches for accounting for the effects of non-stationarity in peak flows and uncertainties in the assessment of hydrologic risk under non-stationary conditions vis-à-vis traditional stationary analysis are discussed.

Keywords Extreme value analysis • Floods • Non-stationarity • Climate change • Uncertainty • Hydrologic risk • Non-stationary design levels • Return period • Probability of failure under non-stationarity

---

A. Mondal (✉)
Department of Civil Engineering, Indian Institute of Technology Bombay, Powai, Mumbai 400076, India
e-mail: arpita567@gmail.com

P.P. Mujumdar
Department of Civil Engineering, Indian Institute of Science, Bangalore 560012, India
e-mail: pradeep@civil.iisc.ernet.in

E. Kolokytha et al. (eds.), *Sustainable Water Resources Planning and Management Under Climate Change*, DOI 10.1007/978-981-10-2051-3_2

2.1 Introduction

Assessment and evaluation of environmental risk are major challenges faced by the society. In particular, hydrologic extremes such as floods and droughts affect people's lives and properties. While such extremes are difficult to understand due to their complexity and rarity, their impact is reportedly increasing as the environment changes and population and industrialization grow, exposing more human settlements to natural hazards. Changes in hydrologic extremes can be caused by rapid modifications in land use/land cover, human interventions, and climate change (Sivapalan and Samuel 2009). Such changes imply that the past can no longer be used as a guide to the future and has led to the belief that "stationarity is dead" (Milly et al. 2008). In particular, climate change is believed to possibly increase the frequency and magnitude of extreme events (IPCC 2012). Global warming may lead to an intensification of the hydrologic cycle owing to the greater moisture holding capacity of the atmosphere (Meehl et al. 2007). This increases the likelihood of more frequent high-precipitation events that influence the risk of pluvial flooding. On the other hand, increasing variability of precipitation and increased evapotranspiration because of rising temperatures may lead to more frequent droughts. Indeed, the risk of both floods (Hirabayashi et al. 2013; Kundzewicz et al. 2010) and droughts (Dai 2013) is reported to be increasing under climate change projections in the twenty-first century. An increasing number of studies in hydrologic literature therefore use future forcing-based scenarios to replace the assumption of stationarity in order to assess future risk of hydrologic extremes. While human-induced climate change has reportedly led to intensification of extreme temperature or precipitation events for the present as well as the future (Coumou and Rahmstorf 2012), only low to medium confidence exists in the projections of floods and droughts, owing to modeling limitations and lack of good quality records at local or regional scales (IPCC 2012). Moreover, understanding and attributing observed changes in hydroclimatic extremes to anthropogenic climate change are particularly challenging due to large natural variability and poor simulation of human-induced signals by the climate models at regional scales (Mondal and Mujumdar 2012, 2015a). Therefore, a pragmatic and holistic approach is necessary to quantify changes in the environment and how that influences designs (Montanari and Koutsoyiannis 2014). Indeed, in recognizing the caveats in hydrologic frequency analysis under non-stationarity, some recent studies caution researchers by mentioning that "stationarity is immortal" (Montanari and Koutsoyiannis 2014) or "stationarity is undead" (Serinaldi and Kilsby 2015).

Traditionally, hydrologic designs rely on tail quantiles of extremes, such as the "100-year flood." The concept of return period and return level has been put into practice widely in hydrology. For example, dike design in the Netherlands is based on the 10,000-year flood level that represents the magnitude of flood that is expected to occur on an average once every 10,000 years. However, the relationship between the probability of exceedance and the return period is no longer constant in the presence of non-stationarity. For example, if the floods increase (decrease) with time because of increasing extreme rainfall associated with climate change, the 100-

year flood in a 2015 climate may be expected to occur on an average more (less) frequently in a 2050 climate. Non-stationarity influences not only the magnitudes of tail quantiles but also the uncertainties associated with their estimation. Given such uncertainties and the assumptions involved, therefore, it remains to be seen whether non-stationary models are indeed more suitable for characterizing transient extremes as compared to the traditional concepts based on stationarity and whether that leads to more robust assessment of risk of a hazardous event.

This chapter discusses the applications of statistical extreme value theory to characterize non-stationarity in hydrologic extremes, notes the limitations of such methods, presents the different definitions of hydrologic risk under non-stationarity and the uncertainties associated with them, and finally synthesizes the collective know-how on this topic with a view to aid practitioners and water managers.

2.2 Extreme Value Theory for Hydrologic Frequency Analysis

For hydrologic designs, one needs to look at the frequency distributions of extremes such as the peak flows or low flows. Extremes behave differently from the bulk of the distribution and can be modeled separately within the extreme value theory (EVT) framework that follows from stabilization of the distribution of the extremes such as the maximum of a block of values. The family of extreme value distributions is typically characterized by three parameters, the location μ, the scale σ, and the shape ξ, with the shape parameter determining the tail behavior. Under non-stationarity, these parameters can vary as a function of a physical covariate or time. Though some studies transform hydrologic extremes data to normal distribution and then apply standard regression techniques to model transient extremes (Vogel et al. 2011), EVT provides a more meaningful approach (Katz et al. 2002) to account for the tail behavior of extremes. Several studies in the recent times employ EVT techniques to model non-stationarity in hydroclimatic extremes (Katz 2013; Mondal and Mujumdar 2015b, c; Towler et al. 2010; Westra et al. 2013, etc.). However, with the exception of Serinaldi and Kilsby (2015), a comprehensive analysis bringing into focus the issues and uncertainties in modeling non-stationary extremes vis-à-vis traditional approaches based on stationarity is lacking in hydrologic literature.

In the following subsections, a background is provided on applications of EVT to model hydrologic data first for the stationary conditions. The next section discusses the extension of these models and methods for the non-stationary conditions.

2.2.1 Extreme Value Models for Hydrologic Data

Statistical extreme value models for hydrologic data can be broadly divided into two groups – the block maxima approach that considers the maximum of a sufficiently large block of values as extreme and the peak-over-threshold (POT) approach which defines as extremes the values that lie above a sufficiently high threshold. For example, if one starts with daily rainfall data and considers an annual block, the annual maximum daily rainfall denotes a block maximum, while the rainfall values in the entire record that are above the 99th percentile represents the peak-over-threshold values. It may be noted that the POT approach allows characterization of the intensity, duration, as well as frequency of extreme events (Mondal and Mujumdar 2015b). An illustrative example for defining hydrologic extremes under the block maxima and the POT approach is given in Fig. 2.1. In this example, starting with daily precipitation values, a block is represented by a month, and two consecutive blocks are shown in the figure. The maximum value for each month constitutes the block maximum, as shown. The POT approach, on the other hand, considers all precipitation values that are above the threshold. An extreme precipitation spell is defined as a time period during which precipitation values are above the threshold till at least one daily precipitation value goes below the threshold. Therefore, in the first block shown in Fig. 2.1, there are three extreme precipitation spells (frequency = 3). The durations of these three spells are one day, one day, and four days, respectively. However, consecutive precipitation values

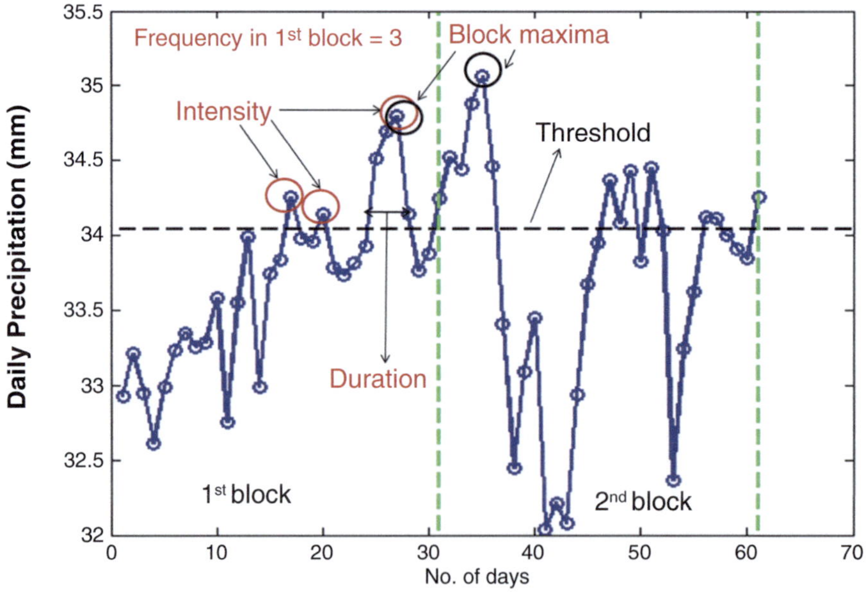

Fig. 2.1 A typical example of characterization of extreme daily precipitation. Two blocks of duration 1 month each is considered. The block maxima consists of two values for each block as shown. In the POT approach, an extreme precipitation spell is defined as a period when one or more consecutive daily precipitation values are above the threshold. Intensity or threshold exceedance is the maximum of each spell. Frequency is given by the number of spells in each block, while duration represents the length of the spell (refer to text)

above the threshold represent the same spell and are dependent on each other. Therefore, following declustering (Coles 2001), only the maximum value in each spell is considered as the intensity or the "threshold exceedance" value.

The threshold exceedances follow the generalized Pareto (GP) distribution, while Poisson and geometric distributions are used to model the frequency and duration of extreme precipitation spells. It is to be noted that the POT method is sensitive to the choice of threshold. In addition to the block maxima and POT approaches, the point process approach that models the threshold exceedances and the rate of their occurrence (frequency) simultaneously can also be used to characterize extremes; however, its application is limited in hydrologic literature. In this chapter, subsequent discussions are focused only on the block maxima approach for brevity, though the concepts can be extended to the POT approach.

For the block maxima approach, the extremal type theorem (Coles 2001) states that a sample block maxima (e.g., annual maximum daily precipitation) can be rescaled to follow one of the distributions constituting the family of extreme value distributions, termed the generalized extreme value (GEV) distributions, with three types widely known as the Gumbel (unbounded light tail), Frechet (heavy tail), and Weibull (bounded tail) families that can be given by

$$G(x;\ \mu,\ \sigma,\xi) = \begin{cases} \exp\left[-\exp\left(-\frac{x-\mu}{\sigma}\right)\right],\ \xi=0 \\ \exp\left[-\left(1+\xi\frac{x-\mu}{\sigma}\right)^{-\xi^{-1}}\right],\ \xi\neq 0,\ 1+\xi\frac{x-\mu}{\sigma}>0\,. \end{cases} \quad (2.1)$$

where μ, $\sigma>0$, and ξ are the location, scale, and shape parameters, respectively, as mentioned earlier. Regardless of the distribution of the original random variable (e.g., daily precipitation), this formulation holds. As is well known, the quantiles of interest or return levels, $z_{(1-p)}$, corresponding to the probability of exceedance p and the return period m ($m=1/p$), can be obtained by inversion of the cumulative distribution function $G(x)$.

$$z_{(1-p)} = \begin{cases} \mu-\frac{\sigma}{\xi}\left[1-(-\log(1-p))^{-\xi}\right]\ for\ \xi\neq 0 \\ \mu-\sigma\ \log(-\log(1-p))\ for\ \xi=0 \end{cases} \quad (2.2)$$

The performance of the fitted models, in particular, the ability of the fitted GEV distribution to capture the tail behavior of the observed extreme series, can be checked by diagnostics such as the quantile or probability plots.

2.2.2 Parameter Estimation and Uncertainty in Design Levels

Several methods can be used for estimation of the parameters $\beta=(\mu,\ \sigma,\ \xi)$ of the GEV distribution. These include the method of least squares, the method of moments, L-moment-based methods, the method of maximum likelihood estimation (MLE), and Bayesian techniques, out of which only the last two methods allow extensions to the non-stationary conditions. MLE is popular because of its simplicity in implementation (Katz 2013) and will be discussed further in this chapter. If g is the probability density function of the GEV distribution, i.e., the derivative of $G(x)$ with respect to x, then the likelihood function is given by

$$L(\beta) = \prod_{i=1}^{k} g(x_i;\ \beta) \tag{2.3}$$

where k is the total number of maxima in the record. The point estimates of the parameters are obtained by maximizing this likelihood function. Because no analytic solution to the optimization problem exists, the MLE approach uses numerical routines. Defining $y_+ = \max\{y, 0\}$, and letting $x_{1,} \ldots,\ x_k$ represent the block maxima consisting of k blocks each of length n, the log-likelihood for the GEV distribution is given by

$$l(\mu,\ \sigma,\ \xi;\ x_{1,} \ldots,\ x_k) = -k \ln \sigma - \left(1 + \frac{1}{\xi}\right)$$

$$\sum_{i=1}^{k} \ln\left[1 + \xi\, \frac{x_i - \mu}{\sigma}\right]_+ - \sum_{i=1}^{k} \left[1 + \xi\, \frac{x_i - \mu}{\sigma}\right]_+ \tag{2.4}$$

For the Gumbel case, a distribution (a separate log-likelihood function can be derived for that case) is first fitted and then the hypothesis $\xi = 0$ is tested against the alternative $\xi \neq 0$. The point estimate of $\beta = (\widehat{\mu},\ \widehat{\sigma},\ \widehat{\xi})$ is then used to arrive at the point estimates of the quantiles, $\widehat{z}_{(1-p)}$. Clearly, these are point estimates and should be communicated along with their standard errors of estimation resulting from sampling uncertainty.

The Bayesian method for parameter estimation yields uncertainty bounds of the quantiles directly. Within the frequentist framework, various methods can be used for estimating the uncertainty associated with the quantiles or return levels. They include the profile likelihood method that is less popular in hydrology (Obeysekera and Salas 2014), bootstrap resampling-based methods that are computationally expensive (Serinaldi and Kilsby 2015), and the relatively simple delta method (Oehlert 1992) that can provide reasonable uncertainty bounds (Serinaldi and Kilsby 2015). By the delta method (Oehlert 1992), the variance of this return level is given by

$$\mathrm{Var}\left(\widehat{z}_{(1-p)}\right) \approx \nabla z_{(1-p)}{}^{T}\ V\ \nabla z_{(1-p)} \tag{2.5}$$

where V is the variance-covariance matrix of $(\mu, \ \widehat{\sigma}, \widehat{\xi})$ and using $y_p = -\log(1-p)$ and superscript T to denote transpose of matrix

$$\nabla z_{(1-p)}{}^{T} = \left[\frac{\partial z_{(1-p)}}{\partial \mu}, \frac{\partial z_{(1-p)}}{\partial \sigma}, \frac{\partial z_{(1-p)}}{\partial \xi} \right]$$
$$= \left[1, -\xi^{-1}\left(1 - {y_p}^{-\xi}\right), \sigma\xi^{-2}\left(1 - {y_p}^{-\xi}\right) - \sigma\xi^{-1}{y_p}^{-\xi} \log y_p\right] \quad (2.6)$$

evaluated at $(\mu, \ \widehat{\sigma}, \ \widehat{\xi})$. This variance can be used to compute the confidence interval of the return levels, assuming that the distribution of the quantiles is described by a Gaussian distribution. The 100(1−α) % confidence limits are therefore given by $\widehat{z}_{(1-p)} \pm \varphi_{1-\alpha/2}\widehat{\sigma}$ where φ represents standard Gaussian quantile.

2.3 Non-stationarity in Hydrologic Extremes

Typically, hydrologic data is separated into deterministic components such as trends and periodicities and random stochastic components. The presence of trends and change points is often considered to be rendering the time series non-stationary (Salas 1993). For extremes, therefore, several studies employ nonparametric tests such as the Mann-Kendall trend test (Clarke 2013; Westra et al. 2013) or the Pettitt change point detection test to investigate whether significant changes are discernible in the observed record. The presence of non-stationarity in hydrologic extremes can also be detected by building a non-stationary extreme value model and testing for its significance.

2.3.1 Non-stationary Extreme Value Models

Under non-stationarity, one or more parameters of the extreme value distribution can vary with a covariate such as time. For example, in the case of annual maxima, the location, scale, and shape parameters of the GEV distribution can be expressed as functions of time t in the following form:

$$\mu(t) = \mu_0 + \mu_1\, t, \ \sigma(t) = \exp\left(\sigma_0 + \sigma_1\, t\right), \ \xi(t) = \xi \quad (2.7)$$

where t denotes the year, and the exponential in the scale parameter ensures positive values. The shape parameter is kept constant in most studies as it is difficult to estimate and unlikely to vary with time (Coles 2001). Physically, for annual peak flows, increase in the location parameter might result from increasing extreme precipitation because of climate change or land-use processes such as urbanization,

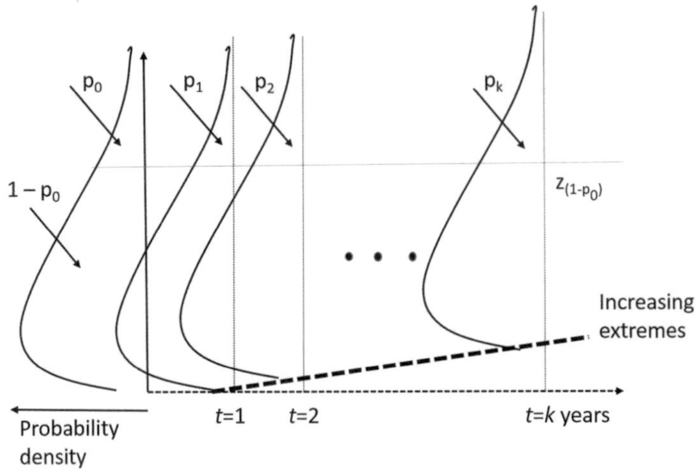

Fig. 2.2 Changing probability density function for increasing extremes. The probability of exceedance corresponding to a fixed initial quantile is shown to vary with time

while the scale parameter may vary with the variability of precipitation. An illustrative example of increasing extremes is provided in Fig. 2.2. The probability of exceedance corresponding to a fixed initial quantile $z_{(1-p_0)}$ is shown to increase with time. Alternately, the quantiles corresponding to a fixed probability of exceedance increase with time.

The likelihood ratio test is commonly used to test the suitability of the non-stationary GEV distribution. If M_o represents the stationary sub-model with constant $\mu,\ \sigma,$ and ξ, M_1 is the non-stationary model with parameters $\mu(t),\ \sigma(t),$ and ξ, and $l_o(M_o)$ and $l_1(M_1)$ are the maximized log-likelihoods for M_o and M_1, respectively, then the null hypothesis of M_o can be rejected in favor of validity of M_1 at a significance level of α if $2\{l_1(M_1)-l_o(M_o)\}> c_\alpha$ where c_α is the $(1-\alpha)$ quantile of the chi-square distribution whose degree of freedom equals the number of additional parameters in M_1. This test can also be used to formally establish trends in hydrologic extremes. For example, if the model $M_1 = \mathrm{GEV}\ (\mu = \mu_0 + \mu_1\ t,\ \sigma,\ \xi)$ is found significant against the stationary model $M_o = \mathrm{GEV}\ (\mu = \mu_0,\ \sigma,\ \xi)$, it would imply that the trend parameter μ_1 is significant, thereby leading to positive detection of trends (Clarke 2013).

For multiple candidate models (all nested), model selection criteria such as Akaike's information criterion (AIC) or the Bayesian information criterion (BIC) can be used. Through these criteria, the likelihoods are penalized for the number of parameters estimated. If any model M_i involves d parameters and the maximized log-likelihood for M_i is $l_i(M_i)$, then

$$\mathrm{AIC}\ (M_i) = 2\ li(Mi) + 2d,\ \ \mathrm{BIC}\ (M_i) = 2\ li(Mi) + d\ \ln\ k \qquad (2.8)$$

where k is the sample size (total number of years for this example on block maxima).

Under non-stationary conditions, diagnostics plots such the quantile and probability plots have to be modified to account for the non-homogeneity in distribution assumptions. Such a modification is achieved by transformation of the original variable $X_t \sim \text{GEV}\left(\widehat{\mu}(t), \widehat{\sigma}(t), \widehat{\xi}\right)$ to a stationary standardized variable given by (Coles 2001)

$$\check{X}_t = \frac{1}{\widehat{\xi}} \ln\left\{1 + \widehat{\xi}\left(\frac{X_t - \widehat{\mu}(t)}{\widehat{\sigma}(t)}\right)\right\} \tag{2.9}$$

that follows the standard Gumbel distribution with $\mu = 0$, $\sigma = 1$. The estimates of the parameters can then be substituted in Eq. (2.9), and a conventional quantile-quantile plot can be generated in terms of the transformed variable $\check{X}_t$ as compared to a standard Gumbel distribution. For estimation of the parameters, the likelihood function of (2.3) can be re-written for $\beta = (\mu_o, \mu_1, \sigma_o, \sigma_1, \xi)$ and maximized to arrive at the estimates of the parameters.

2.3.2 *Computational Packages for Analysis of Hydrologic Extremes Under Non-stationarity*

Many known computational packages for the analysis of hydrologic extremes under non-stationarity are available for the free and open-course statistical environment R. Some products were originally written for a similar commercial environment, S-plus (Coles 2001). A recent review on the available packages for analysis of extremes can be found in Gilleland et al. (2013), though it has to be kept in mind that this is a continuously evolving field. Table 2.1 provides the details (not exhaustive) on recent packages that allow incorporating non-stationarity into modeling of extremes. Among them, the package extRemes (Gilleland and Katz 2011) has a growing popularity because of its graphical user interface. The first package for non-stationary extreme value analysis for Matlab environment is NEVA (Cheng et al. 2014).

2.3.3 *Limitations of Non-stationary Models for Hydrologic Extremes*

With the above well-established background, it is worth discussing here some alternate views in hydrologic literature that criticize many aspects of the mentioned methods. Such fundamental critical points are listed below:

– First and foremost, the two fundamental assumptions necessary for any hydrologic frequency analysis are as follows (Klemes 2000): (1) the hydrologic

Table 2.1 Summary of computational packages for non-stationary extreme value analysis

Name of package	Environment	Estimation method	Multivariate capability
Ismev	R	MLE, LM	No
extRemes	R	MLE, LM, Bayesian (limited)	No
VGAM	R	MLE, BFA	Yes
Evd	R	MLE	Yes (limited)
Evdbayes	R	Bayesian	No
SpatialExtremes	R	MLE, MCLE, Bayesian	Yes
GenStat	Independent commercial	MLE	No
NEVA	Matlab	Bayesian	No
GAMLSS	R	PLE	No

MLE maximum likelihood estimation, LM L-moment method, BFA backfitting algorithm, PLE penalized likelihood estimation, MCLE maximum composite likelihood estimation

variable is an independent and identically distributed (2) random variable fully described by a continuous cumulative distribution function, and (3) the k-year observed record is a random sample from the distribution. Both these assumptions are believed to be not applicable to real-world complex hydrologic phenomena (Klemes 2000; 1986). It has in fact been argued that the frequency analysis techniques, with theoretically derived distributions, may not actually lead to any more robust estimation of hydrologic risk than a simple extrapolation of the age-old duration curves (Klemes 2000).

– Next, for the extreme value theory to hold for block maxima (say, annual maximum daily flows), it is assumed that the original (daily) data are iid distributed which definitely do not apply to daily flows. However, as counterargument, it is noted that the asymptotic extreme value distributions do not change even if the parent data are dependent, as long as certain relatively weak mixing conditions hold (Leadbetter et al. 1983). This is applicable to the stationary condition. Whether such mixing conditions hold under non-stationary conditions and whether at every time step the time-varying GEV distribution conforms to the family given by Eq. (2.1) are most often left unchecked.
– The non-stationary model M_1 always contains additional parameters as compared to the stationary model M_o. If the model M_1 captures the non-stationary behavior of hydrologic extremes well enough, it will lead to less biased simulations of future conditions for extrapolation beyond the range of observed data. However, the variance of the estimates will increase owing to larger number of parameters (Montanari and Koutsoyiannis 2014). Effective and robust ways to address this bias-variance tradeoff are conspicuously lacking in hydrologic literature.
– In stationary conditions, the probability law has to be extrapolated beyond the range of observed data, by the very definition of the problem. Non-stationary conditions involve additional extrapolation of the law relating the parameters of the distribution and the covariates (e.g., Eq. (2.7)), introducing a further source

of uncertainty (Serinaldi and Kilsby 2015). The subjective choices of covariates and structural forms of the relations between parameters and covariates also add to the uncertainty of non-stationary extreme value models (Serinaldi and Kilsby 2015).

- More fundamentally, some authors emphasize that change in observed data records is not the same as non-stationarity and that non stationarity may be only applicable for the modeling world (Koutsoyiannis 2006; Montanari and Koutsoyiannis 2014). Indeed, the observed time series is only a realization of the stochastic process, and inferring ensemble statistical properties from the temporal statistical properties would involve the assumption of ergodicity. However, if the process is non-stationary, it cannot be ergodic (Serinaldi and Kilsby 2015).
- The trends in hydrologic extremes (and subsequently in the parameters) are assumed to be deterministic in nature, though a random component is always there that is stationary. Typically, large temporal scale variations (such as trend) in hydrologic data are treated as deterministic, whereas small temporal scale variations are modeled as random components though they are clearly not measurement errors and do not come from repeatable experiments (Koutsoyiannis 2006). Observed changes at both temporal scales may be a part of larger natural variability (Koutsoyiannis 2013), and therefore, projecting the deterministic trends in the future may lead to unrealistic values. A suggested possible modeling alternative is to consider regime-switching dynamics under global stationarity (Serinaldi and Kilsby 2015).
- Last but not the least, some studies (von Storch 1995) argue that the null hypothesis of stationarity is tested against an alternative of non-stationarity which is formulated only after examining the data and *observing* the trends. Therefore, the hypothesis tested is not independent of the data. Moreover, the null hypothesis assumes "random" nature of the variables; therefore, any method to detect and model non-stationarity that ignores serial dependence in the variables may lead to incorrect results.

Before further analysis, it is therefore necessary to duly acknowledge the above caveats and finally incorporate them in the recommendations on design protocols under non-stationarity.

2.4 Hydrologic Risk Communication Under Climate Change

As established previously, the unique relationship between the probability of exceedance p and the return period m does not hold in non-stationary conditions such as that due to climate change. Therefore, there is a growing interest in the scientific community as well as practitioners on assessment and communication of hydrologic risk under non-stationarity.

2.4.1 Fixed Probability of Exceedance-Based Approach

The most straightforward way of communicating the changing nature of risk in non-stationary conditions is through the time-varying quantiles or "effective return levels" (Katz et al. 2002) corresponding to a fixed probability of exceedance. The transient $(1-p)$th quantile of the non-stationary GEV distribution is thus a function of the covariate and is obtained by year-wise inverting equation the non-stationary GEV cumulative distribution function, as given by

$$z_{(1-p)}(t) = \begin{cases} \mu(t) - \frac{\sigma(t)}{\xi}\left[1 - (-log\,(1-p))^{-\xi}\right] \; for\ \xi \neq 0 \\ \mu(t) - \sigma(t)\,[log\,(-log\,(1-p))] \; for\ \xi = 0 \end{cases} \tag{2.10}$$

for the non-stationary GEV model given by Eq. (2.7). Thus, the effective return levels are akin to a risk measure that assumes that the flood plain is changing at each time step. For conservative risk estimates, one may consider the highest effective return level during the design life of a structure. This quantity is termed "minimax design level" by Rootzén and Katz (2013).

The variance of effective return levels, which is also a function of the covariate, can also be obtained by the delta method, as follows:

$$\mathrm{Var}\left(z_{(1-p)}(t)\right) \approx \nabla z_{(1-p)}{}^{T}\, V\, \nabla z_{(1-p)} \tag{2.11}$$

where V is the variance-covariance matrix of $(\mu_{0,}\ \mu_{1},\ \sigma_{0},\ \sigma_{1},\ \xi)$ and

$$\nabla z_{(1-p)}{}^{T} = \left[\frac{\partial z_{(1-p)}}{\partial \mu_0}, \frac{\partial z_{(1-p)}}{\partial \mu_1}, \frac{\partial z_{(1-p)}}{\partial \sigma_0}, \frac{\partial z_{(1-p)}}{\partial \sigma_1}, \frac{\partial z_{(1-p)}}{\partial \xi}\right] \tag{2.12}$$

evaluated at the estimated values $(\widehat{\mu_o},\ \widehat{\mu_1}, \widehat{\sigma_o},\ \widehat{\sigma_1},\ \widehat{\xi})$. As in the stationary case, variance-covariance matrix V is obtained by inversion of the observed information matrix evaluated at the estimated value of the parameters. The derivatives in Eq. (2.12) can be computed numerically when their analytical forms are cumbersome. The confidence intervals of the effective return levels at each time step t can be computed similar to the stationary condition.

2.4.2 Waiting Time-Based Approach

The traditional definition of return period in stationary conditions can have two interpretations both leading to the same expression. The first interpretation is based on expected waiting time until the next exceedance of the hazardous event, while the second is based on the expected number of events in any time period. These interpretations can be extended to the non-stationary conditions (Cooley 2013).

Based on the expected waiting time (Cooley 2013; Salas and Obeysekera 2013), the m-year return levelr_m is defined as that value of the design level for which the expected value of the waiting time is m years. If the random variable T represents the time at which a hazardous event exceeding the level r will occur for the first time, the probability that the first event that exceeds r occurs at time t is given by

$$\begin{aligned}\Pr(T=t) &= \Pr(X_1 \le r)\Pr(X_2 \le r)\ldots\ldots\Pr(X_{t-1} \le r)\Pr(X_t > r)\\ &= \prod_{y=1}^{t-1} G_y(r)\,(1-G_t(r)) \end{aligned} \tag{2.13}$$

where X_i represents the annual maximum at year i. Thus, the waiting time follows a generalized geometric distribution whose parameter varies in time. The m-year return levelr_m can be obtained by equating the expected value of this waiting time T with m years as follows:

$$\mathrm{E}(T) = \sum_{t=1}^{\infty} t \prod_{y=1}^{t-1} G_y(r_m)\,(1-G_t(r_m)) = 1 + \sum_{t=1}^{\infty}\prod_{y=1}^{t} G_y(r_m) = m \tag{2.14}$$

where the right-hand side comes from algebraic manipulations given in Cooley (2013). The solution of this equation is computationally challenging as it involves an infinite summation. However, since $G_y(x)$ is a monotonic decreasing function as $y \to \infty$, it is possible to bind the right-hand side by choosing an arbitrary, large number (Cooley 2013). Salas and Obeysekera (2013) present the first hydrologic application of this definition of return level under non-stationarity.

2.4.3 Frequency-Based Approach

The expected number of event-based definition aims to find the design quantile r_m such that the expected number of events in any given m years is one (Cooley 2013). If Q represents the random variable that denotes the number of exceedances occurring in m years, beginning with the year $y = 1$ and ending with $y = m$, the distribution of Q is binomial in the stationary case, whereas it varied with time as the probability of exceedance changes. I is chosen as an indicator variable which takes a value of one if the design level r is exceeded, and zero otherwise. Therefore,

$$\begin{aligned} Q = \sum_{y=1}^{m} I\,(Z_y > r) &\Rightarrow E(Q) = \sum_{y=1}^{m} E\left[I\,(Z_y > r_m)\right] = \sum_{y=1}^{m}\Pr\left(Z_y > r_m\right)\\ &= \sum_{y=1}^{m}\left(1 - G_y(r_m)\right) = 1 \end{aligned} \tag{2.15}$$

Therefore, the computation of the m-year return level based on the expected number of events is much more convenient than that based on the expected waiting time.

The implicit delta method (Cooley 2013) can be used to obtain the uncertainties associated with the return levels under non-stationarity based on these two definitions. If $\widehat{\beta}$ represents the MLE estimates of the parameter vector β, and $V\left(\widehat{\beta}\right)$ is the variance-covariance matrix associated with $\widehat{\beta}$, then the m-year return level r_m can be expressed as a function of the parameter vector, $r_m = f(\beta)$, whose gradient is given by ∇r_m. According to the delta method, the variance of $r_m$is approximately given by

$$\mathrm{Var}\left(r_m\right) \approx \nabla {r_m}^T \, V\left(\widehat{\beta}\right) \nabla r_m \tag{2.16}$$

In the absence of analytical expressions for f, a generalized expression of the form $m = s\left(\beta,\ r_m\right)$ can be obtained. If m_0 is the desired return period and $r_{m0}$is the corresponding return level, then the numerical derivatives of gradient of $r_m$can be obtained if s is known. Applying chain rule of differentiation (Cooley 2013)

$$\begin{aligned}\frac{\partial r}{\partial \beta_i}\bigg|_{\beta=\widehat{\beta};m=m_0} &= \frac{\partial r}{\partial m}\bigg|_{\beta=\widehat{\beta};m=m_0} \frac{\partial m}{\partial \beta_i}\bigg|_{\beta=\widehat{\beta};m=m_0} \\ &= \left(\frac{\partial m}{\partial r}\bigg|_{\beta=\widehat{\beta};r=r_{m0}}\right)^{-1} \frac{\partial m}{\partial \beta_i}\bigg|_{\beta=\widehat{\beta};m=m_0}\end{aligned} \tag{2.17}$$

the necessary gradients can be computed.

2.4.4 Probability of Failure-Based Approach

Recent studies (Serinaldi and Kilsby 2015) argue that the concept of return period and return levels may misrepresent hydrologic risk by virtue of being derived quantities and implicitly hiding the associated *annual* probability of exceedance $p = 1/m$. The probability of failure or risk, on the other hand, presents the probability of observing at least one critical event in the lifetime of M years and allows one to compute a design value r_d for the given probability. It can be showed that when design life M years = return period m years = 100 years, assuming the data to be iid, the probability of at least one critical event occurring during the lifetime of the structure is 63 % (Serinaldi 2014) which is quite high. Therefore, the probability of failure may not always be intuitively understood based only on return periods.

The generic representation of probability of failure is given by

$$\begin{aligned} p_M &= 1 - \Pr\left(X_1 \le r_d \cap X_2 \le r_d \ldots\ldots \cap X_M \le r_d\right) \\ &= 1 - H_M\left(X_1 \le r_d, X_2 \le r_d, \ldots .X_M \le r_d\right) \\ &= 1 \quad C_M\left\{G_1\left(r_d\right),\ G_2\left(r_d\right), \ldots\ldots G_M\left(r_d\right)\right\} \end{aligned} \tag{2.18}$$

where H_M denotes the joint distribution of the set of random variables X_i for $i = 1, \ldots M$, G_i denotes the marginal univariate distribution of X_i, and C_M is the copula describing the margin-free dependence structure (Serinaldi 2014). Equation (2.18) therefore can be applicable for any condition and not just for iid data. Under iid conditions, $G_i = G\ \forall i = 1,\ 2, \ldots,\ M$, and $C_M = \prod$. For the independent but not identically distributed data,$C_M = \prod$, but G_i changes at each year.

On the other hand, Rootzén and Katz (2013) independently suggest that the necessary information for hydrologic design should include (1) the design life of the structure (e.g., say, the next 100 years) and (2) the probability of a critical event occurring within that design life. They propose a new quantity as a measure of hydrologic risk, the "design life level," that can be thought of as a special case of Eq. (2.18), for independent but not identically distributed data. For example, for highest minimum winter temperatures at Fort Collins, Colorado, USA, that reportedly witnessed an increasing trend in warming winters, "the 2021–2100 10 % design life level is 24 °F" (Rootzén and Katz 2013). This means that the maximum probability of exceedance of any winter day having temperature below 24 °F, in any given year in the design life period 2021–2100, is at most 10 %. Note here that this is different from a design value based on 10 % *annual* exceedance probability which actually corresponds to a 10-year storm, whose magnitude will be much higher than 24 °F. In stationary conditions, a 10 % risk of failure over the design life of 80 years corresponds to a return period of about 760 years.

The delta method can be used here also to compute the uncertainties in the design life level (Rootzén and Katz 2013). For the design life of M years, as explained above, the estimate of joint distribution of annual maxima, assuming independence, is given by

$$\widehat{H_M}(x) = \prod_{i=1}^{M} \widehat{G}_i(x) \tag{2.19}$$

Let $\widehat{q}(y)$ represent an estimate of the quantile function that can be obtained by inversion of $\widehat{H_M}(x)$. The design life levels are obtained from this quantile function. For example, if $M = 80$ years, $\widehat{q}_{2021-2100}(0.9)$ represents an estimate of the 2021–2100 10 % design life level. Therefore, we are interested in the uncertainty of $\widehat{q}(y) = \widehat{H_M}^{-1}(y)$. The variance of $\widehat{q}(y)$, by the delta method, is given by (Rootzén and Katz 2013)

$$\mathrm{Var}\ (\widehat{q}(y)) \approx \left(\frac{\partial}{\partial \beta_i} q\,(y;\beta) \Big|_{\beta=\widehat{\beta}_i};\ 1 < i < d \right)$$

$$V\left(\widehat{\beta}\right) \mathrm{x} \left(\frac{\partial}{\partial \beta_i} q\,(y;\beta) \Big|_{\beta=\widehat{\beta}_i};\ 1 < i < d \right)^T \tag{2.20}$$

where d is the length of the vector β (=5 for the model considered in Eq. (2.7)). The confidence intervals of the design life level can then be easily computed assuming asymptotic Gaussian distribution of $\widehat{q}(y)$.

2.4.5 Example

For illustrative purpose with a view to synthesize the concepts discussed so far, a case study on floods is discussed here that has been studied twice in earlier studies, with different perspectives, by Cooley (2013) and Serinaldi and Kilsby (2015). These studies consider annual peak flows in the Red River of the North at Halstad, Minnesota, USA, as this region had experienced damaging flood events in recent times. The data can be obtained from the United States Geological Survey's National Water Information System (http://nwis.waterdata.usgs.gov/). The observed peak flows are shown in Fig. 2.3, while the details of the fitted stationary and non-stationary models, following Cooley (2013), are provided in Table 2.2. The non-stationary model used here considers a trend only in the location parameter of the GEV distribution. Though there is some indication that the floods may have increased in this region (Cooley 2013), they have not been formally attributed to

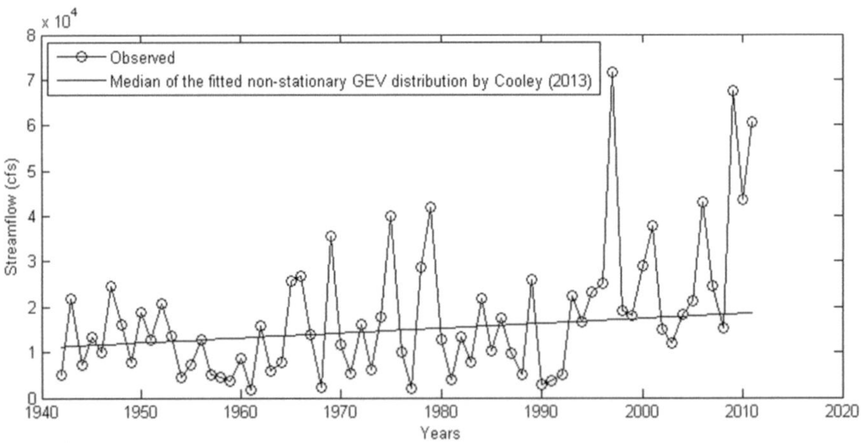

Fig. 2.3 Observed annual maximum daily flows in the Red River at Halstad. The data is publicly available at http://nwis.waterdata.usgs.gov/. The median of the fitted non-stationary GEV distribution, following Cooley (2013), is also shown

Table 2.2 Details of the fitted stationary and non-stationary models following Cooley (2013). The results of the Mann-Kendall trend test and the likelihood ratio test are also shown

Models	Parameters	Estimate value (Cooley 2013)	Maximized log-likelihood values (Cooley 2013)	Mann-Kendall trend test	Likelihood ratio test
Stationary model M_0	μ (SE)	10392 (1095)	−751.59	p-value = 0.00089	Deviance statistics = 3.38
	σ (SE)	7924 (944)			
	ξ (SE)	0.323 (0.137)			
Non-stationary model M_1	μ_0 (SE)	7976 (2247)	−749.90		Chi-square critical value at 95 % confidence, 1 degree of freedom = 3.84
	μ_1 (SE)	106 (65)			
	σ (SE)	8630 (1024)			
	ξ (SE)	0.216 (0.132)			

Table 2.3 Estimates of hydrologic risk for the Red River at Halstad for the design life of 2012–2061

Design level	Magnitude (cfs)
Stationary return level	72,306 (Cooley 2013)
Highest historical (1942–2011) effective return level	68,134
Expected waiting time-based return level (trend continued)	73,150 (Cooley 2013)
Expected waiting time-based return level (trend stopped at end of design life, 2061)	71,650
Expected number of event-based return level	70,950 (Cooley 2013)
Minimax design level	73,434
Design life level: 2012–2061 10 % level	1,29,301
Design life level: 2012–2061 5 % level	1,54,735

climate change. The various hydrologic design levels for this example are presented in Table 2.3. Serinaldi and Kilsby (2015) also perform a thorough exploratory data analysis on the peak flows in this river, in its and surrounding locations, and explore the effects of human regulation on such flows. Cooley (2013) cautions on several caveats of the non-stationary extreme value models which are later discussed in more details by Serinaldi and Kilsby (2015). Based on their findings and the analysis presented here, the following are some critical points to be noted for this example. Though these issues are discussed here with respect to a particular case study, they are in general consequential for analysis of hydrologic extremes:

– Although the Mann-Kendall trend test shows a significant increasing positive trend in peak flows (Table 2.2, Column 5) in this region and the AIC favors selection of the non-stationary model (Cooley 2013), the likelihood ratio test shows that the null hypothesis of a stationary model cannot be rejected (last column of Table 2.2). Also, the 95 % confidence interval of the trend parameter μ_1, assuming approximate Gaussian distribution is given by $106 \pm 1.96 \times 65 = [-21.4, 233.4]$ which not only represents a huge uncertainty band but also includes zero, thereby questioning the correctness of the non-stationary model. Indeed, such high ranges of uncertainty are very common in hydrologic data that is often limited by short record lengths.

- Serinaldi and Kilsby (2015) note that the entire observed record length was affected by human regulations and therefore emphasize caution while using this dataset for future hydrologic risk assessment. They advocate a thorough exploratory data analysis including borrowing information from nearby sites and local experts.
- The uncertainty bounds of the design levels reported by Cooley (2013) are very high (e.g., [33324, 111286]). Moreover, the shape parameter differs by a significant amount in the stationary and the non-stationary models though there is no apparent reason for that, except possible improper estimation of parameters.
- All the non-stationary design levels presented in Table 2.3 involve extrapolation of the fitted model into the future using its deterministic form. Given the presence of uncertainty and limitations of the assumed model, such extrapolations may not be appropriate.
- Taking into consideration the uncertainties in the non-stationary design levels, a comparison between such levels and the stationary 50-year return level is necessary to conclude whether the non-stationary model and the sophisticated non-stationary design levels have led to improved understanding and robust assessment of hydrologic risk.
- The non-stationary design levels are not completely agreeing with each other and are dependent on subjective choices such as how long the assumed trend should be projected into the future. In general, though recent studies (Rootzén and Katz 2013; Serinaldi 2014) favor the probability of failure- and risk-based hydrologic designs, the design life levels are found to lead to overly conservative risk estimates, questioning their feasibility in practice.
- Duly acknowledging the limitations of data-driven techniques, it is advised that the actual dynamics that affects non-stationarity needs to be first identified (Serinaldi and Kilsby 2015) without "looking at" the peak flow dataset and performing a statistical analysis.

2.5 Summary and Conclusions

2.5.1 Summary

In this chapter, the applications of statistical extreme value theory to characterize non-stationarity in hydrologic extremes are discussed, along with the limitations of such methods. Different definitions of hydrologic risk under non-stationarity and the uncertainties associated with them are also presented comprehensively. Through an example, we discuss various caveats associated with non-stationary modeling of extremes that is usually followed in recent literature, given there are

Table 2.4 Advantages and disadvantages of hydrologic risk measures under non-stationarity

Name of design level	Advantages	Disadvantages
Effective return level/minimax design level	Communicates changing risk clearly	Akin to changing flood plains every year, thus infeasible
	Simple to compute	A derived quantity representing *average annual* change – can mislead perception
Expected waiting time-based return level	Easy and sensible interpretation	Computationally challenging. Sensitive to how long the parameter-covariate relationship is projected
	More closely linked with life span calculation	A derived quantity representing *average annual* chance – can mislead perception
Expected number of event-based return level	Easy interpretation	Still involves extrapolation of the parameter-covariate relationship
	Easy computation, though iterative	A derived quantity representing *average annual* chance – can mislead perception
	Extrapolation of parameter-covariate relationship is only for the design life	
Design life level	Does not necessitate the iid assumption	Conservative estimates
	Represents collective risk across design life and not average annual probability of exceedance	Leads to very high equivalent return period, therefore requires large dataset for estimation
		Uncertainties are higher
		Sensitive to how long the parameter-covariate relationship is projected

statistically significant trends and/or change points in the data. The advantages and disadvantages of the various hydrologic risk measures under non-stationarity are presented in Table 2.4. It can be seen that they do not constitute a unifying framework for defining hydrologic risk under transient conditions, neither is any of them devoid of flaws or assumptions. Therefore, whether they can replace traditional design concepts still remains an open question. Further research efforts are therefore likely to spawn in this area, and perhaps a physics-guided statistical approach can improve our understanding of hydrologic extremes, how they vary with climate change, and how they can affect societies.

2.5.2 Concluding Remarks and Recommendations for Sustainable Planning

With increasing awareness on anthropogenic climate change and its possible links with hydro-climatic extremes that continue to cause unprecedented loss and damage to societies, more and more practitioners and policy makers are interested to account climate change in some way in water planning (Milly et al. 2015). Indeed, as growing population gets exposed to the risk of extreme events, hydrologists face the challenge of understanding the spatiotemporal behavior of extremes, especially under rapidly changing conditions. Therefore, extreme value models and their non-stationary extensions or other similar techniques (Read and Vogel 2015) and more sophisticated risk communication approaches based on them offer potential tools for investigating such extremes under climate change. However, considering the issues and caveats discussed in details in this chapter and the assumptions involved, at the current state of knowledge, non-stationary extreme value models cannot be universally claimed to be a suitable replacement of traditional design concepts based on stationarity. In general, it is advisable to make use of all possible information sources including exploratory data analysis and expert opinions, to acknowledge and communicate the uncertainties involved in the analysis, and to retain stationary models as benchmarks until alternate approaches are proved to overcome the current limitations.

Acknowledgments The authors thank Dan Cooley, Rick Katz, Holger Rootzen, and Francesco Serinaldi for their helpful clarifications. The R-code to calculate the expected waiting time and expected number of event-based return levels was shared by Dan Cooley.

References

Cheng L, AghaKouchak A, Gilleland E, Katz RW (2014) Non-stationary extreme value analysis in a changing climate. Clim Chang 127(2):353–369

Clarke RT (2013) How should trends in hydrological extremes be estimated? Water Resour Res 49(10):6756–6764

Coles S (2001) An introduction to statistical modeling of extreme values. Springer, London

Cooley D (2013) Return periods and return levels under climate change. In: AghaKouchak A, Easterling D, Hsu K, Schubert S, Sorooshian S, AghaKouchak A, Easterling D and Hsu K (eds) Extremes in a changing climate: detection, analysis, and uncertainty (pp 97–114). Springer, New York

Coumou D, Rahmstorf S (2012) A decade of weather extremes. Nat Clim Chang 2:491–496

Dai A (2013) Increasing drought under global warming in observations and models. Nat Clim Chang 3:52–58

Gilleland E, Katz RW (2011) New software to analyze how extremes change over time. Eos Trans Am Geophys Union 92(2):13–14

Gilleland E, Ribatet M, Stephenson AG (2013) A software review for extreme value analysis. Extremes 16:103–119

Hirabayashi Y, Mahendran R, Koirala S, Konoshima L, Yamazaki D, Watanabe S, . . . Kanae S (2013) Global flood risk under climate change. Nat Clim Change 3:816–821

IPCC (2012) Summary for policymakers. In: Field CB, Barros V, Stocker TF, Qin D, Dokken DJ, Ebi KL, . . . Midgley PM (eds) Managing the risks of extreme events and disasters to advance climate change adaptation. A special report of Working Groups I and II of the Intergovernmental Panel on Climate Change (pp 1–19). Cambridge University Press, Cambridge, UK/New York

Katz RW (2013) Statistical methods for nonstationary extremes. In: AghaKouchak A, Easterling D, Hsu K (eds) Extremes in a changing climate: detection, analysis and uncertainty. Springer, Dordrecht, pp 15–37

Katz RW, Parlange MB, Naveau P (2002) Statistics of extremes in hydrology. Adv Water Resour 25(8):1287–1304

Klemes V (1986) Dilettantism in hydrology: transition or destiny? Water Resour Res 22(9S):177S–188S

Klemes V (2000) Tall tales about tails of hydrological distributions. I. J Hydrol Eng 5(3):227–231

Koutsoyiannis D (2006) Nonstationarity versus scaling in hydrology. J Hydrol 324(1):239–254

Koutsoyiannis D (2013) Hydrology and change. Hydrol Sci J 58(6):1177–1197

Kundzewicz ZW, Hirabayashi Y, Kanae S (2010) River floods in the changing climate observations and projections. Water Resour Manag 24(11):2633–2646

Leadbetter M, Lindgren G, Rootzén H (1983) Extremes and related properties of random sequences and processes. Springer, New York

Meehl GA, Stocker TF, Collins WD, Friedlingstein P, Gaye AT, Gregory JM, . . . Raper SC (2007) Global climate projections. In: Solomon S, Qin D, Manning M, Chen Z, Marquis M, Averyt KB, . . . MHL (eds) Climate change 2007: the physical science basis. Contribution of Working Group I to the Fourth Assessment Report of the Intergovernmental Panel on Climate Change. Cambridge University Press, Cambridge, UK/New York

Milly PC, Betancourt J, Falkenmark M, Hirsch RM, Kundzewicz ZW, Lettenmaier DP, Stouffer RJ (2008) Stationarity is dead: whither water management? Science 319:573–574

Milly PC, Betancourt J, Falkenmark M, Hirsch RM, Kundzewicz ZW, Lettenmaier DP, Krysanova V (2015) On critiques of "Stationarity is dead: whither water management?". Water Resour Res 51(9)

Mondal A, Mujumdar P (2012) On the basin-scale detection and attribution of human-induced climate change in monsoon precipitation and streamflow. Water Resour Res 48(10)

Mondal A, Mujumdar PP (2015a) On the detection of human influence in extreme precipitation over India. J Hydrol 529(3):1161–1172

Mondal A, Mujumdar PP (2015b) Modeling non-stationarity in intensity, duration and frequency of extreme rainfall over India. J Hydrol 521:217–231

Mondal A, Mujumdar PP (2015c) Return levels of hydrologic droughts under climate change. Adv Water Resour 75:67–79

Montanari A, Koutsoyiannis D (2014) Modeling and mitigating natural hazards: stationarity is immortal! Water Resour Res 50:9748–9756

Obeysekera J, Salas JD (2014) Quantifying the uncertainty of design floods under nonstationary conditions. J Hydrol Eng 19(7):1438–1446

Oehlert GW (1992) A note on the delta method. Am Stat 46(1):27–29

Read LK, Vogel RM (2015) Reliability, return periods, and risk under nonstationarity. Water Resour Res 51(8):6381–6398

Rootzén H, Katz RW (2013) Design life level: quantifying risk in a changing climate. Water Resour Res 49(9):5964–5972

Salas JD (1993) Analysis and modeling of hydrologic time series. In: Maidment D (ed) Handbook of hydrology. McGraw-Hill, New York, pp 19.1–19.72

Salas JD, Obeysekera J (2013) Revisiting the concepts of return period and risk for nonstationary hydrologic extreme events. J Hydrol Eng 19(3):554–568

Serinaldi F (2014) Dismissing return periods! Stoch Env Res Risk A 29(4):1–11

Serinaldi F, Kilsby CG (2015) Stationarity is undead: uncertainty dominates the distribution of extremes. Adv Water Resour 77:17–36

Sivapalan M, Samuel JM (2009) Transcending limitations of stationarity and the return period: process-based approach to flood estimation and risk assessment. Hydrol Process 23:1671–1675

Towler E, Rajagopalan B, Gilleland E, Summers RS, Yates D, Katz RW (2010) Modeling hydrologic and water quality extremes in a changing climate: a statistical approach based on extreme value theory. Water Resour Res 46(11)

Vogel RM, Yaindl C, Walter M (2011) Nonstationarity: flood magnification and recurrence reduction factors in the United States. J Am Water Resour Assoc 47:464–474

von Storch H (1995) Misuses of statistical analysis in climate. In: von Storch H, Navarra A (eds) Analysis of climate variability: applications of statistical techniques. Springer, Berlin, pp 11–26

Westra S, Alexander LV, Zwiers FW (2013) Global increasing trends in annual maximum daily precipitation. J Clim 26(11):3904–3918

Chapter 3
Impact of Climatic and Land Use Changes on River Flows in the Southern Alps

Roberto Ranzi, Paolo Caronna, and Massimo Tomirotti

Abstract River flow time series are far from being stationary and always experienced changes in the past, also dramatic in long time horizons. In recent years it seems that both climatic and anthropogenic factors are accelerating the variability of hydrological processes. It is not clear, however, whether climatic or anthropic factors represent the major forcing to the hydrological cycle. Long-term statistics, lasting over 150 years, of annual runoff for the five major Italian rivers in the Central Alps are presented and compared with precipitation, temperature and land use changes. A homogeneous decreasing trend of annual runoff is observed, and the significance of such a trend at the local and regional scale is tested with Mann-Kendall, Sen-Theil and Sen-Adichie statistical tests. It is shown that for some rivers, the increased agricultural water demand and land use changes are a likely major source of non-stationarity, possibly more relevant than meteorological ones. A natural feedback which is being observed also at the global scale is discussed on the basis of land use in the Adige river basin by comparing cadastral maps of the mid-nineteenth century with recent aerial photographs in four sample areas. Results are consistent with the reduced speed of deforestation observed at the global scale and the natural afforestation observed in Europe occurring over the last decades. This process can play a major role in regulating the hydrological cycle and mitigating flood and drought extremes, but also enhancing evapotranspiration losses and thus reducing runoff volumes.

Keywords River flow regime • Runoff • Trends • Climatic change • Anthropogenic changes

3.1 Introduction

In the scientific debate on the recent and projected climate at the global scale (IPCC 2013), a wide consensus is achieved on the fact that the Earth is experiencing a fast climate warming which will affect natural and human systems. Evidences on

---

R. Ranzi (✉) • P. Caronna • M. Tomirotti
DICATAM, University of Brescia, Via Branze 43, 25123 Brescia, Italy
e-mail: roberto.ranzi@unibs.it; paolo.caronna@unibs.it; massimo.tomirotti@unibs.it

E. Kolokytha et al. (eds.), *Sustainable Water Resources Planning and Management Under Climate Change*, DOI 10.1007/978-981-10-2051-3_3

the atmospheric temperature and CO_2 concentration increase, sea level rise and glaciers' retreat were fundamental in convincing 195 countries at the Paris climate conference COP21 to adopt a global action plan with the objective to reduce the global greenhouse gases emissions (UNFCCC 2015). However, in the scientific community, there still exist a few exceptions to this position (Gerhard 2006) and some criticism to the credibility of climate projections for hydrological purposes (Koutsoyiannis et al. 2008; Blöschl and Montanari 2010). In terms of impact on the hydrological cycle, the outcomes of the IPCC 5th Assessment Report (IPCC 2013) and the report by Bates et al. (2008) depict low confidence or unclear scenarios when dealing with soil moisture and mean river runoff. Quoting IPCC (2013) "The most recent and most comprehensive analyses of river runoff do not support the IPCC Fourth Assessment Report (AR4) conclusion that global runoff has increased during the 20th century. New results also indicate that the AR4 conclusions regarding global increasing trends in droughts since the 1970s are no longer supported," and "Projected changes in runoff show decreases in northern Africa, western Australia, southern Europe and southwestern USA and increases larger than the internal variability in northwestern Africa, southern Arabia and southeastern South America associated to the projected changes in precipitation. Owing to the simplified hydrological models in many CMIP5 climate models, the projections of soil moisture and runoff have large model uncertainties."

In spite of these uncertainties on future runoff changes also in the water engineering community, structural and nonstructural adaption measures are being taken to face the problems that could arise from an unbalance between water resources availability and demand in a changing planet (Ranzi et al. 2015). In fact stream flow changes could have relevant feedbacks, in the near future, on the water management policies and engineering design, particularly in those regions which are deeply exploited for hydropower production and where irrigation and domestic water demand are in conflict with industrial uses. The scientific hydrology community is paying more and more attention on several aspects, related to both physical and societal factors, of the changing dynamics of the water cycle in connection with rapidly changing human systems (Montanari et al. 2013). Changes and their reasons are now at the core of research and its technical applications in the water sector, and this chapter aims at contributing to this streamline.

Because of the high regional variability of precipitation projections, it is important to monitor recent changes of river runoff at the regional scale in order to detect changes, if any significant occurs, and to identify the possible natural or anthropogenic reasons for these changes. Several studies at the regional scale report about significant variations on a statistical basis in the twentieth century, which can be linked to temperature changes in a very weak sense (Labat et al. 2004) and are often controversial (Legates et al. 2005). Some statistical analyses indicate a statistically significant decrease of runoff and a temperature and evapotranspiration losses increase, for instance, for the Xiaoxi river close to the sea coast (Zhang et al. 2014), but others, even in a similar geographic context, show a stationary or slightly increasing annual runoff (Liu et al. 2010; Kumar et al. 2009).

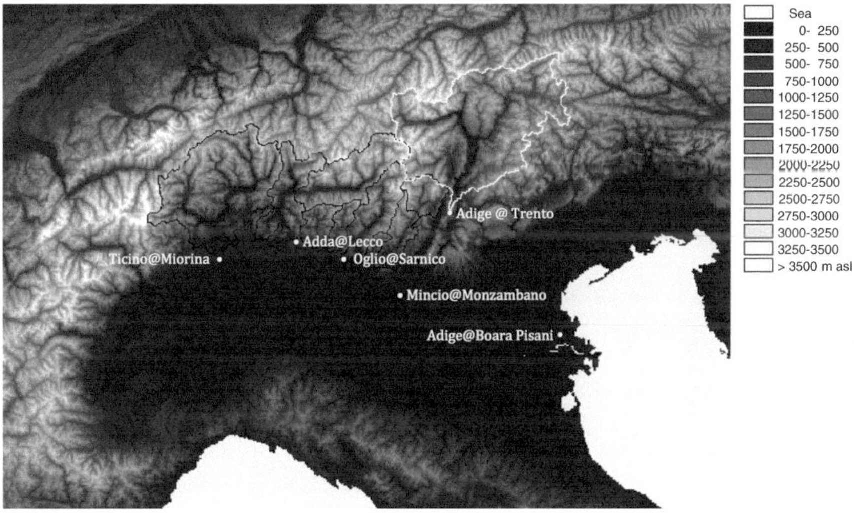

Fig. 3.1 The five investigated river basins in the Central Southern Alps: from *left* to *right*, the Adige, Sarca-Mincio, Oglio, Adda and Ticino rivers. Downstream the Adige river in Trento surface water and groundwater resources are heavily exploited for agricultural uses

In this study we update with more recent and extended data a previous investigation conducted by Ranzi (2012) on runoff in the Central Southern Alps. Annual runoff series with a duration ranging from 62 to 170 years, one starting from year 1845, for five major rivers and six hydrometric stations are analysed to detect trends (Fig. 3.1 and Table 3.1). Because similar patterns of the decreasing trends were observed for the time series of annual runoff of all five basins the null hypothesis they exhibit a common constant trend was tested and results are presented in this chapter together with statistical tests on precipitation and temperature changes in the Adda and Adige basins.

3.2 The Investigated Basins

The first basin investigated is the Adige river, which drains the eastern part of the Central Italian Alps (Fig. 3.1). Because of the lower precipitation and the increase of evapotranspiration losses which are likely to occur also as a consequence of the observed increase of the forested areas due to afforestation (Ranzi et al. 2002), runoff decreased more rapidly than precipitation already in the twentieth century. Focusing on the 1923–2011 period when data are available for two stations of the river (Fig. 3.2), with a mean of 664 mm for the Adige at Trento (with a basin area of 9763 km^2), a 163 mm/century decrease of runoff was observed, but a more pronounced decrease (−270 mm/century) was observed at the Adige basin outlet

Table 3.1 Statistics of the annual runoff, precipitation and temperature for the investigated basins

River basin	Station	Area (km^2)	Observation years (N. data)	Mean	St. dev.	Trend slope	Z_{MK}	Z_{SR}	Theil-Sen slope	TS' slope conf. limits
			Runoff	(mm)	(mm)	(mm year^{-1})	(−)	(−)	(mm year^{-1})	(mm year^{-1})
Adige	Boara Pisani	11954	1923–1978, 1980–1986, 1988–2011 (87)	573	145	**−2.70 ± 0.28**	**−4.46**	**−4.31**	**−2.39**	**(−3.45, −1.36)**
Adige	Trento	9763	1862–2011 (150)	710	140	**−1.42 ± 0.24**	**−5.54**	**−5.31**	**−1.33**	**(−1.83, −0.90)**
Mincio	Monzambano	2380	1950–2011 (62)	709	182	**−3.25 ± 1.23**	**−2.78**	**−2.52**	**−3.33**	**(−5.35, −0.94)**
Oglio	Sarnico	1840	1933–2011 (79)	979	227	−1.62 ± 1.11	−0.99	−0.99	−1.16	(−3.28, 0.75)
Adda	Lecco	4508	1845–2014 (170)	1150	237	**−0.90 ± 0.36**	**−2.47**	**−2.50**	**−0.90**	**(−1.64, −0.19)**
Ticino	Miorina	6599	1921–2012 (92)	1378	351	−2.02 ± 1.37	−1.11	−1.20	−1.67	(−4.47, 1.36)
Central Alps		25070	1950–2011 (62)	946	201	−1.86 ± 1.42	0.00	−0.03	0.00	(−3.25, 2.39)
Adige + Adda	Trento and Lecco	14271	1862–2011 (150)	843	160	**−1.21 ± 0.29**	**−3.95**	**−4.02**	**−1.17**	**(−1.72, −0.60)**
			Precipitation	(mm)	(mm)	(mm year^{-1})	(−)	(−)	(mm year^{-1})	(mm year^{-1})
Adige	Trento	9763	1864–2011 (148)	890	159	−0.40 ± 0.30	−1.32	−1.27	−0.40	(−1.00, 0.20)
Adda	Lecco	4508	1864–2009 (146)	1240	264	−0.95 ± 0.51	**−1.99**	−1.95	**−0.92**	**(−1.84, −0.02)**
Adige + Adda	Trento and Lecco	14271	1864–2009 (146)	1000	185	−0.61 ± 0.36	−1.70	−1.69	−0.58	(−1.31, 0.10)
			Temperature	(°C)	(°C)	(°C 10 years^{-1})	(−)	(−)	(°C 10 years^{-1})	(°C 10 years^{-1})
Adige	Trento	9763	1862–2011 (150)	2.4	0.7 °C	**+0.10 ± 0.1**	**7.59**	**6.93**	**+0.11**	**(0.08, 0.13)**

Results of Mann-Kendall (MK), Spearman (SR), Theil-Sen (TS) slope tests and parameters of linear regression line trend (slope and its standard deviation) for the considered stations. In bold statistics of time series with negative trends at 5 % significance level

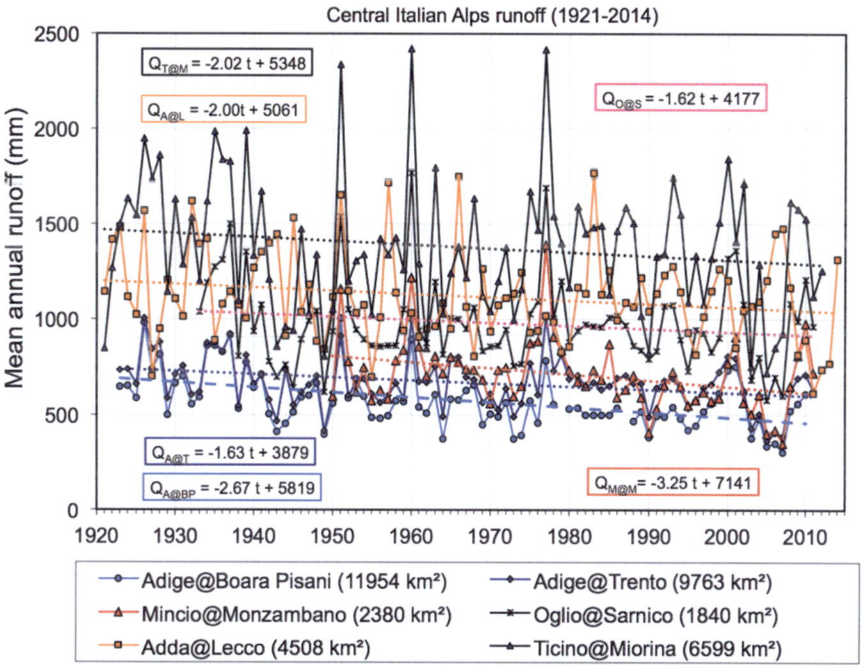

Fig. 3.2 Time series of annual runoff for four major river basins in the Central Southern Alps

into the Adriatic sea, close to the Boara Pisani station which gauges a basin area of 11 964 km^2. It is worth noting that runoff measured in the mountain part of the catchment gauged at Trento was quite close to that measured at the outlet of Boara Pisani until 1950 (697 vs. 655). After that period, runoff at the downstream station of Boara Pisani, accounting for 536 mm in the 1950–2011 period (with the exception of the years 1979 and 1987 when the missing daily runoff data were more than 10 %), was much lower than that measured at Trento, averaging 650 mm. Considering a longer time period of 150 years from 1862 to 2011, the mean annual runoff of the Adige river reached 710 mm $\pm$ 140 mm, and the slope of the linear regression trend line decreases to -142 mm/century. It has to be mentioned that data for the Adige river in Trento for the 1943 to 1950 period were reconstructed as described in Zolezzi et al. (2009) through a linear regression with the downstream stations of Serravalle and Boara Pisani.

A possible explanation of the specific runoff decrease in the downstream part of the river in the second half of the last century is the increase of water pumped from the river and the surrounding groundwater for irrigation purposes in the downstream agricultural areas and increased evapotranspiration losses. Because the exploitation of water resources increased significantly in the second half of the century, this anthropogenic factor is a likely reason of the runoff decrease in the downstream areas, more than climate change. Another factor of minor importance is the diversion of a maximum discharge of 500 m^3/s from the Adige river to the Garda lake and the Mincio river which occurred during about ten flood events starting with the year 1960 when a by-pass gallery was constructed.

A marked decrease of mean annual runoff (−325 mm/century) is observed also for the shortest time series investigated, the one of the Mincio river at Monzambano (709 mm is the mean value of annual runoff over the years 1950–2011), at the outlet of the Garda lake and for the Oglio at Sarnico (see Figs. 3.1 and 3.2 and Table 3.1), accounting for a mean annual runoff of 979 mm. A new time series lasting for 170 years and spanning the 1845–2014 period was investigated here. It is the series of annual outflow of the Adda river from the Como lake at Lecco, which was previously investigated by Moisello and Vullo (2009) for the trends in annual maxima. Also the Adda river exhibits a less pronounced but statistically significant decreasing trend of 90 mm/century over the 1845–2014 period, when the mean annual runoff is 1150 ± 237 mm. For the Ticino river at Miorina, at the outlet of the Maggiore lake, a 202 mm/century runoff decrease of the lake's inflow is observed over the 1921–2012 period, when the mean annual runoff is 1378 ± 351 mm. It has to be pointed out that trend lines are very sensitive to anomalies in the time series. For instance, an increasing trend of 69 mm/century was previously observed for the Ticino river over the 1943–2012 period, characterised by a dry decade in the 1940s, possibly because of the recent precipitation increase observed in the NW Alps and the retreat of glacierised areas, which accelerated in the last decades.

3.3 Statistical Tests on Trends

In order to test the significance level of trends in the time series, several statistical tests are available in the literature (e.g. McCuen 2003); among them, Mann-Kendall (MK), Spearman's rho (SR) and run tests are widely used. These tests are nonparametric, one of their primary merits being that they do not assume that the data under analysis were drawn from a given distribution. MK (Mann 1945; Kendall 1970) and SR (Spearman 1904) tests were selected for our analysis since they have been extensively used in hydrology; on the other hand, their performances are almost identical in terms of power (Yue et al. 2002).

Both MK and SR tests are based on the null hypothesis that a sample of data $\{x_1, \ldots, x_n\}$ is independent and identically distributed, which means that there is no trend or serial correlation among the data. The alternative hypothesis of the two-sided tests is that the distributions of x_j and x_k are not identical for all $k, j \leq n$ and $k \neq j$. The tests can be used to detect a monotonically increasing or decreasing trend according to the sign of the test statistic Z_{MK} or Z_{SR} (positive for an increasing trend, negative for a decreasing one), which we computed as in Appendix A in Yue et al. (2002) (thus taking into account the so-called correction for continuity for Z_{MK}). The trend is significant if the null hypothesis cannot be accepted for the given significance level α. The results obtained by MK and SR tests for the considered gauging stations selecting a significance level $\alpha = 0.05$ are shown in

Table 3.1. Since both Z_{MK} and Z_{SR} for large sample size (>10 for Z_{MK}, >20 for Z_{SR}) can be considered distributed according to a standard normal distribution (Kendall 1970), setting $Z = Z_{MK}$ or Z_{SR}, the acceptance region of MK and SR tests is given by the interval $-1.96 \leq Z \leq 1.96$, while the rejection region is given by $Z < -1.96$ (significant decreasing trend) and $Z > 1.96$ (significant increasing trend). It is worth noting that the values obtained for Z_{MK} and Z_{SR} are very close together and the conclusions about the significance of the trends are in agreement for all the series, except for the annual precipitation over the Adda river basin; however, in this case the values of Z_{MK} and Z_{SR} are very close to the boundary of the non-rejection region.

The trend slopes indicated in Table 3.1 were obtained by linear regression. To this purpose, the nonparametric Theil-Sen estimator (TS in the following) was used as described in Appendix A, while the corresponding confidence interval was obtained according to Appendix B. With respect to the traditional (parametric) least squares approach, there are several important advantages in using TS estimator (e.g. Wilcox 2010). First, it is more robust in the presence of outliers. On the other hand, while it is substantially comparable with least squares estimator in the presence of both normality and homoscedasticity of the response variable Y (where least squares method exhibits the best behaviour), it is superior in terms of standard error even preserving normality hypothesis, if only homoscedasticity condition is removed.

In addition to the tests for trend detection of single series, a test of homogeneity of trends at regional scale was performed. To test the hypothesis that the runoff and precipitation series exhibit a common trend slope, the Sen-Adichie test (SA in the following) for the parallelism of regression lines was performed as described in Appendix C. The test was applied to three sets of series. First, the null hypothesis was that the $k = 5$ regression lines for the annual runoff time series $Y(x)$ of Adige at Trento, Mincio, Oglio, Adda and Ticino have a common, unspecified β slope, i.e. they are parallel. The above hypothesis was tested both on the full extension of the available data and on the 1950–2011 period, which was monitored more homogeneously. Finally it was assumed that the overall moisture fluxes in the region, including both the five runoff time series and the annual precipitation averaged over the Adige and Adda river basins, exhibit a long-term trend with the same slope (in this case the number of regression lines is $k = 6$).

As anticipated, the test was performed first assuming, as null hypothesis, that the five time series of the runoff data have the common slope β_A. The pooled least squares SA estimator (C.2) of trend provided the value $\beta_A = -1.26$ mm year^{-1} for the long-term Adige, Mincio, Oglio, Adda and Ticino runoff time series. As for the V statistic (C.6) that resulted $V = 3.05$ and the upper tail chi-square limit $\chi^2_{4,0.05} = 9.49$, the null hypothesis of a common slope is not rejected.

The observation period of the series is not homogeneous, and therefore the test was repeated for the same time series but selecting a common time interval lasting from the year 1950 until the year 2011, for which the time series are complete. In this case the pooled least square estimator resulted $\beta_B = -2.00$ mm year^{-1} and the SA statistic V =5.43, well below the same chi-square limit $\chi^2_{4,0.05} = 9.49$. The region of non-rejection of the null hypothesis is wider, also because of the minor size of the data sample.

Finally, the hypothesis was tested that also the rainfall time series has the same trend slope of the runoff series, assuming that the moisture fluxes in the whole hydrosphere exhibit a common decrease. The annual rainfall averaged spatially over the Adige and Adda river basins for the 1864–2009 time period was added to the five complete runoff time series. In this case the pooled least square slope, because of the milder decreasing trend of rainfall, results $\beta_C = -1.10$ mm year^{-1}. The null hypothesis of this value for a unique slope is not rejected as the V statistics is 6.02 and $\chi^2_{5,0.05} = 11.1$.

3.4 Climate, Anthropogenic and Land Use Changes

In order to infer the possible factors influencing runoff trends in the Central Southern Alps, the major changes that occurred in that area over the last 150 years which we consider as more relevant are the following:

- Precipitation changes
- Temperature changes
- Land use changes including afforestation and urbanisation
- Retreat of glacierised areas

In this chapter some of these factors were investigated, starting from temperature, precipitation and land use changes in the Adige basin. From the analyses described in the following, some conjectures on the factors influencing runoff changes in the investigated area can be drawn.

3.4.1 Climatic Changes in the Adige Basin

The uncertainty on the response of precipitation to a climate warming is confirmed by recent climatological analyses in the Alps (Auer et al. 2007). A tendency to an increase of precipitation in the NW part of the Alps and to a decrease in the SE part is observed, but only the former trend was considered as statistically significant by Brunetti et al. (2009), with an 11 % increase over the 1901–2000 period. In the Adige river basin gauged at the Trento station, mean annual precipitation, estimated as 832 mm in the 1923–1996 period using a network of 14 stations, showed a decrease of 105 mm/century, a tendency which is consistent with the precipitation decrease in the southeastern Alps (Ranzi and Barontini 2010). Extending the observation period to the years 1864–2011 based on the HISTALP database (Auer et al. 2007), a mean annual precipitation of 890 mm at the catchment scale was estimated.

The observed −40 mm/century linear trend slope, a value which was consistent with the estimate using the Theil-Sen slope, as shown in Table 3.1, is not statistically significant at the 5 % significance level. This result is consistent with the statistics of mean areal precipitation estimated over the Adda basin and the combinations of the Adige and Adda basins, where annual precipitation decreases with a rate of 95 and 61 mm/century, respectively, a value which, however, is not sufficient to reject the null hypothesis of no trend, at the 5 % significance level.

As it has been pointed out by Auer et al. (2007), the whole Alpine region is characterised by uniform trends of temperature. Over the 1886–2005 time period, in this area the rate of increase of mean annual temperature was 1.4 ± 0.1 °C/century (Brunetti et al. 2009) with significance higher than 99 % with Mann-Kendall test, thus confirming the global trend to a fastening warming. In the second half of the twentieth century, trends are higher than those observed in the same regions in the first half of the century. This is a possible reason why the temperature change for the Adige basin estimated on the HISTALP data base depicts a slower decrease of −1 °C/century over the 1862–2011 period. These values are, however, being investigated more in detail to complete the series with more stations, and comments of results are under preparation.

The impact of the combined effect of increased temperature and enhanced precipitation on runoff is not trivial although the runoff trend analysis is consistent with the above-mentioned "dipole" effect of precipitation changes over the Alps. An area-weighted sum of runoff in the five rivers, considering only the Trento station for the Adige river, provides a time series of runoff in a fictitious station draining an overall area of 25 070 km^2. Because of the short time series for the Oglio river at Sarnico, corresponding to the inflow in the Iseo Lake, the fictitious station data covers a 62-year period only, and the resulting mean runoff of 946 mm shows a negative trend of −186 mm/century which is, however, not significant at 5 % level. Therefore, for the Central Southern Alps, the null hypothesis of the existence of no trend cannot be rejected at 5 % significance level. This conclusion changes when a smaller hydrological region is investigated, including only the Adige and Adda basins, but with a time series 150 years long (Fig. 3.3). Although the runoff decrease is less marked, because of the availability of more data, the null hypothesis of no trend cannot be rejected with the same significance level (Table 3.1). In Fig. 3.3 annual runoff is compared with annual precipitation which exhibits a milder negative slope of 61 mm/century which confirms that precipitation decrease rate is less marked than that of runoff and is not statistically significant.

3.4.2 Land Use Changes in the Adige Basin

To explain the runoff decrease, land use changes for the Adige basin were investigated. A land use change which can be partially ascribed to anthropogenic factors is the increase of the forested areas in Europe, including Italy, in the last decades, as reported in the official reports by FAO (2011). Forests have a positive effect on the hydrological cycle by mitigating runoff extremes and storing moisture in soils and in the canopy, but they enhance evapotranspiration losses and therefore they have impact on runoff as well.

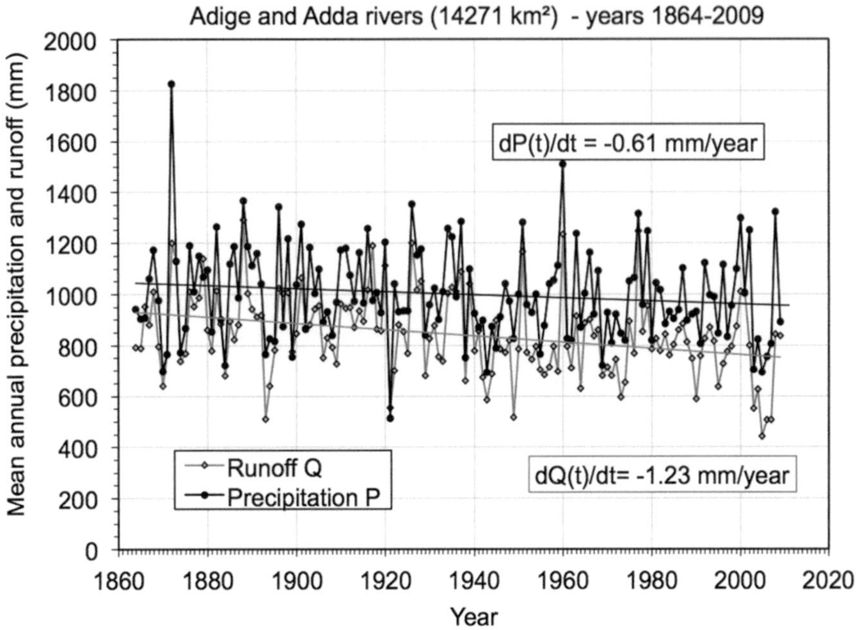

Fig. 3.3 The area-weighed runoff for a fictitious river basin in the Central Southern Alps draining an area of 14,271 km^2 including the Adige at Trento and Adda at Lecco river basins observed over the 1864–2009 period when precipitation data are available as well

Because of the combined effect of natural afforestation due to the decreased use of wood as a source of energy and the abandoning of agricultural areas (see, for instance, Ranzi et al. 2002 who described the effect of these land use changes on floods) and good management practices including protection of forests for the conservation of biodiversity and forests plantations, forested areas are growing in Europe, in the Russian Federation, North America and East Asia and in other subregions of our planet.

A study focused on four different areas of the Adige watershed (Fig. 3.4) was conducted to investigate the land use variations in order to link them with the runoff changes in the same time period.

The Adige River sources in Italy at an altitude of 1586 m a.s.l near the Resia Lake, close to the borders with Austria and Switzerland, 410 km in length, it is the second longest river in Italy after the Po River and it flows into the Adriatic Sea. The watershed is included in the Italian provinces of Bolzano and Trento, and, in the Veneto Region, it flows in the province of Verona and (for a small part) in the province of Vicenza. A small portion of the basin is located in Switzerland (Canton of Grisons). The valley bottom ranges from 1300 to 1500 m a.s.l. in the northern part of the watershed, while the river flows at an altitude of 240 m a.s.l. in the Bolzano plain and 190 m a.s.l. near Trento. In Fig. 3.4 the Adige catchment with the Trento's outlet is represented: this drainage area has an extension of 9763 km^2.

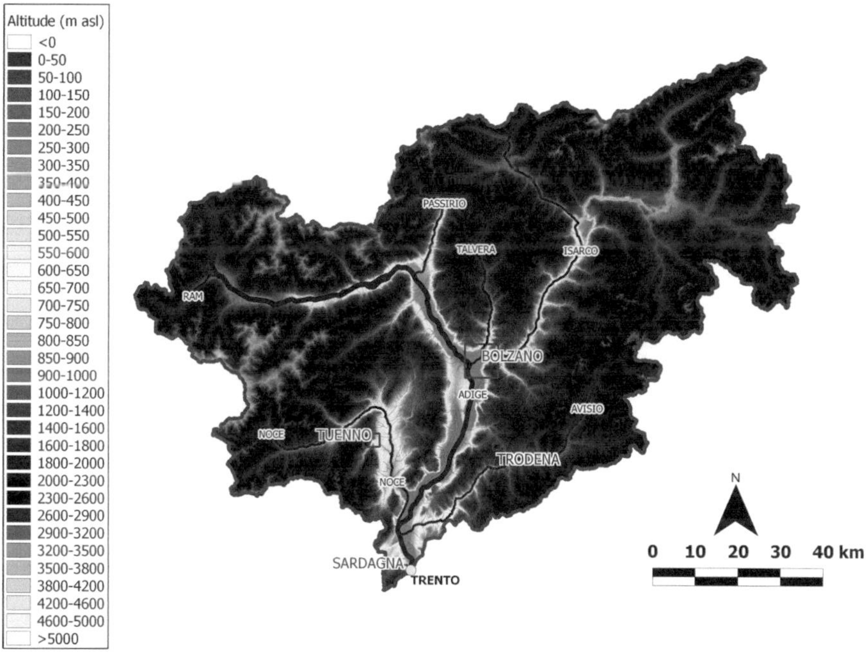

Fig. 3.4 Adige Watershed (digital elevation model), main hydrographic network and investigated areas (Tuenno, Bolzano, Sardagna, Trodena) (Image created with Quantum GIS Open Source Geospatial Foundation Project. http://qgis.osgeo.org)

The land use changes that occurred between mid-nineteenth century and present days were analysed on the basis of up-to-date GIS techniques. One of the most evident findings in this investigation was the strong increase of forest extension in the basin. This fact is not surprising since the increase of amount of land covered in forest is well known in Europe. FAO (2011) reports that this increase accounts for more than 15,000 thousands of ha out of 989,000 thousands estimated in 1990. Considering the global situation, FAO (2015) stated that planted forest area has increased by over 110 million ha since 1990 (7% of the world's forest area), with peak growing rates from 2000 to 2010. FAO observations started in 1990, but such a change is observed also in the investigated area, where land use close to Trento and Bolzano in the Adige river basin is compared considering the cadastral maps of the mid-nineteenth century with recent aerial photographs in a sample area and the Corine Land Cover 2006 maps. The comparison between natural and planted forests is an interesting topic and matter of debate among many stakeholders (natural forests play an important role in maintaining biodiversity; planted forests are important for wood production and environmental protection purposes), but the data available for the study were not sufficient to point out this demarcation.

The analysis was based on three main sources: the mid-nineteenth-century land use maps were obtained from the Province of Trento cadastral office and from the web GIS service of the Autonomous Province of Bolzano (South Tyrol), http://

Fig. 3.5 Mid-nineteenth-century cadastral map of the area close to Tuenno (Courtesy of Ufficio Catasto of Provincia Autonoma di Trento)

gis2.provinz.bz.it/geobrowser/. The current land use maps were generated from aerial photographs provided by Google Earth (Image Landsat,© 2009 GeoBasis - DE/BKG, © 2015 Google). To complete the current observable information, the Corine Land Cover 2006 map was used. All the mentioned sources were imported in the Quantum GIS open-source software, redrawn and transformed from the original format to vector polygonal files. Hence, for each watershed subregion, it was possible to create the past and present land use maps and overlap them to precisely quantify the land use variation in more than 150 years. In Figs. 3.5, 3.6, 3.7 and 3.8, the start and end points of this process are shown. The Tuenno area (1150 ha) is represented as drawn in the digitalised cadastral map in Fig. 3.5. In Fig. 3.6 the same area is shown, illustrating the present-day situation. Both images were then manually converted into vector shape files: Figs. 3.7 and 3.8 can be observed to appreciate this transformation. The same colours were used for identical land uses in the two ages, to appreciate the similarities and the changes occurred.

Considering the Tuenno area, which covers more than 1150 ha, forested areas changed from 45.52 to 60.67 %. The most part of the gained area derived from the abandoned pastures that totally disappeared during the considered period. The percentage increase of urbanised area was significant but negligible in absolute terms considering the whole extent. The grasslands halved, one part contributing to the crops' increase and the other part to the forest's increase. Also the noncultivated areas are more than doubled. The past and present uses of the land in percentage, together with the land use migrations, are shown in Table 3.2. In the second column,

Fig. 3.6 Current aerial photograph of the area close to Tuenno (Trento, Italy). The territory included in the *light grey box* corresponds to the same area in Fig. 3.5. © 2015 Cnes/Spot Image, © 2009 GeoBasis – DE/BKG, © 2015 Google

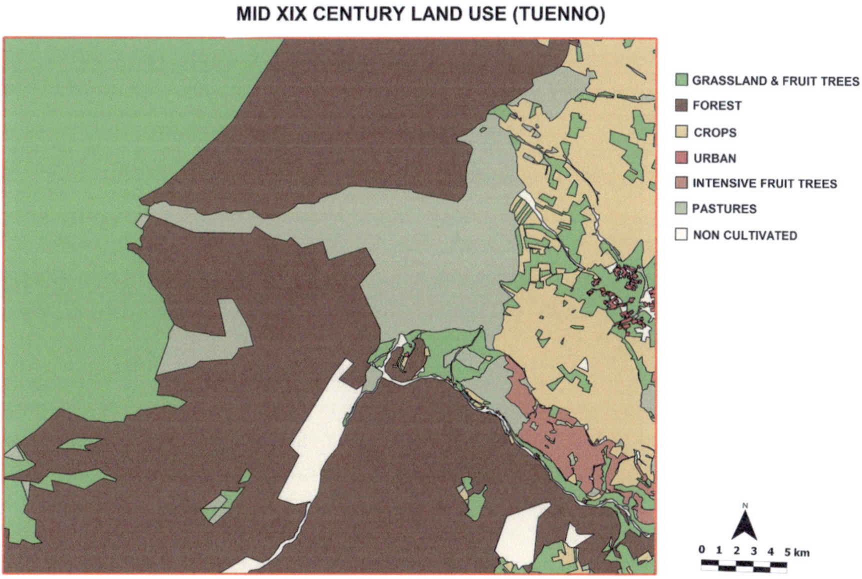

Fig. 3.7 Mid-nineteenth-century cadastral map of the area close to Tuenno (Trento, Italy) converted in a polygonal vector file (Image created with Quantum GIS (Quantum GIS Development Team, 2015). Open Source Geospatial Foundation Project. http://qgis.osgeo.org)

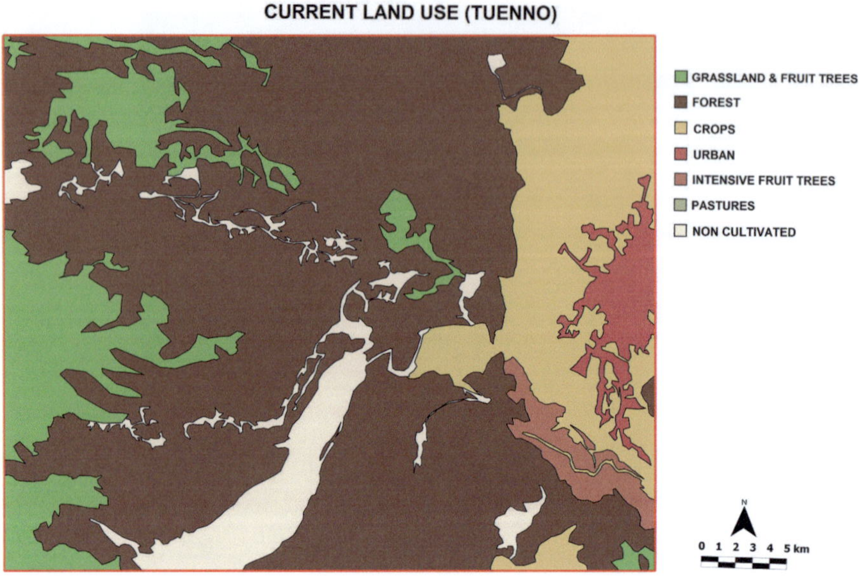

Fig. 3.8 Current aerial photograph of the area close to Tuenno (Trento, Italy) converted in a polygonal vector file (Image created with Quantum GIS (Quantum GIS Development Team 2015). Open Source Geospatial Foundation Project. http://qgis.osgeo.org)

the mid-nineteenth-century land use is listed; in the second row, the current land use is represented. In the mid-cells, each variation is quantified. In the last column and row, the total lost and gained extents for each land use category are shown. For example, in the mid-nineteenth century, forests covered 45.52 % of the total area. Reading the corresponding row, we can see that 0.16 % of the total area is now occupied by crops, 2.23 % by grassland (with minor presence of fruit trees) and 4.20 % by noncultivated lands; current urbanised (and, obviously, pasture) areas didn't gain territories from the forests. On the other hand, we can observe the column corresponding to the current extension of the forests (60.67 %) observing that 0.03 % was gained from previously urbanised areas, 0.15 % from crops, 11.67 % from grassland (with minor presence of fruit trees), 0.06 % from intensive cultivation of fruit trees, 9.14 % from pastures and 0.79 from noncultivated areas. The comparison between past and present land use is shown in the chart of Fig. 3.9.

The same analysis was conducted in three other sample areas representative of different geographical conditions. The land use alterations for the area close to Sardagna are shown in Fig. 3.10. Here the percentage land use changes over the past 150 year are computed for a mountain area of 256 ha (with altitude ranging between 600 and 1200 m a.s.l): it is evident how significant was the land use change from grassland, agricultural and fruit crops to forests. Forested areas increased from 52.07 % in the mid-nineteenth century to the present 77.80 %, with a growth of urbanised area from 1.02 to 10.95 % and a reduction of agricultural crops from 19.89 to 9.46 %.

Table 3.2 Percentage land use changes in the area close to Tuenno (Trento, Italy)

			Present land use (% out of 11.50 km²)								
			2.85	14.14	13.33	60.67	1.92	0.00	7.10		
	Total area 100 %		Urban	Crops	Grass-land & fruit trees	Forest	Intensive fruit trees	Pastures	Non cultiated		
Mid nineteenth century land use (% out 11.50 km²)	0.23	Urban	-	0.00	0.00	0.03	0.00	0.00	0.00	Lost area (%)	0.03
	11.18	Crops	1.46	-	0.03	0.15	0.14	0.00	0.03		1.81
	26.62	Grass-land & fruit trees	0.88	3.25	-	11.67	0.38	0.00	0.49		16.67
	45.42	Forest	0.00	0.16	2.23	-	0.00	0.00	4.20		6.59
	1.64	Intensive fruit trees	0.00	0.37	0.00	0.06	-	0.00	0.00		0.43
	11.93	Pastures	0.03	0.77	1.12	9.14	0.16	-	0.71		11.93
	2.98	Non cultivated	0.27	0.23	0.00	0.79	0.03	0.00	-		1.32
			Gained area (%)								
			2.64	4.77	3.38	21.84	0.71	0.00	5.43		

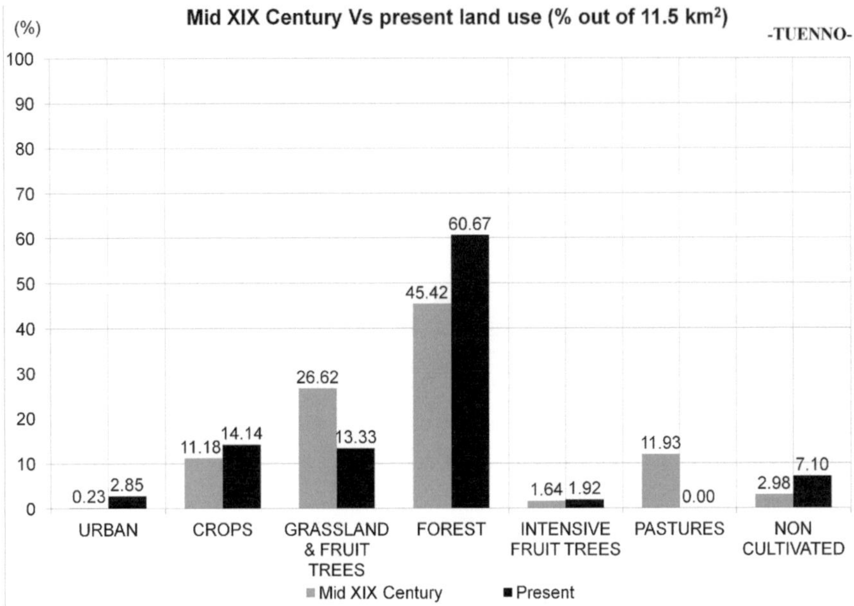

Fig. 3.9 Comparison between the mid-nineteenth century and the present land use in the area close to Tuenno (Trento, Italy)

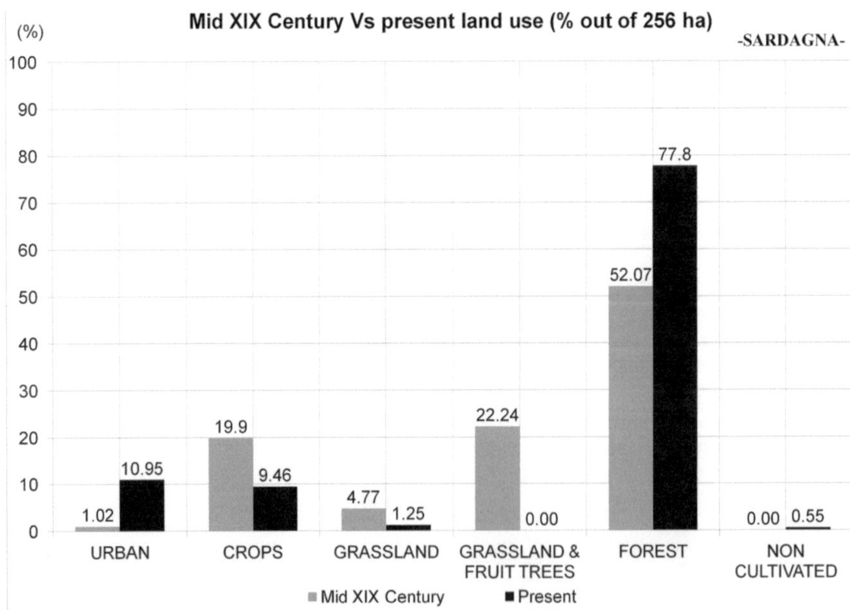

Fig. 3.10 Comparison between the mid-nineteenth century and the present land use in the area close to Sardagna (Trento, Italy)

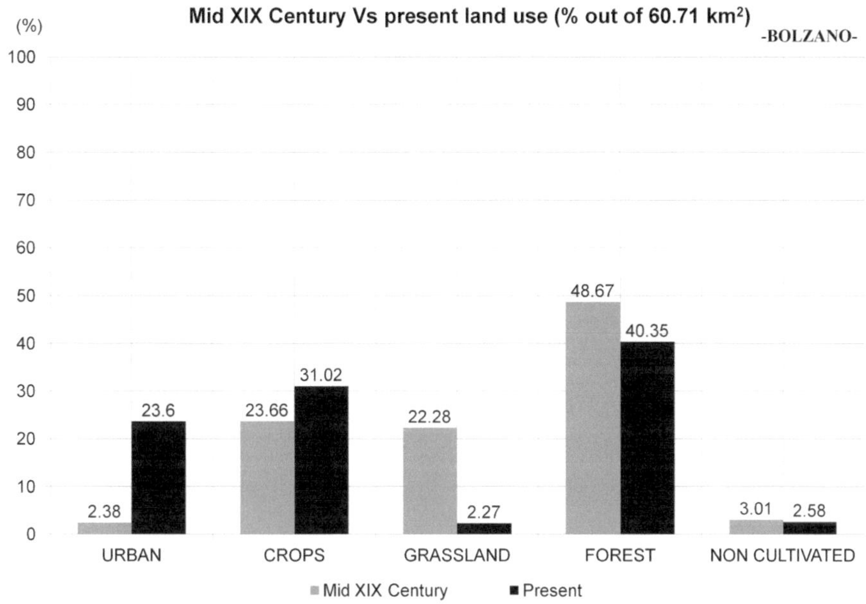

Fig. 3.11 Comparison between the mid-nineteenth century and the present land use in the area close to Bolzano (Italy)

Considering an old rural district, now transformed into a medium-size city (Bolzano), forested areas only changed from 48.67 % in the mid-nineteenth century to the present 40.35 %. The increase of urbanised area was substantial, instead: from 2.38 to 23.60 %. The agricultural crops increased from 23.66 to 31.02 % in the analysed area, while grasslands and pastures strongly decreased (see Fig. 3.11).

The land use in the mid-nineteenth-century cadastral maps was then compared with the present situation in the mid-mountain area of the Adige river basin close to Trodena (see Fig. 3.12). Here forested areas changed radically from 11.50 % in the mid-nineteenth century to a surprising current value of 86.90 %.

On the basis of the analysis conducted over the Adige watershed, the forest expansion occurring in the last decades in Europe can be confirmed. Under the hydrologic point of view, the resulting increase of evapotranspiration and interception losses from forests is one of the possible reasons of the runoff decrease.

3.5 Closing Remarks

Time series of annual runoff for the five major river basins in the Central Italian Alps were collected and analysed for trend detection applying three statistical tests. Trend analyses show a decreasing trend of mean annual runoff for all the investigated basins, significant at 5 % level according to a Mann-Kendall test for some of the

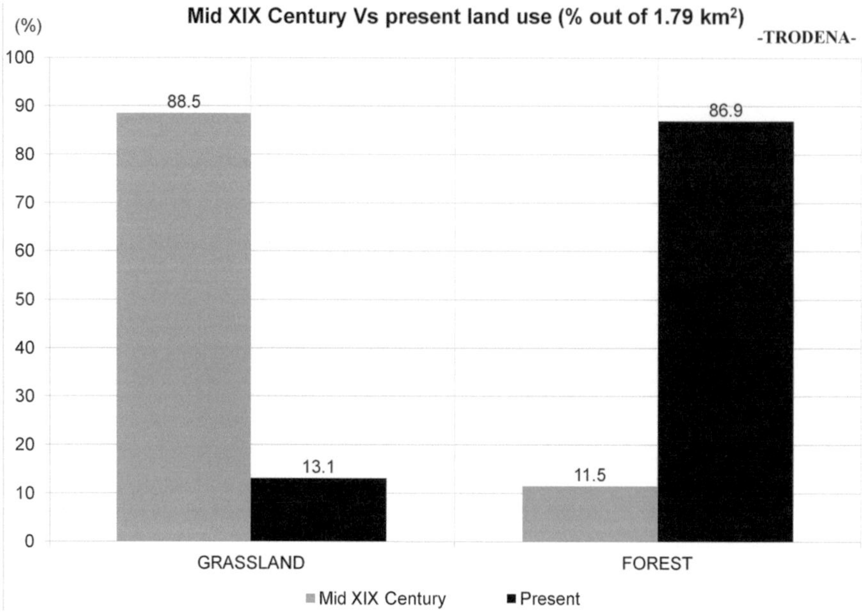

Fig. 3.12 Comparison between the mid-nineteenth century and the present land use in the area close to Trodena (Bolzano, Italy)

basins only. As the pattern of runoff decrease is similar for the Adige, Mincio, Oglio, Adda and Ticino rivers, the null hypothesis that a constant decreasing trend of $\beta = -1.26$ mm year^{-1} is exhibited by all the series was tested with the Sen-Adichie test. The hypothesis is not rejected with a 5 % significance level and the confidence limits of $\beta_1 = -1.79$ mm year^{-1} to $\beta_2 = -0.84$ mm year^{-1} were estimated. To look for possible reasons of such a decrease of runoff, the climatology of precipitation and temperature was analysed resulting in a slight, not statistically significant decrease of precipitation in the Adige and Adda river basins and a +1 °C/century increase of air temperature over the 1862–2011 period, which can be a reason of major evapotranspiration losses.

These losses can be enhanced also by the afforestation of areas previously dedicated to agriculture and pasture. An example of the analysis conducted on four representative areas with different altitudes and levels of urbanisation was presented. Considering the mountain area near Trodena, forested areas changed from 11.50 % in the mid-nineteenth century to a surprising current value of 86.90 %, with grassland having an opposite land use change. In other mid-altitude areas, the pattern is similar. In another area with altitude ranging between 600 and 1200 m.a.s.l., forested areas changed from 52.07 % in the mid-nineteenth century to the present 77.80 %, with an increase of urbanised area from 1.02 to 10.95 % and a reduction of agricultural crops from 19.89 to 9.46 % of the analysed area. The resulting increase of evapotranspiration and interception losses from forests is one of the possible reasons of the runoff decrease observed in the Adige basin and in the other investigated basins as well.

Thus anthropogenic effects, as land use changes, are factors to be included in both analyses of trends and development of adaptation and mitigation measures to climatc change.

Appendices

Appendix A: Estimation of the Regression Line

According to Theil-Sen estimator (Theil 1950a, b, c; Sen 1968; Hollander et al. 2014), trend slope can be computed as

$$\widehat{\beta} = median\left(S_{ij}, 1 \leq i < j \leq n\right), \tag{A.1}$$

where S_{ij} denote the $N = n(n\text{-}1)$ individual slopes between values Y_i and Y_j of the series, corresponding to time positions x_i and x_j:

$$S_{ij} = \frac{Y_j - Y_i}{x_j - x_i}. \tag{A.2}$$

To calculate (A.1) the sample slopes S_{ij} are ranked in ascending order $S^{(1)} \leq \ldots \leq S^{(N)}$; then

$$\widehat{\beta} = S^{(k+1)}, \quad N \text{ odd} \quad (N = 2k + 1) \tag{A.3}$$

$$\widehat{\beta} = \frac{S^{(k)} + S^{(k+1)}}{2}, \quad N \text{ even} \quad (N = 2k). \tag{A.4}$$

Accordingly, the intercept of the regression line can be estimated as (Hollander et al. 2014)

$$\widehat{\alpha} = median\ (A_1, \ldots, A_n), \quad A_i = Y_i - \widehat{\beta}x_i. \tag{A.5}$$

In this way, if $A^{(1)} \leq \ldots \leq A^{(n)}$ denote the ordered A_i values,

$$\widehat{\alpha} = A^{(k+1)}, \quad n \text{ odd} \quad (n = 2k + 1) \tag{A.6}$$

$$\widehat{\alpha} = \frac{A^{(k)} + A^{(k+1)}}{2}, \quad n \text{ even} \quad (n = 2k). \tag{A.7}$$

Appendix B: Confidence Interval for the Trend Slope

A symmetric two-sided confidence interval for the trend slope can be obtained in terms of the upper (α/2)th percentile $k_{\alpha/2}$ of the null distribution of the Kendall statistic (Hollander et al. 2014):

$$C = \sum_{i=1}^{n-1} \sum_{j=1+1}^{n} sign\left(D_j - D_i\right), \quad D_s = Y_s - \widehat{\beta} x_s. \tag{B.1}$$

Setting $C_\alpha = k_{\alpha/2}-2$ and

$$M = \frac{N - C_\alpha}{2}, \quad Q = \frac{N + C_\alpha}{2} = M + C_\alpha \quad \text{with } N = \frac{n(n-1)}{2} \tag{B.2}$$

the (1–α) confidence interval (β_L, β_U) is given by

$$\beta_L = S^{(M)}, \quad \beta_U = S^{(Q+1)} \tag{B.3}$$

according to the notation of Appendix A.

For large n, the following approximation can be used:

$$C_\alpha \approx z_{\alpha/2}\left[\frac{n(n-1)(2n+5)}{18}\right]^{1/2}, \tag{B.4}$$

where $z_{\alpha/2}$ denotes the upper (α/2)th percentile of the standard normal distribution. In general, Eq. (B.4) provides a noninteger value; in this case the integer part of the right-hand side can be used.

Appendix C: Test for the Parallelism of Several Regression Lines

The null hypothesis H_0 is that k regression lines have a common, but unspecified slope β:

$$H_0: \quad [\beta_1 = \cdots = \beta_k = \beta, \quad \text{with } \beta \text{ unspecified}]. \tag{C.1}$$

The first step is to obtain an estimate of the common slope β under the null hypothesis (C.1); to this purpose, the following pooled least squares estimator is used:

$$\overline{\beta} = \frac{\sum_{i=1}^{k}\sum_{j=1}^{n_i}\left(x_{ij}-\overline{x}_i\right)Y_{ij}}{\sum_{i=1}^{k}\sum_{j=1}^{n_i}\left(x_{ij}-\overline{x}_i\right)^2}, \qquad \overline{x}_i = \frac{1}{n_i}\sum_{j=1}^{n_i} x_{ij}, \tag{C.2}$$

where Y_{ij} is the value of the i-th response variable Y_i corresponding to the value x_{ij} and the ith independent variable x_i (for $i = 1, \ldots, k$).

Then for each of the k regression lines, the aligned observations can be computed as

$$Y_{ij}^* = Y_{ij} - \overline{\beta}x_{ij} \quad i = 1, \ldots, k; \quad j = 1, \ldots, n_i, \tag{C.3}$$

whose rank in the ith regression sample is denoted by r_{ij}^*.

Setting

$$T_i^* = \frac{\sum_{j=1}^{n_i}\left(x_{ij}-\overline{x}_i\right)r_{ij}^*}{n_i+1}; \quad i = 1, \ldots, k \tag{C.4}$$

and

$$C_i^2 = \sum_{j=1}^{n_i}\left(x_{ij}-\overline{x}_i\right)^2; \quad i = 1, \ldots, k, \tag{C.5}$$

the Sen-Adichie statistic V is given by

$$V = 12\sum_{i=1}^{k}\left[\frac{T_i^*}{C_i}\right]^2. \tag{C.6}$$

Under the null hypothesis (C.1), V is asymptotically distributed as a chi-square with k–1 Degrees of freedom (Sen 1969). This implies that hypothesis (C.1) versus the alternative

$$H_1: \quad [\beta_1, \ldots, \beta_k \quad \textit{not all equal}] \tag{C.7}$$

has to be rejected at the level of significance α if

$$V \geq \chi_{k-1,\alpha}^2 \tag{C.8}$$

where $\chi^2_{k\text{-}1,\alpha}$ is the upper αth percentile of a chi-square distribution with k–1 degrees of freedom (Adichie 1976, 1984; Hollander et al. 2014).

References

Adichie JN (1976) Testing parallelism of regression lines against ordered alternatives. Commun Stat Theory Methods 5(11):985–997

Adichie JN (1984) Rank tests in linear model. In: Krishnaiah PR, Sen PK (eds) Handbook of statistics 4: nonparametric methods. North Holland, Amsterdam, pp 229–257

Auer I, Böhm R, Jurkovic A, Lipa W, Orlik A, Potzmann R, Schoener W, Ungersboeck M, Matulla C, Briffa K, Jones P, Efthymiadis D, Brunetti M, Nanni T, Maugeri M, Mercalli L, Mestre O, Moisselin J-M, Begert M, Mueller-Westermeier G, Kveton V, Bochnicek O, Stastny P, Lapin M, Szalai S, Szentimrey T, Cegnar T, Dolinar M, Gajic-Capka M, Zaninovic K, Majstorovic Z, Nieplova E (2007) HISTALP – historical instrumental climatological surface time series of the Greater Alpine Region. Int J Climatol 27:17–46. doi:10.1002/joc.1377

Bates BC, Kundzewicz ZW, Wu S, Palutikof JP (eds) (2008) Climate change and water. Technical paper VI of the Intergovernmental Panel on Climate Change, IPCC Secretariat, Geneva

Blöschl G, Montanari A (2010) Climate change impacts: throwing the dice? Hydrol Process 24(3):374–381, 6693. doi: 10.1002/hyp.7574

Brunetti M, Lentini G, Maugeri M, Nanni T, Auer I, Böhm R, Schöner W (2009) Climate variability and change in the Greater Alpine Region over the last two centuries based on multi-variable analysis. Int J Climatol 29(15):2197–2225. doi:10.1002/joc.1857

FAO (2011) State of the world's forests, 2011. Food and Agriculture Organization of the United Nations, Rome (available at http://www.fao.org/docrep/013/i2000e/i2000e00.htm)

FAO (2015) Global forest resources assessment 2015: how are the world's forests changing? Food and Agriculture Organization of the United Nations, Rome (available at http://www.fao.org/3/a-i4793e.pdf)

Gerhard LC (2006) Climate change: conflict of observational science, theory, and politics: reply. AAPG Bull 90(3):409–412. doi:10.1306/11030505107

Hollander M, Wolfe DA, Chicken E (2014) Nonparametric statistical methods, 3rd edn. Wiley & Sons, New York

IPCC (2013) Climate change 2013: the physical science basis. In: Stocker TF, Qin D, Plattner G-K, Tignor M, Allen SK, Boschung J, Nauels A, Xia Y, Bex V, Midgley PM (eds) Contribution of Working Group I to the Fifth Assessment Report of the Intergovernmental Panel on Climate Change. Cambridge University Press, Cambridge, UK/New York, 1535 pp

Kendall MG (1970) Rank correlation methods, 4th edn. Griffin, London

Koutsoyiannis D, Efstratiadis A, Mamassis N, Christofides A (2008) On the credibility of climate predictions. Hydrol Sci J 53(4):671–684. doi:10.1623/hysj.53.4.671

Kumar S, Merwade V, Kam J, Thurner K (2009) Streamflow trends in Indiana: effects of long term persistence, precipitation and subsurface drains. J Hydrol 374:171–183. doi:10.1016/j.jhydrol.2009.06.012

Labat D, Godderis Y, Probst J, Guyot J (2004) Evidence for global runoff increase related to climate warming. Adv Water Resour 27(6):631–642. doi:10.1016/j.advwatres.2004.02.020

Legates DR, Lins HF, McCabe GJ (2005) Comments on "Evidence for global runoff increase related to climate warming". Adv Water Resour 28:1310–1315. doi:10.1016/j.advwatres.2005.04.006

Liu D, Chen X, Lian Y, Liu Z (2010) Impacts of climate change and human activities on surface runoff in the Dongjiang River basin of China. Hydrol Process 24:1487–1495. doi:10.1002/hyp.7609

Mann HB (1945) Nonparametric tests against trend. Econometrica 13(3):245–259

McCuen RH (2003) Modeling hydrologic change. Lewis Publishers, Boca Raton

Moisello U, Vullo F (2009) I massimi di portata dell'Adda a Lecco. L'Acqua 6:9–27

Montanari A, Young G, Savenije HHG, Hughes D, Wagener T, Ren LL, Koutsoyiannis D, Cudennec C, Toth E, Grimaldi S, Blöschl G, Sivapalan M, Beven K, Gupta H, Hipsey M, Schaefli B, Arheimer B, Boegh E, Schymanski SJ, Di Baldassarre G, Yu B, Hubert P, Huang Y, Schumann A, Post DA, Srinivasan V, Harman C, Thompson S, Rogger M, Viglione A,

McMillan H, Characklis G, Pang Z, Belyaev V (2013) "Panta Rhei–Everything Flows": change in hydrology and society–The IAHS Scientific Decade 2013–2022. Hydrol Sci J 58(6):1256–1275. doi:10.1080/02626667.2013.809088

Ranzi R (2012) Climate and anthropogenic change impact on the hydrological cycle, water management and engineering. In: Proceedings of 4th International Conference on Estuaries and Coasts, Hanoi, 8–11 October 2012, Science and Technics Publishing House, Hanoi (Vietnam), vol 2, pp. 439–446

Ranzi R, Barontini S (2010) Are there evidences of the impacts of global warming on runoff regimes in the Southern Alps. In: Proceedings of the international workshop on impacts of global warming from hydrological and hydraulics issues, Kyoto, 16 March 2010, pp 1–6

Ranzi R, Bochicchio M, Bacchi B (2002) Effects on floods of recent afforestation and urbanisation in the Mella River (Italian Alps). Hydrol Earth Syst Sci 6(2):239–253

Ranzi R, Nalder G, Abdalla AA, Ball J, De Costa GS, Galvão C, Jia Y, Kim YO, Kolokytha E, Lee SI, Nakakita E, Nguyen VTV, Paquier A, Patel PL, Peviani MA, Teegavarapu R (2015) Summary of recommendations for policymakers on adaption to climate change in water engineering. Hydrolink 3:93–95

Sen PK (1968) Estimates of the regression coefficient based on Kendall's tau. J Am Stat Assoc 63(324):1379–1389

Sen PK (1969) On a class of rank order tests for the parallelism of several regression lines. Ann Math Stat 40(5):1668–1683

Spearman C (1904) The proof and measurement of association between two things. Am J Psychol 15(1):72–101

Theil H (1950a) A rank-invariant method of linear and polynomial regression analysis, I. Proceedings of the Koninklijke Nederlandse Akademie van Wetenschappen Series A 53:386–392

Theil H (1950b) A rank-invariant method of linear and polynomial regression analysis, II. Proceedings of the Koninklijke Nederlandse Akademie van Wetenschappen Series A 53:521–525

Theil H (1950c) A rank-invariant method of linear and polynomial regression analysis, III. Proceedings of the Koninklijke Nederlandse Akademie van Wetenschappen Series A 53:1397–1412

UNFCCC-United Nations Framework Convention on Climate Change (2015) Adoption of the Paris Agreement. http://unfccc.int/resource/docs/2015/cop21/eng/l09r01.pdf, 32 pp

Wilcox RR (2010) Fundamentals of modern statistical methods, 2nd edn. Springer, New York

Yue S, Pilon P, Cavadias G (2002) Power of the Mann-Kendall and Spearman's rho tests for detecting monotonic trends in hydrological series. J Hydrol 259(1–4):254–271. doi:10.1016/S0022-1694(01)00594-7

Zhang C, Zhang B, Li W, Liu M (2014) Response of streamflow to climate change and human activity in Xiaoxi river basin in China. Hydrol Process 28:43–50. doi:10.1002/hyp.9539

Zolezzi G, Bellin A, Bruno MC, Maiolini B, Siviglia A (2009) Assessing hydrological alterations at multiple temporal scales: Adige River, Italy. Water Resour Res 45(12): W12421-1-W12421-15. doi: 10.1029/2008WR007266

Part II
Decision Making for Managing Water Resources Systems Under Climate Change

Chapter 4
Ranking of Global Climate Models for Godavari and Krishna River Basins, India, Using Compromise Programming

K. Srinivasa Raju, D. Nagesh Kumar, and Naga Babu I

Abstract Forty-five Coupled Model Intercomparison Project 5 (CMIP5)-based global climate models are assessed for two major river basins in India, namely, Krishna and Godavari on precipitation rate. About 44 and 59 grids of $1° \times 1°$ size are studied for this purpose, respectively. Metrics employed for evaluation of GCMs are average absolute relative anomaly (AARA), normalised root mean square anomaly (NRMSA), absolute normalised mean bias anomaly (ANMBA) and correlation coefficient (CoC). Entropy method is applied to obtain weights of AARA, NRMSA, ANMBA and CoC. Ranks of 45 GCMs are obtained through a Euclidean distance-based technique, the compromise programming. Additive ranking is employed to aggregate the grid-wise ranking pattern.

Keywords Additive ranking • Compromise programming • Entropy • Global climate models • Performance metrics

4.1 Introduction

Global warming is playing a major role in water resources planning, effecting changes in climate. Climate change impact on precipitation and temperature was studied by numerous authors (Mujumdar and Nagesh Kumar 2012; Knutti 2008). Global climate models (GCMs), simulation-based mathematical models that assess climate change, are becoming key components for studying the hydrological implications on water resources at the river basin level. This gives raise to the view that selection of suitable GCMs among the available is necessary for reliable

K.S. Raju • Naga Babu I
Department of Civil Engineering, Birla Institute of Technology and Science-Pilani, Hyderabad Campus, India
e-mail: ksraju@hyderabad.bits-pilani.ac.in; nagababuimmadi@gmail.com

D.N. Kumar (✉)
Department of Civil Engineering, Indian Institute of Science, Bangalore, India
e-mail: nagesh@civil.iisc.ernet.in

E. Kolokytha et al. (eds.), *Sustainable Water Resources Planning and Management Under Climate Change*, DOI 10.1007/978-981-10-2051-3_4

and practical implementation at the basin level. These discussed aspects lead to the necessity for evaluating GCMs based on relevant performance metrics for possible ranking of the same (Tebaldi and Knutti 2007; Raju and Nagesh Kumar 2014). Concise literature review is presented to highlight relevant recent studies.

Alexander and Arblaster (2009) analysed multiple simulations from nine GCMs of Coupled Model Intercomparison Project 3 (CMIP3) for predicting extremes of precipitation and temperature over Australia. Significant variations between individual models were observed even though most of the GCMs indicated reasonable trend. Reifen and Toumi (2009) analysed 17 GCMs for surface temperature. The emanated results suggested that the average of all GCMs can be used as the basis for predicting climate change. Kundzewicz and Stakhiv (2010) discussed the ability of GCMs for their use in infrastructure and water resources planning and management problems. Knutti et al. (2010) discussed model diversity, potential challenges, independence and structural errors. They stressed the importance of weighting of models. Wilby (2010) analysed the utility of GCMs for adaptation and long-term planning and suggested guidelines for selecting GCMs for hydrological applications.

Crosbie et al. (2011) suggested to use multiple GCMs for adaptation studies and cautioned regarding the choice of downscaling methods. Schaller et al. (2011) analysed 24 CMIP3-based GCMs for precipitation and temperature simulations. They concluded that multimodel average outperforms the individual models and investigated causes there for. Anandhi et al. (2011) evaluated the efficacy of 19 Coupled Model Intercomparison Project 5 (CMIP5) GCMs, based on skill score, to simulate variations in the snow water equivalent in water supply watersheds of New York City. GFDL 2.0 and CCSM models were found to be the high- and low-performing models, respectively, and grouped the GCMs into low, medium and high performing based on skill score. Overland et al. (2011) suggested not to use single model evaluation for regional projection and provided guidelines for model credibility. Chen et al. (2012) analysed nine GCMs for their efficacy to simulate the near-surface wind over China. All GCMs exhibited lower interannual variability than the reanalysis data. Chen et al. (2012) investigated climate change impact on run-off and opined that single GCM outputs would not be robust.

Fu et al. (2013) assessed the performance of 25 GCMs for monthly and annual rainfall, monthly mean sea level pressure and air temperature over the Southeastern Australia region on 11 assessment criteria. Model for Interdisciplinary Research on Climate (MIROC) 3.2_hires and Beijing Climate Center (BCC)-CM1 are the high- and low-performing models for monthly and annual temperature, whereas Institute for Numerical Mathematics (INM)-CM30 and BCC-CM1 are high and low preferred for monthly and annual mean sea level pressure. Sillmann et al. (2013) analysed CMIP5-based GCMs in simulating daily precipitation and temperature and concluded that the models were able to simulate climate extremes quite well. Yao et al. (2013) analysed 28 CMIP5- and 24 CMIP3-based GCMs for simulating extreme monthly average temperatures and inferred that both groups of models performed well.

Shili et al. (2014) analysed 12 simulations from CMIP5 models over China and concluded that CanESM2, CCSM4 and MIROC-ESM were found to be suitable in replicating the average index values and ascertained that extreme temperature indices were better simulated than the extreme indices of precipitation. Elguindi et al. (2014) assessed 45 CMIP5 GCMs and observed that model resolution improved the skill of the models in most of the cases and improvement was significantly visible for temperature. Grose et al. (2014) compared the performance of 27 CMIP5 GCMs with that of CMIP3 GCMs for western tropical Pacific and found that CMIP5 models showed improvement than CMIP3 models. Ludwig et al. (2014) discussed various perspectives of climate change adaptation and integrated water resources planning and management and their relationship in the present planning situations. Perez et al. (2014) evaluated the skill of GCMs of CMIP3 and CMIP5 for North-east Atlantic Ocean region. ECHAM5/MPI-OM, UKMO-HadGEM2 and MIROC3.2 (hires) from CMIP3 and CMCC-CM, EC-EARTH, ACCESS1.0, HadGEM2-ES and HadGEM2-CC from CMIP5 were preferred. Lee and Wang (2014) studied future changes by employing 20 CMIP5 GCMs due to anthropogenic global warming and evaluated the performance of the models. Mehrotra et al. (2014) explored the potential of nine CMIP5 GCMs and suggested structural modelling approach.

Wang et al. (2015) evaluated the skill of 23 GCMs of CMIP3 and 25 of CMIP5 in simulating the spatial structure of sea surface temperature variability and found that CMIP5 ensemble performed better than the CMIP3 ensemble. Chen and Sun (2015) analysed the performance of CMIP5 GCMs in simulating climate extreme events in China, and comparison was made with that of CMIP3 GCMs. Shamir et al. (2015) in their study of eight dynamically downscaled GCMs for upper Santa Cruz River inferred that climate change projections increased the uncertainty in precipitation. Gulizia and Camilloni (2015) evaluated the ability of CMIP5 and CMIP3 GCMs over South America, south of the equator and in three particular subregions. CMIP3 and CMIP5 models' ensembles exhibited some differences in precipitation simulation. Moise et al. (2012), DelSole and Shukla (2012) and Su et al. (2013) studied similar aspects.

It is evident from the studied literature that most of the researchers ranked GCMs with individual criterion. No significant studies are reported to consider the problem in multiobjective perspective for precipitation rate in CMIP5 environment. Keeping this aspect in view, the proposed objectives of the study are (i) performance metrics identification and computation of entropy approach-based weights of performance metrics and (ii) application of Euclidean distance-based technique, a compromise programming for grid-wise ranking of GCMs for two major Indian river basins, namely, Krishna and Godavari. Metrics employed for evaluation of GCMs are average absolute relative anomaly (AARA), normalised root mean square anomaly (NRMSA), absolute normalised mean bias anomaly (ANMBA) and correlation coefficient (CoC). Additive ranking approach (Bui 1987) is applied to aggregate the ranking pattern for Krishna and Godavari river basins.

In total 45 GCMs, namely, ACCESS1.3, ACCESS1.0, BCC-CSM 1.1 m, BCC-CSM 1.1, BNU-ESM, CCSM4, CANESM2, CESM1(WACCM), CESM1

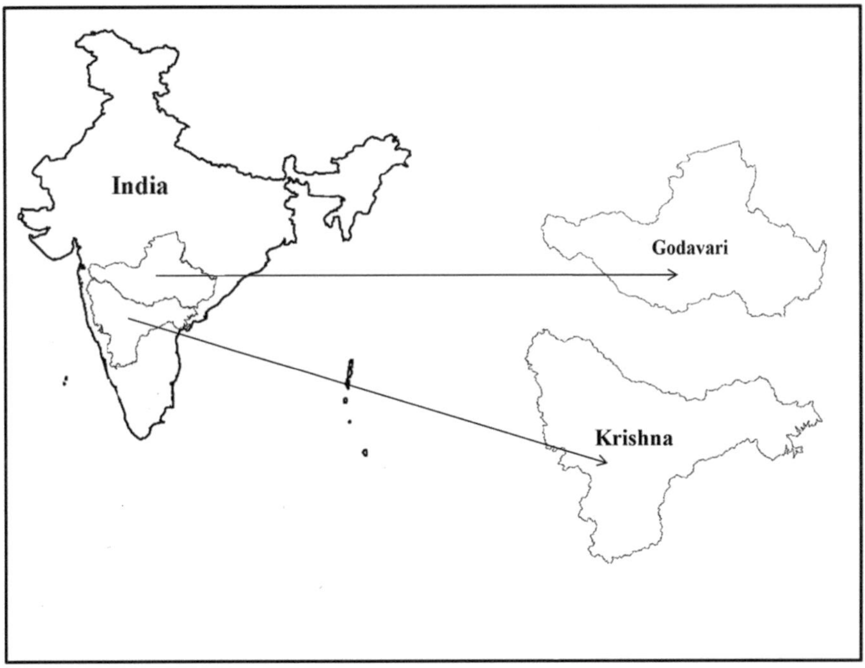

Fig. 4.1 Location of Godavari and Krishna river basins

(FASTCHEM), CESM1(CAM5), CESM1(BGC), CMCC-CMS, CMCC-CM, CMCC-CESM, CNRM-CM5, CNRM-CM5-2, CSIRO-Mk3.6.0, EC-EARTH, FIO-ESM, FGOALS-g2, GFDL-CM3, GFDL-ESM2M, GFDL-ESM2G, GISS-E2-R, GISS-E2-H-CC, GISS-E2-H, GISS-E2-R-CC, HadCM3, HadGEM2-ES, HadGEM2-CC, HadGEM2-AO, INM-CM4, IPSL-CM5A-MR, IPSL-CM5A-LR, IPSL-CM5B-LR, MIROC5, MIROC-ESM-CHEM, MIROC-ESM, MIROC4h, MPI-ESM-P, MPI-ESM-MR, MPI-ESM-LR, MRI-ESM1, MRI-CGCM3 and NorESM1-M, are assessed for predicting precipitation rate for Krishna and Godavari river basins. Location map of Krishna and Godavari river basins is presented in Fig. 4.1. Introduction after which details of the literature review, performance metrics and techniques employed, and case study area, results are presented followed by analysis and conclusions.

4.2 Performance Metrics and Techniques

The metrics presented in Table 4.1 are employed for evaluating the GCMs.

Entropy method is used to compute the weights of the performance metrics based on entropy value (E) and degree of diversification (1-E). Normalisation of degree of diversification values yields the required weights (Pomerol and Romero 2000).

Table 4.1 Metrics employed in the present study

Metric	Mathematical representation	Indication
Correlation coefficient (CoC)	$CoC=\frac{\sum_{i=1}^{m}(x_i-\bar{x})(y_i-\bar{y})}{(n-1)\sigma_x\sigma_y}$ x_i & y_i, $\bar{x}$ & $\bar{y}$, σ_x & σ_y are observed, average and standard deviation values of x and y; m is number of observations	Strength of the relationship which is linear can be assessed between the simulated and observed. Range of CoC is −1 to +1
Absolute normalised mean bias anomaly (ANMBA)	$ANMBA=\left\vert\frac{\frac{1}{m}\left(\sum_{i=1}^{m}(y_i-x_i)\right)}{\frac{1}{m}\sum_{i=1}^{m}x_i}\right\vert$	Smaller values of ANMBA indicate better performance
Normalised root mean square anomaly (NRMSA)	$NRMSA=\frac{\left[\left(\frac{1}{m}\right)\sum_{i=1}^{m}(x_i-y_i)^2\right]^{1/2}}{\left(\frac{1}{m}\right)\sum_{i=1}^{m}x_i}$	Smaller values of NRMSA indicate better performance
Average absolute relative anomaly (AARA)	$AARA=\frac{1}{m}\sum_{i=1}^{m}\left\vert\frac{(y_i-x_i)}{x_i}\right\vert$	Smaller values of AARA indicate better performance

Table 4.2 Description of compromise programming technique

Step	Description	Mathematical expression/remark
1	Normalise the payoff matrix if required	Choose suitable method of normalisation
2	Determine ideal value for each metric j among available GCMs	$f_j^*()$ $j=1,2,\ldots..J$ where J is number of metrics
3	Calculate the Lp value for each GCM a	$L_{pa}=\left[\sum_{j=1}^{J}w_j^p\left\vert f_j^*-f_j(a)\right\vert^p\right]^{\frac{1}{p}}$ $f_j(a)$ = Metric value of j for GCM a; w_j = Weight assigned to the metric j; p = Parameter ($p=1,2,..$)
4	Rank the GCMs based on the Lp	The lower the Lp value, the better the GCM

Compromise programming, a decision-making technique based on Euclidean distance measure, is analysed in this study. It involves steps like payoff matrix normalisation, identification of ideal values, computation of L_p metric values of GCMs and ranking accordingly. Table 4.2 presents details of compromise programming.

Additive ranking is arithmetic addition of ranks of each GCM obtained by compromise programming across grid points and is used in group decision-making approach for aggregating the ranks.

4.3 Case Study and Data

Godavari river basin lies between 16° 16′ N to 22° 36′ N and 73° 26′ E and 83° 07′ E with catchment area of 312,812 km^2. The catchment of river Godavari extends in the states of Maharashtra, Madhya Pradesh, Andhra Pradesh and Telangana, Chhattisgarh, Karnataka and Orissa to the extent of 48.6 %, 10.0 %, 23.40 %, 10.9 %, 1.4 % and 5.7 % of the total catchment area, respectively. The number of grid points covered for this area with 1° × 1° grid is 59. Krishna river basin lies between 13° 07′ N to 19° 25′N and 73° 21′ E to 81° 09′ with catchment area of 258,948 km^2. The catchment of river Krishna extends in the states of Maharashtra, Karnataka, Andhra Pradesh and Telangana to the extent of 26.8 %, 43.7 % and 29.5 % of the total catchment area, respectively. The number of grid points covered for this area with 1° × 1° grid is 44. The location of these basins is presented in Fig. 4.1. Details of these basins are available in Integrated Hydrological Data Book (2012).

Indian Meteorological Department (IMD) precipitation data for the period 1951–2005 is used as observed data at a grid resolution of 1° × 1°. All the GCM data which were available at various grid sizes are interpolated using the nearest neighbourhood approach to 1° × 1° for comparison with IMD data (Raju and Nagesh Kumar 2014).

4.4 Results and Analysis

4.4.1 Krishna River Basin

Weights were computed using entropy-based approach for all the grids of Krishna river basin. It is observed that weights are varying for both metric and grid. CoC is varying from 32.34 to 98.96 %, NRMSA from 0.12 to 21.00 %, ANMBA from 0.17 to 37.60 % and AARA from 0.42 to 12.96 %. CoC occupied the first position all the 44 times (i.e. grid points), and AARA, NRMSA and ANMBA have not attained the first position at all. Out of the four metrics, CoC is thus found to be the preferred metric. Weight distribution of metrics in different ranges over the 44 grid points obtained by entropy method is presented in Table 4.3. It is noticed that the numbers of grid points within 10 % weightage are 24, 22, 32 and 0 in the case of NRMSA, ANMBA, AARA and CoC, whereas CoC is spread out across almost all weight ranges above 30 % barring one weight range.

Compromise programming is applied to compute L_p metric value of each GCM for each grid point, to compute the rank of each GCM. MATLAB-based computer program as shown in Fig. 4.2 is developed for compromise programming. Analysis is conducted for all the 44 grid points of Krishna river basin. Parameter p is chosen as 2. Occupation of GCMs in the first, second, third and last positions (among 44 grids) is presented in columns 2, 3, 4 and 5 of Table 4.4, respectively.

Table 4.3 Distribution of ranges of weights over 44 grid points for Krishna basin obtained by entropy method

Range of weights (in %)	CoC	NRMSA	ANMBA	AARA
≤10	–	24	22	32
>10 and ≤20	–	19	–	12
>20 and ≤30	–	1	9	–
>30 and ≤40	8	–	13	–
>40 and ≤50	5	–	–	–
>50 and ≤60	8	–	–	–
>60 and ≤70	1	–	–	–
>70 and ≤80	–	–	–	–
>80 and ≤90	1	–	–	–
>90 and ≤100	21	–	–	–

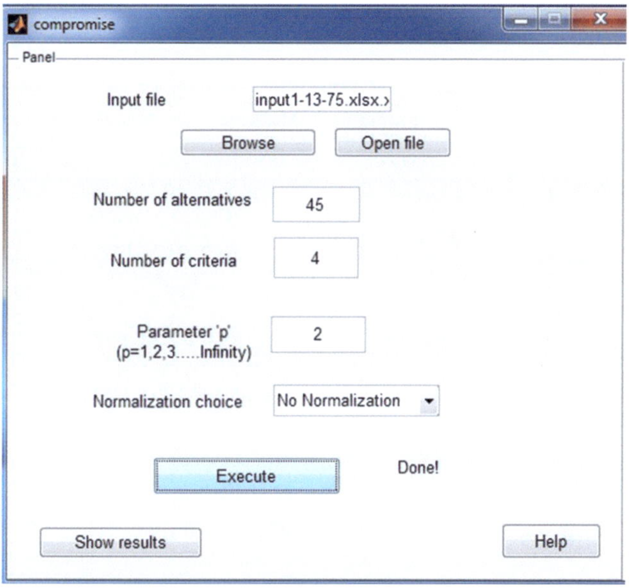

Fig. 4.2 Overview of compromise programming module

- CANESM2, CCSM4, CMCC-CESM, GISS-E2-H-CC, GISS-E2-H, GISS-E2-R-CC, HadGEM2-ES and IPSL-CM5A-MR never occupied the first three or last positions over 44 grid points in Krishna river basin. Hence, these are not included in Table 4.4.
- EC-EARTH, GFDL-CM3, INM-CM4 and NorESM1-M occupied the first position 14, 8, 4 and 6 times (32/44, i.e. 72.72 %) (column 2, Table 4.4).
- CESM1(WACCM), EC-EARTH, GFDL-ESM2G, MIROC-ESM and MPI-ESM-P and MPI-ESM-MR occupied the second position 3, 4, 3, 3, 6 and 3 times (22/44, i.e. 50.00 %) (column 3, Table 4.4).

Table 4.4 Number of times the GCMs occupied the first, second, third and last positions in Krishna river basin

GCM	First position	Second position	Third position	Last position
ACCESS1.3	–	–	–	2
ACCESS1.0	–	2	1	3
BCC-CSM 1.1m	–	1	–	–
BCC-CSM 1.1	–	–	–	2
BNU-ESM	–	2	3	–
CESM1(WACCM)	2	3	3	4
CESM1(FASTCHEM)	1	1	–	–
CESM1(CAM5)	1	1	2	1
CESM1(BGC)	1	1	4	–
CMCC-CMS	1	1	1	–
CMCC-CM	–	–	1	1
CNRM-CM5	–	2	–	–
CNRM-CM5-2	–	1	2	–
CSIRO-Mk3.6.0	–	–	1	–
EC-EARTH	14	4	3	–
FIO-ESM	–	2	–	–
FGOALS-g2	–	–	–	1
GFDL-CM3	8	1	1	–
GFDL-ESM2M	–	–	–	1
GFDL-ESM2G	–	3	1	3
GISS-E2-R	–	1	–	1
HadCM3	–	–	1	–
HadGEM2-CC	–	–	1	–
HadGEM2-AO	–	–	–	1
INM-CM4	4	2	2	–
IPSL-CM5A-LR	1	–	–	4
IPSL-CM5B-LR	–	–	1	10
MIROC5	–	–	5	1
MIROC-ESM-CHEM	1	1	1	–
MIROC-ESM	–	3	–	–
MIROC4h	–	–	–	6
MPI-ESM-P	1	6	2	–
MPI-ESM-MR	1	3	1	–
MPI-ESM-LR	2	2	5	–
MRI-ESM1	–	–	1	1
MRI-CGCM3	–	–	–	2
NorESM1-M	6	1	1	–

Table 4.5 Distribution of ranges of weights over 59 grid points for Godavari basin obtained by entropy method

Range of weights (in %)	CoC	NRMSA	ANMBA	AARA
≤10	–	32	30	41
>10 and ≤20	–	26	1	18
>20 and ≤30		1	13	–
>30 and ≤40	7	–	15	–
>40 and ≤50	9	–	–	–
>50 and ≤60	13	–	–	–
>60 and ≤70	–	–	–	–
>70 and ≤80	–	–	–	–
>80 and ≤90	7	–	–	–
>90 and ≤100	23	–	–	–

- BNU-ESM, CESM1(WACCM), CESM1(BGC), EC-EARTH, MIROC5 and MPI-ESM-LR occupied the third position 3, 3, 3, 5 and 5 times (23/44, i.e. 52.27 %) (column 4, Table 4.4).
- ACCESS1.0, CESM1(WACCM), GFDL-ESM2G, IPSL-CM5B-LR, IPSL-CM5A-LR and MIROC4h occupied the last position 3, 4, 3, 10, 4 and 6 times (30/44, i.e. 68.18 %) (column 5, Table 4.4).
- Additive ranking rule is applied to aggregate the ranking over 44 grid points of Krishna river basin. It is found that EC-EARTH, CNRM-CM5 and CNRM-CM5-2 occupied first three positions based on their aggregate ranking points, i.e. ranking of each GCM over all grid points (368, 668, 710), whereas IPSL-CM5B-LR occupied the last position (ranking points 1377).

4.4.2 Godavari River Basin

CoC is varying from 32.34 to 98.86 %, NRMSA from 0.06 to 21.00 %, ANMBA from 0.09 to 37.60 % and AARA from 0.79 to 16.99 %. Out of the 59 evaluations, CoC occupied the first position all the 59 times (i.e. 59 grid points), and AARA, NRMSA and ANMBA have not attained the first position at all. Out of the four metrics, CoC is thus found to be the preferred metric. It is observed from Table 4.5 that the numbers of grid points within 10 % weightage are 32, 30, 41 and 0 in the case of NRMSA, ANMBA, AARA and CoC.

Similar procedure as explained in the previous section is repeated for Godavari river basin for computing ranking pattern. Table 4.6 presents the occupation of GCMs in first, second, third and last positions (among chosen grids) which are presented in columns 2, 3, 4 and 5, respectively.

- BCC-CSM 1.1, BCC-CSM 1.1 m, CANESM2, GISS-E2-H, FGOALS-g2 and MRI-CGCM3 never occupied the first three or the last positions over 59 grid points in Godavari river basin. Hence, these are not included in Table 4.6.
- For one grid point 22° N 80° E, both CCSM4 and CESM1 (FASTCHEM) occupied the third position making the number of times of occupation 60 against 59 grid points.
- EC-EARTH, GFDL-CM3, HadGEM2-ES, INM-CM4, IPSL-CM5A-LR, MIROC-ESM-CHEM and NorESM1-M occupied the first position 12, 9, 3, 11, 4, 3 and 3 times (45/59, i.e. 76.27 %) (column 2, Table 4.6).
- ACCESS1.0, CESM1(CAM5), CMCC-CESM, EC-EARTH, GFDL-CM3, INM-CM4, IPSL-CM5B-LR, MIROC-ESM-CHEM, MPI-ESM-P and NorESM1-M occupied the second position 4, 3, 3, 3, 4, 3, 4, 4, 6 and 3 times (37/59, i.e. 62.71 %) (column 3, Table 4.6).
- CMCC-CMS, EC-EARTH, HadGEM2-AO, MIROC5, MPI-ESM-P, MPI-ESM-LR and NorESM1-M occupied the third position 4, 4, 4, 5, 3, 4 and 4 times (28/59, i.e. 47.46 %) (column 4, Table 4.6).
- ACCESS1.3, ACCESS1.0, CESM1(WACCM), CMCC-CMS, GFDL-ESM2G, GISS-E2-R, HadCM3, IPSL-CM5B-LR, MIROC5 and MIROC4h occupied the last position 4, 5, 5, 3, 3, 7, 5, 6, 3 and 7 times (48/59, i.e. 81.35 %) (column 5, Table 4.6).
- Additive ranking rule is applied to aggregate the ranking over 59 grid points of Godavari river basin. It is found that EC-EARTH, MPI-ESM-LR, INM-CM4 and GFDL-ESM-2 M occupied the first three positions and last position, respectively, based on their aggregate ranking points (588, 777, 846 and 1857).

4.5 Summary and Conclusions

Four performance metrics – correlation coefficient (CoC), normalised root mean square anomaly (NRMSA), absolute normalised mean bias anomaly (ANMBA) and average absolute relative anomaly (AARA) – are used to rank 45 GCMs of CMIP5 for Krishna and Godavari basins for the climate variable, the precipitation rate. Weights of the metrics are computed by entropy method. Euclidean distance-based method, compromise programming, is employed to rank the 45 CMIP5 GCMs.

Table 4.6 Number of times the GCMs occupied the first, second, third and last positions in Godavari river basin

GCM	First position	Second position	Third position	Last position
ACCESS1.3	1	–	–	4
ACCESS1.0	1	4	2	5
BNU-ESM	1	1	2	–
CCSM4	–	1	1	–
CESM1(WACCM)	–	–	1	5
CESM1(FASTCHEM)	1	1	1	2
CESM1(CAM5)	–	3	2	–
CESM1(BGC)	1	1	2	1
CMCC-CMS	–	–	4	3
CMCC-CM	–	1	2	1
CMCC-CESM	–	3	2	–
CNRM-CM5	1	2	2	–
CNRM-CM5-2	–	2	2	–
CSIRO-Mk3.6.0	–	–	1	–
EC-EARTH	12	3	4	–
FIO-ESM	–	1	–	–
GFDL-CM3	9	4	–	–
GFDL-ESM2M	–	–	–	2
GFDL-ESM2G	–	–	1	3
GISS-E2-R	–	1	2	7
GISS-E2-H-CC	1	1	–	–
GISS-E2-R-CC	–	–	1	–
HadCM3	–	–	–	5
HadGEM2-ES	3	1	–	–
HadGEM2-CC	2	1	2	–
HadGEM2-AO	–	2	4	–
INM-CM4	11	3	–	–
IPSL-CM5A-MR	1	1	1	2
IPSL-CM5A-LR	4	1	–	1
IPSL-CM5B-LR	–	4	2	6
MIROC5	–	–	5	3
MIROC-ESM-CHEM	3	4	–	–
MIROC-ESM	–	1	2	–
MIROC4h	1	–	–	7
MPI-ESM-P	–	6	3	–
MPI-ESM-MR	1	2	1	–
MPI-ESM-LR	2	1	4	–
MRI-ESM1	–	–	–	2
NorESM1-M	3	3	4	–

Conclusions emanated from the study are:

1. EC-EARTH, MPI-ESM-LR and INM-CM4 occupied the first three positions for Godavari basin whereas EC-EARTH, CNRM-CM5 and CNRM-CM5-2 in the case of Krishna river basin.
2. EC-EARTH, GFDL-CM3, INM-CM4 and NorESM1-M occupied the first position 14, 8, 4 and 6 times for Krishna river basin contributing to 72.72 %.
3. EC-EARTH, GFDL-CM3, HadGEM2-ES, INM-CM4, IPSL-CM5A-LR, MIROC-ESM-CHEM and NorESM1-M occupied the first position 12, 9, 3, 11, 4, 3 and 3 times for Godavari river basin contributing to 76.27 %.
4. Correlation coefficient is found to be a significant metric to assess the association between the ranking patterns obtained.

The present study in our opinion helps to formulate sustainable water resources planning strategies in the changing climate environment. This can be possible through the downscaling approaches that can be based on suitable GCMs suggested in this case study.

Acknowledgement This work is supported by the Council of Scientific and Industrial Research, New Delhi, through project no. 23(0023)/12/EMR-II dated 15 October 2012. Authors acknowledge the modelling groups for making their simulations available for analysis, the Program for Climate Model Diagnosis and Intercomparison (PCMDI) for collecting and archiving the CMIP5 model output and the WCRP's Working Group on Coupled Modelling (WGCM) for organising the model data analysis activity.

References

Alexander LV, Arblaster JM (2009) Assessing trends in observed and modelled climate extremes over Australia in relation to future projections. Int J Climatol 29:417–435

Anandhi A, Frei A, Pradhanang SM, Zion MS, Pierson DC, Schneiderman EM (2011) AR4 climate model performance in simulating snow water equivalent over Catskill Mountain watersheds, New York, USA. Hydrol Process 25:3302–3311

Bui TX (1987) Coop: a group decision support system for cooperative multiple criteria group decision making. Springer, Berlin

Chen H, Sun J (2015) Assessing model performance of climate extremes in China: an intercomparison between CMIP5 and CMIP3. Clim Chan 129:197–211

Chen H, Xu CY, Guo S (2012) Comparison and evaluation of multiple GCMs, statistical downscaling and hydrological models in the study of climate change impacts on runoff. J Hydrol 434–435:36–45

Chen L, Pryor SC, Li D (2012) Assessing the performance of Intergovernmental Panel on Climate Change AR5 climate models in simulating and projecting wind speeds over China. J Geophys Res 117:D24102

Crosbie RS, Dawes WR, Charles SP, Mpelasoka FS, Aryal S, Barron O, Summerell GK (2011) Differences in future recharge estimates due to GCMs, downscaling methods and hydrological models. Geophys Res Lett 38:L11406

DelSole T, Shukla J (2012) Climate models produce skillful predictions of Indian summer monsoon rainfall. Geophys Res Lett 39:L09703

Elguindi N, Giorgi F, Turuncoglu U (2014) Assessment of CMIP5 global model simulations over the subset of CORDEX domains used in the Phase I CREMA. Clim Chan 125:7–21
Fu G, Liu Z, Charles SP, Xu Z, Yao Z (2013) A score-based method for assessing the performance of GCMs: a case study of southeastern Australia. J Geophys Res 118:4154–4167
Grose MR, Brown JN, Narsey S, Brown JR, Murphy BF, Langlais C, Gupta AS, Moise AF, Irving DB (2014) Assessment of the CMIP5 global climate model simulations of the western tropical Pacific climate system and comparison to CMIP3. Int J Climatol 34:3382–3399
Gulizia C, Camilloni I (2015) Comparative analysis of the ability of a set of CMIP3 and CMIP5 global climate models to represent precipitation in South America. Int J Climatol 35:583–595
Integrated Hydrological Data Book: Non-classified River Basins (2012) Central Water Commission, New Delhi
Knutti R (2008) Should we believe model predictions of future climate change? Philos Trans R Soc 366:4647–4664
Knutti R, Furrer R, Tebaldi C, Cermak J, Meehl GA (2010) Challenges in combining projections from multiple climate models. J Clim 23:2739–2758
Kundzewicz ZW, Stakhiv EW (2010) Are climate models ready for prime time in water resources management applications, or is more research needed? Hydrol Sci J 55(7):1085–1089
Lee JY, Wang B (2014) Future change of global monsoon in the CMIP5. Clim Dyn 42:101–119
Ludwig F, Slobbe EV, Cofino W (2014) Climate change adaptation and Integrated Water Resource Management in the water sector. J Hydrol 518:235–242
Mehrotra R, Sharma A, Bari M, Tuteja N, Amirthanathan G (2014) An assessment of CMIP5 multi-model decadal hindcasts over Australia from a hydrological viewpoint. J Hydrol 519:2932–2951
Moise F, Colman RA, Brown JR (2012) Behind uncertainties in projections of Australian tropical climate: analysis of 19 CMIP3 models. J Geophys Res 117:D10103
Mujumdar PP, Nagesh Kumar D (2012) Floods in a changing climate: hydrologic modeling international hydrology series. Cambridge University Press, Cambridge
Overland JE, Wang M, Bond NA, Walsh JE, Kattsov VM, Chapman WL (2011) Considerations in the selection of global climate models for regional climate projections: the arctic as a case study. J Clim 24:1583–1597
Perez J, Menendez M, Mendez FJ, Losada LJ (2014) Evaluating the performance of CMIP3 and CMIP5 global climate models over the north-east Atlantic region. Clim Dyn 43:2663–2680
Pomerol JC, Romero SB (2000) Multicriterion decision in management: principles and practice. Kluwer, Boston
Raju KS, Nagesh Kumar D (2014) Ranking of global climatic models for India using multicriterion analysis. Clim Res 60(2):103–117
Reifen C, Toumi R (2009) Climate projections: past performance no guarantee of future skill? Geophys Res Lett 36:L13704
Schaller N, Mahlstein I, Cermak J, Knutti R (2011) Analyzing precipitation projections: a comparison of different approaches to climate model evaluation. J Geophys Res 116:D10118
Shamir E, Megdal SB, Carrillo C, Castro CL, Chang HI, Chief K, Corkhill FE, Eden S, Georgakakos KP, Nelson KM, Prietto J (2015) Climate change and water resources management in the Upper Santa Cruz River, Arizona. J Hydrol 521:18–33
Shili Y, Jinming F, Wenjie D, Jieming C (2014) Analyses of extreme climate events over China based on CMIP5 historical and future simulations. Adv Atmos Sci 31(5):1209–1220
Sillmann J, Kharin VV, Zhang X, Zwiers FW, Bronaugh D (2013) Climate extremes indices in the CMIP5 multimodel ensemble: Part 1. Model evaluation in the present climate. J Geophys Res Atmos 118:1716–1733
Su F, Duan X, Chen D, Hao Z, Cuo L (2013) Evaluation of the global climate models in the CMIP5 over the Tibetan Plateau. J Clim 26:3187–3208
Tebaldi C, Knutti R (2007) The use of the multi-model ensemble in probabilistic climate projections. Philos Trans R Soc 365:2053–2075

Wang G, Dommenget D, Frauen C (2015) An evaluation of the CMIP3 and CMIP5 simulations in their skill of simulating the spatial structure of SST variability. Clim Dyn 44:95–114

Wilby RL (2010) Evaluating climate model outputs for hydrological applications – Opinion. Hydrol Sci J 55(7):1090–1093

Yao Y, Luo Y, Huang J, Zhao Z (2013) Comparison of monthly temperature extremes simulated by CMIP3 and CMIP5 models. J Clim 26:7692–7707

Chapter 5
Integrated Reservoir Operation Considering Real-Time Hydrological Prediction for Adaptive Water Resources Management

Daisuke Nohara and Tomoharu Hori

Abstract Changes in hydrological systems associated with climate change may threaten to reliability of water resources management systems. Sophistication of reservoir operation is considered to be important to enhance existing water resources management systems. In this chapter, an adaptive operation method of a multipurpose reservoir considering real-time hydrological prediction is presented for more robust water resources management under climate change. A method to design and assess integrated reservoir operation for both the flood control and water utilization is developed with consideration of imperfect real-time hydrological predictions in order to enhance the capability of an existing multipurpose reservoir.

Keywords Reservoir operation • Prior release • Flood control • Water supply • Power generation • Water resources management system • Multipurpose reservoir • Hydrometeorological prediction • Real-time hydrological prediction • Monte Carlo simulation • Prediction uncertainty • Synthetic generation of inflow prediction • Synthetic generation of long-term inflow regimes • Autoregressive model • Normal distribution • Lognormal distribution • Error growth function • Thomas-Fiering model • Serial correlation

5.1 Introduction

Water-related disasters such as floods or droughts hinder the development of society and economy in the affected areas. Adequate water resources management is therefore needed to establish secure society protected from threats of those disasters to human lives and livelihoods. As reservoirs play a significant role in the water resources management, improved management of reservoirs is considered to enhance existing water resources management systems.

D. Nohara (✉) • T. Hori
Disaster Prevention Research Institute, Kyoto University, Gokasho, Uji 611-0011, Japan
e-mail: nohara.daisuke.2v@kyoto-u.ac.jp; hori.tomoharu.3w@kyoto-u.ac.jp

E. Kolokytha et al. (eds.), *Sustainable Water Resources Planning and Management Under Climate Change*, DOI 10.1007/978-981-10-2051-3_5

On the other hand, scientific reports or studies warn about the possibility that changing climate affects the hydrological cycle. Intergovernmental Panel on Climate Change (IPCC 2014) reported that changing precipitation as well as melting snow and ice are altering hydrological systems in many regions (with medium confidence), affecting water resources in terms of quantity and quality. IPCC (2014) also indicated the possibility that frequency of extreme events of flood or drought may increase in some regions. Although the impact of climate change on hydrological systems is considered to vary from region to region, changes in hydrological systems associated with changing climate may threaten reliability of existing water resources management system developed so as to adapt to the current hydrological system. From this point of view, sophistication of reservoir management is also considered to be important to enhance existing water resources management systems.

Although the ability of reservoir systems to control water-related disasters can be enhanced with many measures including construction of new reservoirs and enlargement of an existing reservoir's capacity to store or release water, sophistication of real-time operation of existing reservoirs would be an attractive option. It usually needs much lower cost and shorter time to apply compared to the structural measures described above, because it does not require large construction works. It is considered that there is especially much potentiality to sophisticate operation of multipurpose reservoirs, which are operated for multiple purposes such as flood management, water supply, or power generation by introducing real-time hydrological predictions. In most multipurpose reservoirs, it is more desirable to keep storage water level as high as possible for water utilization including water supply and power generation, while lower storage water level is preferred for flood management. In order to balance these conflicting requirements, most multipurpose reservoirs have seasonal operation rules or guidelines that indicate reservoir storage level or release rate considering importance of each purpose of reservoir at each period. This is often accomplished by lowering storage water level during the wet season where the reservoir has more chance to have a flood event to secure more empty storage volume in the reservoir while keeping water level high during the dry season. This operation method, however, degrades flood control capability of the reservoir during the dry season as well as its ability for drought management during the wet season.

On the other hand, various operational hydrometeorological predictions have been provided by the authorities in many regions recently. It is expected that reservoir operation can be sophisticated by considering those hydrometeorological predictions which can provide reservoir managers with estimation on future condition of the target river basin for adaptive decision making on water release in real time. Real-time reservoir operation considering hydrometeorological predictions can also be considered effective to integrate operations of a multipurpose reservoir for different purposes. It is expected to maximize the potential of the reservoir for both flood management and water utilization with a limited storage volume if reservoir is operated in an adaptive way considering the estimation on future condition in the target river basin provided by the latest hydrometeorological prediction. In other words, reservoir can be operated in more flexible manner relaxing constraints defined by operation rules or guidelines compared to when no hydrological prediction is considered. From this point of view, integrated operation of a multipurpose reservoir considering real-time hydrometeorological information has been studied in this decade (e.g., Nohara et al. 2005; Hsu and Wei 2007; Chou and Wu 2013).

Hydrometeorological predictions, however, inherently contain uncertainty as there is no operational prediction with perfect foresight. It is therefore crucially important to take their accuracy into account when hydrometeorological predictions are considered in reservoir operation. Impacts of prediction's uncertainty on effectiveness of introducing real-time hydrometeorological predictions into reservoir operation need to be clarified in order to design an effective reservoir operation method considering those predictions without major drawback in flood or drought management due to their uncertainty.

Considering the circumstances described above, a method to design and assess integrated reservoir operation for both the flood control and water utilization considering imperfect real-time hydrological predictions is discussed in this chapter in order to enhance the capability of an existing multipurpose reservoir. Prior release operation, which comes into practice in Japan recently and can combine reservoir operation for both the flood management and water utilization based on real-time hydrological prediction, is considered here as a method for integrated operation of a multipurpose reservoir. Effectiveness of introduction of the hydrological prediction into real-time operation of a multipurpose reservoir is demonstrated analyzing impacts of prediction uncertainty.

5.2 Integrated Reservoir Operation for Flood Control and Water Utilization

Multipurpose reservoirs are operated for more than one purpose, some of which conflict each other. The conflict usually exists between the operations for flood management and for water utilization including water supply and power generation. It is therefore important for more sophisticated operation of a multipurpose reservoir to effectively integrate the operations for the two contradictory purposes.

5.2.1 Prior Release Operation of a Multipurpose Reservoir

As one of promising methods for integrated reservoir operation for flood control and water utilization, prior release operation of multipurpose reservoirs has been studied in Japan recently. The prior release operation is an operation method for flood management, in which water stored in the reservoir is released just before the floodwater arrives to the reservoir to secure empty storage for flood control. This operation is expected to bring out enhanced capability of a multipurpose reservoir for integrated water resources management because the reservoir can keep its water level as high as possible for water utilization in no-flood situation at the same time as safely decreasing water level in advance of flood events for flood management. In other words, the operation has a potentiality to temporarily increase storage capacity of the reservoir for flood management during the flood situation without decreasing its ability for drought management or power generation. This operation method therefore starts being introduced in some of multi-reservoirs in Japan.

However, the operation may disadvantage water utilization if water storage is not recovered to the original level after the flood event even though storage level is lowered by prior release operation. In order to avoid from this problem, real-time predictions of hydrological states such as rainfall or streamflow in the catchment are needed to estimate the scale of the flood in advance to find out how much water the reservoir receives during the flood event. Good estimation on the amount of flood discharge enables the reservoir to release water as prior release to the extent that storage level can be recovered with certainty after the flood event, maximizing the advantage of prior release operation in flood management.

It is, however, difficult to perfectly predict meteorological or hydrological conditions in the future. Predictions essentially contain some degree of error or uncertainty. Therefore, it can rather reduce the functions of the reservoir for water utilization or flood management if prior release operation is conducted based on a hydrological prediction with low accuracy. For example, if prior release operation is conducted based on the significantly overestimated hydrological prediction, water storage is not recovered after the flood, and risks in water utilization increase due to the unnecessary prior release. On the other hand, flood risk would increase if prior release operation is conducted based on the underestimated hydrological prediction not securing empty volume enough to control the flood safely. It is, therefore, needed to consider the effect of uncertainty contained in the prediction when effectiveness of prior release operation for a multipurpose reservoir considering imperfect hydrological prediction is investigated.

5.2.2 *Issues and Approaches to Analyze the Effect of Real-Time Hydrological Prediction and Its Uncertainty*

In order to analyze the effect of advanced reservoir operation considering real-time hydrological predictions provided by a prediction system with an error structure on flood management or water utilization, one needs to investigate the effect by integrating results from a number of reservoir operations, respectively, conducted with consideration of a prediction provided by the system with the same error structure. However, it is often difficult to collect real prediction data enough much to analyze the effect of the advanced reservoir operation in this manner. One of the reasons for this difficulty is because many river basins have not experienced so many flood or drought events since operational hydrological prediction started being provided for the basin. Another reason is that operational hydrological prediction systems are often updated to incorporate new techniques for improvement, which also changes the error structure of the prediction system. It is therefore difficult to have long historical data of real-time hydrological predictions with the same error structure, and this makes it more difficult to analyze the effect of real-time hydrological prediction and its uncertainty on the performance of reservoir operation considering the prediction.

One of the effective approaches to overcome this scarcity of real-time hydrological prediction data is simulated generation of hydrological predictions. In this approach, a number of hydrological prediction sequences are artificially generated with a designed error structure. This approach is also useful for analysis on effects of prediction uncertainty or accuracy because it can generate a number of predictions with a certain degree of accuracy, which enables to employ stochastic approach for the analysis. From this reason, the simulated generation of hydrological predictions has often been applied in the analysis of reservoir operation considering hydrological predictions (e.g., Sivaarthitkul and Takeuchi 1995; Nohara and Hori 2014). This approach is also considered effective to investigate the effectiveness of prior release operation based on real-time hydrological predictions with uncertainty and employed in the model developed in this study.

5.2.3 *Prior Release Operation of the Target Reservoir*

The method described in this chapter is applied to the Kamafusa Reservoir in the Natori River basin in the northeastern Japan (Fig. 5.1). The Kamafusa Reservoir is a multipurpose reservoir, which is operated for flood control, water supply for irrigation and municipal water, and power generation with 39.3 million cubic meters (MCM) of active storage volume. Like most multipurpose reservoirs in

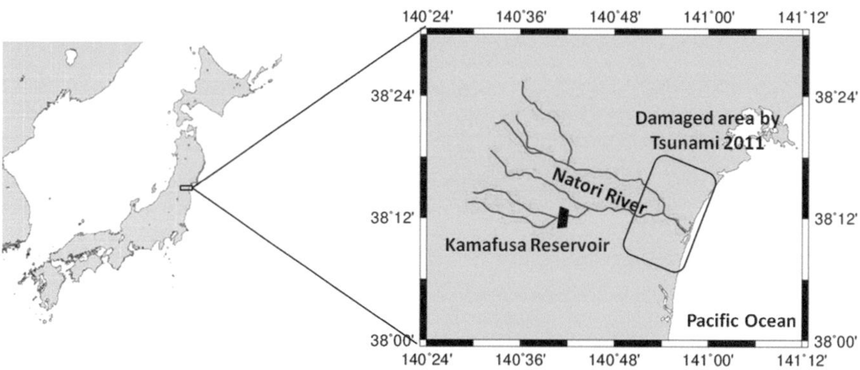

Fig. 5.1 Location of the Kamafusa Reservoir and the Natori River basin

Japan, the storage of the Kamafusa Reservoir is controlled separately between flood control purpose and water utilization purpose (including water supply and power generation) within the storage volumes allocated for those operation purposes. The allocated storage volume for each operation purpose varies according to the season. In the dry season (from October through June), the storage volume for flood control purpose is 2,9 MCM, and the rest of the storage is allocated for water utilization purpose. On the other hand, the storage volume for flood control is enlarged to 21 MCM in the wet season (from July to September) corresponding to increased possibility of flood occurrence. The catchment area of the reservoir is 192.25 km^2, and the most part of the catchment lies in the hilly mountainous area where water runs off very quickly after rainfall in the flood event. Flood control operation of the Kamafusa Reservoir is designed to mitigate the maximal release rate to the 850 m^3/s against the flood at the designed level with the peak inflow of 1,650 m^3/s. The flood control operation is initiated when inflow to the reservoir exceeds 300 m^3/s (called as flood inflow rate), and release rate is decided by the following equation:

$$r(t) = 0.407 \cdot [q(t) - 300] + 300 \tag{5.1}$$

where $r(t)$ is reduced water release rate at time step t [m^3/s] from the reservoir and $q(t)$ is inflow rate to the reservoir at time step t [m^3/s]. The fundamental data of the Kamafusa Reservoir is shown in Table 5.1.

In March in 2011, the great earthquake, which was named by Japan Meteorological Agency as "the 2011 off the Pacific coast of Tohoku Earthquake," hit the Pacific coast areas in the northeastern Japan. The earthquake and following tsunami devastated the Pacific coast areas in the northeastern Japan, leaving more than 18,000 dead victims or missing and damages to many properties and infrastructures. The downstream of the Natori River basin was also damaged by the earthquake and tsunami (Fig. 5.1). They damaged levees and infrastructures for flood management

Table 5.1 Fundamental data of the Kamafusa Reservoir

Items	Values
Catchment area	192.25 km^2
Total active storage	39.3 MCM
Flood control capacity	
Wet season (from July to September)	21 MCM
Dry season (from October to June)	2.9 MCM
Water use capacity	
Wet season (from July to September)	18.3 MCM
Dry season (from October to June)	36.4 MCM
Prior release amount	2.4 MCM
Designed flood inflow rate	1,650 m^3/s
Maximal release rate (w/out prior release)	850 m^3/s
Maximal release rate (w/ prior release)	600 m^3/s
Power generation capacity	1,200 KW

along the most downstream sections of the Natori River, which increased flood risk in the surrounding areas. The earthquake also caused land subsidence in the Pacific coastal areas including the downstream of the Natori River basin, which also reduced flood discharge capacity of the river channel and increased the flood risk.

In order to compensate to the decrease in the flood protection level in the downstream, prior release operation was introduced to the Kamafusa Reservoir. With this prior release operation, 2.4 MCM of water is released from the reservoir in advance of a flood to secure additional empty volume in the reservoir besides the original flood control capacity when a flood is expected to occur considering hydrological predictions. Decision on conducting prior release operation is made considering expected rainfall and inflow for the coming 24 h. The maximal rate of release from the reservoir is reduced to 600 m^3/s by storing inflow water with enlarged empty volume in the reservoir during the flood event in order to lower the water level in the river channel in the downstream where flood protection level was degraded due to damage in the levees and land subsidence. Water storage is recovered at the end of the flood event. However, effectiveness of this prior release operation depends on the accuracy of hydrological predictions considered. This study therefore aims at developing a method to analyze the effectiveness of prior release operation considering hydrological predictions as well as impacts of prediction uncertainty on the effectiveness of prior release operation.

5.2.4 Outline of the Simulation Model to Analyze Effects of Integrated Operation of a Multipurpose Reservoir

The effectiveness of prior release operation is analyzed in this study with the following steps: At first, a model for simulated generation of real-time hydrological prediction with a designed error structure for flood forecast is developed. Inflow to the reservoir is considered as the variable to be predicted in this study. A number of simulations on short-term reservoir operation for flood management including prior release operation are then conducted as a Monte Carlo simulation over the target flood event considering real-time inflow prediction with the designed error structure generated with the simulated generation model of predictions. The effectiveness of prior release operation on flood management can be analyzed through the assessment of the results from these simulations in an integrated manner. Simulations on long-term reservoir operation are then conducted for the following period respectively from the storage water level at the end of the short-term simulations to assess impacts on water utilization including water supply and power generation. The impacts on water utilization are investigated through the Monte Carlo simulation for each simulation of the short-term reservoir operation changing the long-term flow regime assumed to occur after the flood event. Impact of prior release operation on long-term reservoir operation for water utilization is then analyzed considering the results of multiple simulations in an integrated manner.

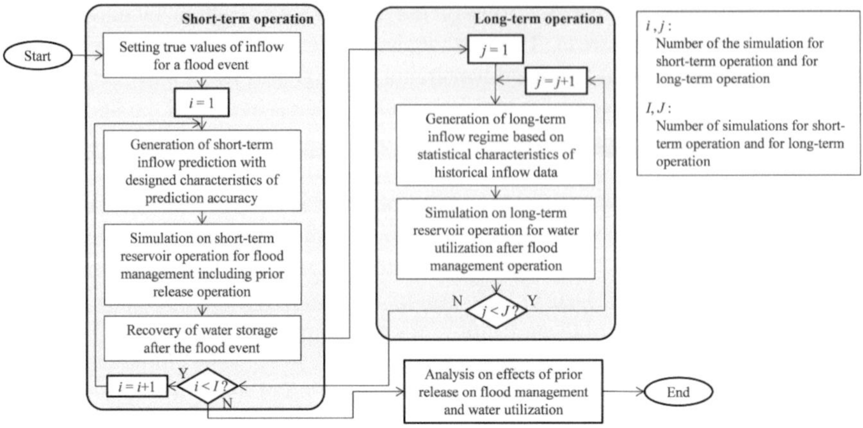

Fig. 5.2 Schematic flow diagram of the simulation for analysis on the effect of prior release operation for a flood event

The effectiveness of real-time inflow prediction with different characteristics of uncertainty can be analyzed by conducting these Monte Carlo simulations changing the settings for the generation model of predictions (described in detail in Sect. 5.3). The schematic flow diagram of the simulation is shown in Fig. 5.2.

5.3 A Method to Analyze Effects of Prior Release Operation on Short-Term Reservoir Operation for Flood Control

A method to analyze effects of prior release operation considering imperfect real-time inflow prediction on short-term reservoir operation for flood control is described in this section. Firstly, a synthetic generation method of inflow prediction with a designed error structure is developed in order to make up a number of prediction data for the Monte Carlo simulation. Deterministic hourly inflow prediction sequences for the coming 24 h are artificially generated in this study. Monte Carlo simulation on short-term reservoir operation for flood control with consideration of generated inflow predictions is then carried out to assess effects of the prior release operation on flood management.

5.3.1 Synthetic Generation Method of Inflow Prediction

Although multiple ways can be considered to represent prediction accuracy when deterministic hydrological predictions are artificially generated, Takeuchi (1990) pointed out two typical methods: One is to represent prediction errors as the degree of difference between a predicted value and the true value (e.g., Lattenmaier 1984; Datta and Burges 1984; Sivaarthitkul and Takeuchi 1995). This method can represent the degradation of prediction accuracy as lead time of the prediction becomes longer, while it is not suitable to represent the mechanism of update of predictions. The other is to represent prediction accuracy as the degree of change in predictions before and after update. This can easily represent the mechanism of prediction update, while it is difficult for this method to mimic the relation between accuracy and lead time of the predictions. In the prior release operation, water is released based on the prediction how much water the reservoir receives in the future during the flood event. Considering the importance of decision making from the early stage whether or not prior release operation should be conducted in order to complete the prior release safely in advance of flood arrival, accuracy of the hydrological prediction for the longer future is considered to influence the effectiveness of prior release operation. It is therefore considered more important to investigate effects of the relation between prediction accuracy and lead time on the effectiveness of the prior release operation. From this reason, the first way is employed to represent prediction errors in synthetic generation of inflow predictions.

In the synthetic generation model of the predictions in this study, a predicted value of inflow is synthetically generated by adding the prediction error value randomly sampled from a probabilistic distribution, which prediction errors are assumed to follow, to the true value (observed value) of inflow. The probabilistic distribution which prediction errors follow is assumed to be a normal distribution with zero mean. Here, the mean of the normal distribution is set to zero assuming that averaged difference between predicted values and true values—or bias of the prediction system—has already been corrected in the prediction. This assumption can be considered reasonable because one can easily correct the bias when the mean error (ME) of the prediction is known based on the statistical analysis of the prediction accuracy with historical data of the prediction. Although normal distributions are employed in this study, one can assume other probabilistic distributions such as lognormal distributions or the Pearson type III distribution which are also often applied for the hydrologic variables based on the analysis of the error characteristics of the target prediction system. On the other hand, it is assumed that the prediction errors in a sequence of inflow prediction have a serial correlation supposing the prediction errors for the subsequent time steps in a prediction sequence provided at the same time step have similar tendency. In order to simulate those characteristics in the prediction errors, a first order autoregressive model (AR(1) model) is employed in this study for the synthetic generation model of the prediction error sequence at each time step of prediction.

The flow of synthetic generation of predictions is as follows. Firstly, a value for the error in the prediction 1 h ahead (next time step) provided at initial time step of the simulation is generated by randomly sampling a value from the probabilistic distribution which errors in one-hour-ahead predictions follow. A sequence of prediction errors in the prediction is provided at the initial time step for the coming

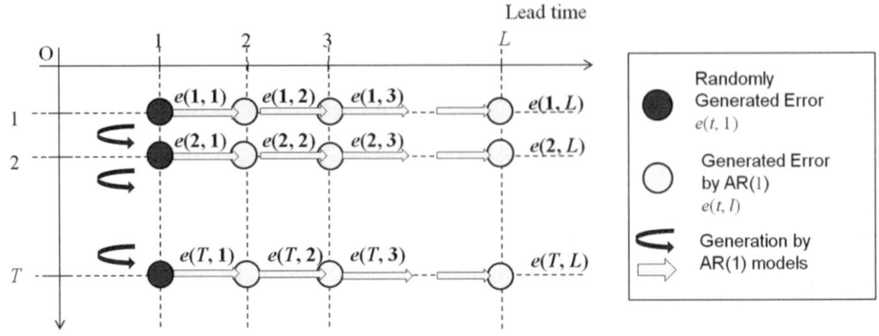

Fig. 5.3 Schematic flow diagram of the synthetic generation method of prediction errors

24 h (23 values for the period from 2 h ahead to 24 h ahead) by the use of an AR(1) model. The errors of 1-h-ahead predictions provided at the following time steps (i.e., updated predictions) are then generated by the use of another AR(1) model, and prediction error series for the following lead times are respectively generated by the use of the first AR(1) model for the prediction error series provided at each time step. The schematic flow diagram of synthetic generation of prediction errors is shown in Fig. 5.3. The AR(1) model for generation of prediction errors in the prediction series provided at each time step can be defined by the following equation:

$$e\,(t,l) = e\,(t,l-1)\cdot\rho_{\mathrm{L}}(1) + w\,(t,l)\,\sqrt{1-[\rho_{\mathrm{L}}(1)]^2}\quad (l\geq 2,\ t\geq 0) \tag{5.2}$$

where $e(t,l)$ is prediction error generated for the prediction for lead time l provided at time step t, $\rho_{\mathrm{L}}(1)$ is serial correlation between the errors in the predictions for successive lead times, and $w(t,l)$ is a random value generated for the prediction for lead time l provided at time step t by randomly sampling from the normal distribution with zero mean. The standard deviation of this normal distribution can be defined to be $[C_{\mathrm{e}}(l)\cdot I_{\mathrm{o}}(t+l)]^2$, where $C_{\mathrm{e}}(l)$ is an error growth function in the lead time direction and $I_{\mathrm{o}}(t+l)$ is the true value of inflow at time step $(t+l)$. This definition of the standard deviation is considered effective to simulate the structure of prediction errors because it can simulate the characteristics that prediction error tends to become greater as lead time of the prediction becomes longer. While various types of function can be considered for the error growth function $C_{\mathrm{e}}(l)$, the following function can be employed when prediction errors are assumed to grow linearly as lead time becomes longer:

$$c_{\mathrm{e}}(l) = \alpha l + \beta \quad (l \geq 1) \tag{5.3}$$

where α and β are parameters. One can employ a nonlinear function such as high-dimensional or exponential functions for the error growth function if the error of the target prediction system grows further greater as lead time of the prediction becomes longer. On the other hand, errors for 1-h-ahead prediction provided at each time step are generated by the use of the following equation:

$$e(t,1) = e(t-1,1)\cdot\rho_{\mathrm{T}}(1) + w(t,1)\sqrt{1-[\rho_{\mathrm{T}}(1)]^2} \quad (t \geq 1) \tag{5.4}$$

where $\rho_{\mathrm{T}}(1)$ is serial correlation between prediction errors provided at the successive time steps. The error in one-hour-ahead prediction provided at the first time step $e(0,1)$, which is the initial value for generation of prediction errors, is generated by sampling a random value from the normal distribution $N(0, [C_e(1)\cdot I_o(1)]^2)$. Inflow prediction sequences can be generated by respectively adding the prediction errors generated by the method described above to the true values of inflow. The value of inflow predicted at time step t for l hours ahead can be described as the following equation:

$$I_p(t,l) = I_o(t+l) + e(t,l) \tag{5.5}$$

where $I_p(t,l)$ is predicted inflow provided at time step t for l hours ahead. With this method, a set of inflow prediction sequences over the target flood event are repeatedly generated under the same setting for the error growth function so as to enable Monte Carlo simulation on the prior release operation of the reservoir considering real-time inflow prediction with the same characteristics of prediction accuracy.

5.3.2 Simulation for Short-Term Reservoir Operation for Flood Management Including Prior Release Operation

For the Monte Carlo simulation for assessment on effects of prior release operation of a multipurpose reservoir considering imperfect hydrological prediction, short-term reservoir operation for prior release and flood control operations of a multipurpose reservoir is modeled. Considering the conditions of the target reservoir, it is supposed in this study that short-term reservoir operations for prior release and flood control are conducted considering real-time inflow prediction for the coming 24 h with temporal resolution of an hour provided (updated) on an hourly basis. Rainfall predictions may also be considered as the hydrological prediction to be considered in short-term reservoir operation, as it is also provided as operational hydrological forecasts in many regions. However, inflow prediction should be calculated from the rainfall prediction in this case in order to estimate the future condition of reservoir storage which needs to be considered in decision of water release strategy. In this case, uncertainty in runoff analysis will also be included in the inflow prediction in addition to that included in the rainfall prediction. In order to make it easy to analyze relationship between prediction uncertainty and effectiveness of prior release operation, inflow is considered as the hydrological variable to be predicted for decision making on prior release operation eliminating effects of uncertainty contained in runoff analysis.

In actual operation of the target reservoir, prior release operation is conducted when a flood is expected to occur within the coming 24 h based on several criteria including predicted situation by rainfall forecasts. On behalf of these criteria, maximum inflow rate in the prediction for the coming 24 h is considered in this study as the equivalent criterion to decide whether or not prior release operation to be conducted in order to make discussion with the results derived by using the generation model of inflow predictions easier. Because the target reservoir must start flood control operation when it receives more than 300 m^3/s of inflow, it is modeled to conduct prior release operation when a value of inflow greater than 300 m^3/s, with which amount of inflow the reservoir must start flood control operation, is included in the prediction for the coming 24 h. However, it usually takes at least 3 h in most reservoirs in Japan for preparatory activities such as informing public and stakeholders for security before prior release is initiated in the actual reservoir management. Therefore, it is modeled in this study that prior release operation is initiated after the situation fulfills the criterion to conduct prior release operation for the first time in the flood event. The storage water level in the reservoir is lowered to the designed level to the extent release rate does not exceed 300 m^3/s before inflow becomes greater than 300 m^3/s, considering maximum increase rate in water release from the previous time step which is applied not to cause rapid increase in river water level in the downstream during prior release operation. The reservoir is operated so as to keep the water level until flood control operation is initiated once the storage water level is lowered to the designed level. These operations for prior release described above are conducted on an hourly basis. The flow of the simulation on prior release operation is shown in Fig. 5.4.

Once the flood control operation is initiated, water release rate is decided according to the Eq. (5.1) except when the reservoir stores water to the full capacity. The flood control operation is also conducted on hourly basis. After the flood situation ends, storage water is recovered to the maximum storage level for water utilization.

5.4 Impact Analysis of Prior Release Operation on Long-Term Reservoir Operation for Water Utilization

A simulation model on long-term reservoir operation for water utilization is developed to investigate impacts of prior release operation on water supply and power generation purposes. The storage volume simulated at the last time step of

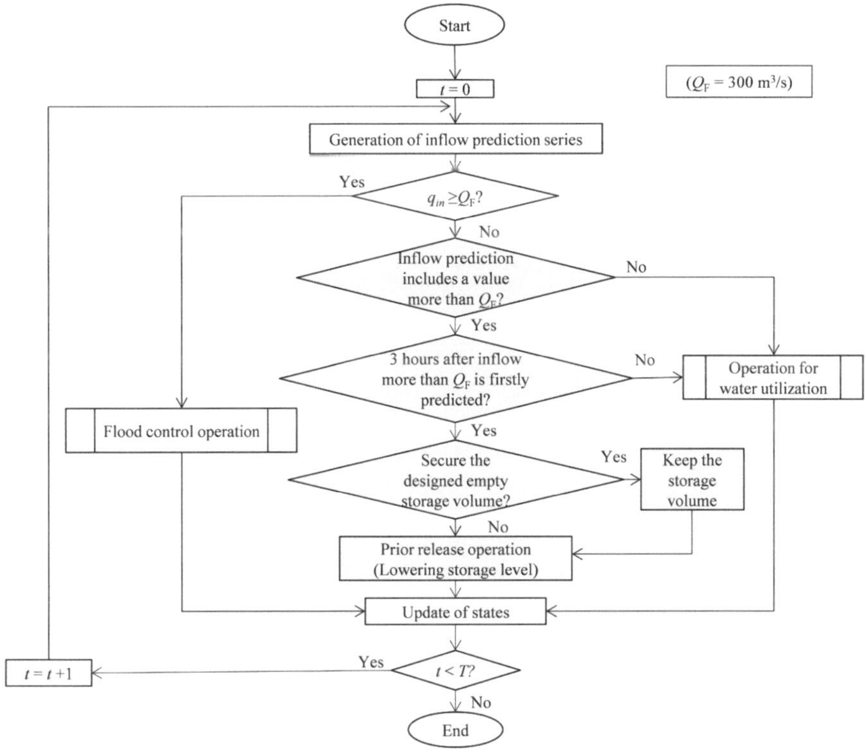

Fig. 5.4 The flow of the simulation on prior release operation

the target flood event in each simulation on short-term reservoir operation for flood management is considered as the initial storage condition in the simulation for long-term reservoir operation.

The outline of the simulation on long-term reservoir operation is as follows. Firstly, a long-term inflow sequence is synthetically generated over the target period based on statistical characteristics of inflow estimated from historical data at the target reservoir. The simulation on long-term reservoir operation for water supply and power generation is then conducted over the target period after the flood event using the long-term inflow regime generated in the previous process. The impacts of reduce in water storage at the end of the flood event due to prior release operation on long-term reservoir operation are evaluated. A number of simulations on long-term reservoir operation are repeatedly carried out in this fashion as a Monte Carlo simulation, which allows statistical analysis on the impact of prior release operation on water supply and power generation purposes in an integrated manner excluding effects arisen by chance due to variance in long-term inflow regimes. A time step for the simulation of long-term reservoir operation is 5 days long.

5.4.1 Synthetic Generation of Long-Term Inflow Regime

In order not to get a biased result by analyzing the impact of prior release operation on long-term reservoir operation with a specific inflow sequence, a number of long-term inflow sequences are artificially generated in this study. A time step of inflow is 5 days long in the simulation on long-term reservoir operation. The type of probabilistic distribution which inflow values follow must be identified in order to randomly generate inflow values simulating their statistical characteristics. If it is annual inflow which is to be generated, it may be possible to assume that the variable follows a normal distribution. However, skewness of the probabilistic distribution is no longer ignorable when inflow of the shorter temporal scale is considered. Generally speaking, a hydrological variable tends to follow a right-skewed probabilistic distribution. These right-skewed probabilistic distributions include lognormal distribution or the Pearson type III distribution. Among them, lognormal distribution with three parameters is employed in this study because the generation method of values following a lognormal distribution is simple compared to the others. The probability density function of the distribution can be described as the following equation:

$$F(x) = \frac{1}{x\zeta\sqrt{2\pi}} \exp\left\{-\frac{1}{2}\left(\frac{\ln(x-c)-\lambda}{\zeta}\right)^2\right\} \tag{5.6}$$

where x is random variables following the lognormal distribution and ζ, c, and λ are parameters.

A synthetic generation model of long-term inflow sequences is then developed so that generated values follow a designed lognormal distribution with three parameters. It is considered important to take sequential characteristics of inflow regime into account when reservoir operation for water utilization including drought management is considered because droughts are caused by low flow situation that persists for the long period. In order to model the persistence of inflow regime, AR(1) model, which can consider serial correlation of the time series, is also employed for synthetic generation of long-term inflow sequences as it was employed in Nohara and Hori (2014). The AR(1) model for generation of long-term inflow sequences can be defined by the following equation:

$$z(t) = \rho_z(t,t-1)\,z(t,t-1) + \sqrt{1-\rho_z(t,t-1)^2}\,\varepsilon(t) \tag{5.7}$$

where $z(t)$ is the normalized value of inflow at time step t, $\rho_z(t, t-1)$ is serial correlation between normalized values of inflow at time step $t-1$ and time step t, and $\varepsilon(t)$ is Gaussian white noise which is randomly sampled from the standard normal distribution N(0,1). This model is also known as Thomas-Fiering Model (Thomas and Fiering 1962) when t is the monthly time step. The serial correlation $\rho_z(t, t-1)$ can be calculated from the serial correlation between the original values (not normalized) by the following equation (Hoshi 1997):

$$\rho_x(t,t-1) = \frac{\exp\left[\lambda(t-1)\lambda(t)\rho_z(t,t-1)\right]-1}{\left\{\exp\left[\lambda(t-1)^2-1\right]-1\right\}^{1/2}\left\{\exp\left[\lambda(t)^2-1\right]-1\right\}^{1/2}} \tag{5.8}$$

where $\rho_x(t, t-1)$ is serial correlation between the original values of inflow at time step $t-1$ and time step t and $\lambda(t)$ is one of the three parameters of the lognormal distribution for inflow values at time step t. Time series of original values of inflow can be calculated from the generated normal value $z(t)$ by the use of the following equation:

$$q(t) = c(t) + \exp\left[\lambda(t) + \zeta(t)z(t)\right] \tag{5.9}$$

where $q(t)$ is generated inflow (original value not normalized) and $c(t)$ and $\zeta(t)$ are parameters of the lognormal distribution for inflow values at time step t. A series of inflow can be synthetically generated over the target period of the simulation on long-term reservoir operation for water utilization by the use of the method described above. A number of long-term inflow series are repeatedly generated for the Monte Carlo simulation on long-term reservoir operation.

5.4.2 Simulation on Long-Term Reservoir Operation for Water Utilization

The simulation on long-term reservoir operation for water utilization is conducted over a long period such as several months using a time step of 5 days. The simulation period can be from the end of the target flood event through the end of the hydrological year, if the target reservoir is not operated to mitigate interannual variability in inflow. In this study, the simulation on long-term reservoir operation is conducted from the end of the flood event through the end of December, where it is in the non-flood season and the variance in inflow is relatively small in the target river basin.

In the simulation on long-term reservoir operation for water utilization in this study, water as much as the target release for each time step is released from the reservoir as far as there is enough much storage water to do so.

For the reservoir where storage water is managed in an integrated manner between water supply and power generation purposes, reservoir operation for water utilization is simulated according to the continuous equation of the reservoir storage described by the following equation:

$$S(t+1) = S(t) + q(t) - r(t) - \kappa(t)$$
$$S_{\min}(t) < S(t) < S_{\max}(t) \tag{5.10}$$

where $S(t)$ is storage volume at the beginning of time step t, $S_{\min}(t)$ and $S_{\max}(t)$ are respectively minimum and maximum storage volumes for water utilization for time step t, and $k(t)$ is water loss due to evaporation from the surface of the reservoir or seepage through the bottom of the reservoir, both of which are ignored in this study (i.e., $k(t) = 0$). Water release for water supply is calculated considering the constraints of the reservoir and water demand by the use of the following equations:

$$r(t) = \max\left\{r_{ws}(t), r_{pg}(t)\right\} \tag{5.11}$$

$$r_{ws}(t) = \max\left\{S(t) + q(t) - S_{\max},\quad \min\left[G_{ws}(t),\ S(t) + q(t) - S_{\min}\right]\right\} \tag{5.12}$$

$$r_{pg}(t) = \max\left\{S(t) + q(t) - S_{\max},\quad \min\left[G_{pg}(t),\ S(t) + q(t) - S_{\min}\right]\right\} \tag{5.13}$$

where $r_{ws}(t)$ and $r_{pg}(t)$ are respectively release rates for water supply and power generation for time step t and $G_{ws}(t)$ and $G_{pg}(t)$ are respectively target release rates (water demand) for water supply and power generation for time step t. Because storage volume is managed in an integrated fashion between water supply and power generation purposes in the target reservoir, Eqs. (5.11), (5.12) and (5.13) are employed to simulate long-term reservoir operation in this study.

5.5 Case Study

5.5.1 Settings for the Simulation

The method described in the previous sections was applied to the Kamafusa Reservoir and the Natori River basin to analyze the effects of prior release operation of the reservoir on flood management and water utilization.

As described in Sect. 5.3, prior release operation is conducted at the Kamafusa Reservoir when a value of inflow greater than 300 m^3/s is included in inflow prediction for the coming 24 h in this study. However, it can change after update of the prediction whether or not the prediction includes a value of inflow more than the criterion where the prediction is updated every time step. Decision on conducting prior release operation needs to be made in a consistent manner considering updates of prediction. Thus, the two options were considered in this case study to reflect predicted inflow condition to decision making on starting prior release operation: the option where prior release operation is immediately initiated if a predicted value of inflow is more than 300 m^3/s and the option where prior release operation is initiated if inflow more than 300 m^3/s is predicted at the successive three time steps. In each option, prior release can be initiated three hours after the predicted condition fulfills the condition due to the time lag needed for communication with public and stakeholders. The two options were also considered here for the way on how to conduct prior release after it is initiated: the option where prior release is completed without interruption and the option where prior release is interrupted if no inflow more than 300 m^3/s is predicted at three successive time steps and restarted if inflow more than 300 m^3/s is predicted again at three successive time steps.

Although the amount to be released from the reservoir is 2.4 MCM for the actual operation of the Kamafusa Reservoir, greater amounts, namely, 4.8 MCM and 7.2 MCM, are also considered in this study as the amount to be released by prior release operation in order to assess the possibility and impacts to enlarge temporal empty storage volume available for flood control for more advanced reservoir operation for flood management. On the other hand, the reservoir needs to release water more than the designed maximum release rate if it is needed to avoid overflow in the reservoir due to extreme large inflow rate more than the designed level. This operation is called as exceptional operation to avoid from overflow in the reservoir and may cause damage more than designed in the downstream due to increase in water release. It can provide enlarged flood control capacity if more water is released in advance of those large flood events considering real-time inflow prediction. From this viewpoint, a method of prior release operation, in which water is released considering real-time inflow prediction so as to secure empty volume much enough to control floodwater during the flood event without conducting the exceptional operation based on predicted total inflow volume during the flood event, can also be considered.

Taking those aspects into account, in this case study, the four methods for prior release operation was simulated: a method where prior release is conducted without interruption once inflow greater than 300 m^3/s is predicted (denoted by Operation I hereafter); a method where prior release is conducted without interruption once inflow greater than 300 m^3/s is predicted for three successive hours (time steps) (denoted by Operation II); a method where prior release is conducted with interruption when inflow more than 300 m^3/s is predicted (denoted by Operation III); and a method where water is released so as to secure empty volume as much as needed for flood control of the flood event considering real-time inflow prediction (denoted by Operation IV).

In Operation I, prior release operation is conducted totally relying on inflow prediction. No countermeasure is considered for prediction uncertainty in this operation. On the other hand, uncertainty of inflow prediction is considered in Operation II, and a wait-and-see option is introduced. In Operation III, prior release operation is initiated relying on inflow prediction without considering prediction uncertainty, but an interruption option is introduced in order to correspond to update of predictions. Operation IV is a kind of most advanced operation methods where the reservoir is required to control more floodwater than designed considering inflow prediction. This operation method is very different from conventional flood control practices in Japan, but is considered to investigate its potential effectiveness. In Operations I to III, three cases were also considered for water amount to be released by prior release operation: (a) 2.4 MCM, (b) 4.8 MCM, and (c) 7.2 MCM.

As for flood events data, 15 flood events in three different scales were considered in the case study: five small-scale historical flood events where peak inflow rate is from 100 m^3/s to 300 m^3/s (from Events 1 to 5), five large-scale historical events where peak inflow rate is more than 300 m^3/s (from Events 6 to 10), and five ideal very large-scale events with a 200-year rainfall (more than the designed level in the target river basin), which rainfall pattern is derived from that in historical flood events (from Events 11 to 15).

The simulation on flood management operation including prior release was conducted for 1000 times as a Monte Carlo simulation, conducting a thousand simulations of reservoir operation for water utilization for the following period with consideration of the result of each simulation on short-term operation as initial condition of storage. The simulation period for long-term reservoir operation was from the end of the target flood event to the end of December. The target release for water supply $G_{ws}(t)$ was estimated based on water demand in the downstream and historical record of discharge from the sub-basins other than the Kamafusa Reservoir catchment. On the other hand, target release for power generation $G_{pg}(t)$ was estimated base on historical record of water release for power generation from the Kamafusa Reservoir.

5.5.2 *Indices for Impacts on Long-Term Reservoir Operation*

Effects of prior release operation on flood management are analyzed based on the results from simulations on short-term reservoir operation, while its impacts on water utilization are investigated from the results of simulations on long-term reservoir operation in this case study. Effects on flood management are evaluated from the following indices: maximal water release rate in the target flood event averaged among simulations and the number of simulations where the exceptional operation was conducted. Feasibility of prior release operation is also assessed in Operations I to III by the number of simulations where a designated empty volume was not secured before flood control was initiated when the reservoir received more than 300 m^3/s of inflow.

On the other hand, the impact of prior release operation on long-term reservoir operation for water supply was assessed by the gap between simulated storage volume and target storage volume, which is defined by the following equations:

$$\frac{1}{I}\frac{1}{J}\sum_i\sum_j\sum_t D_{ws}\left(i,j,t\right) \tag{5.14}$$

$$D_{ws}(t) = \max\left[S_G\left(i,j,t\right) - S\left(i,j,t\right),\ 0\right] \tag{5.15}$$

where $D_{ws}(i, j, t)$ and $S_G(i, j, t)$ are respectively damage and target storage volume in water supply during time step t in simulation j for long-term reservoir operation for simulation i on short-term reservoir operation, and I and J are respectively the number of simulations for short-term reservoir operation and for long-term reservoir operation. Similarly, the impact of prior release on power generation purpose was evaluated by the gap in power generation, which is defined by the following equations:

$$\frac{1}{I}\frac{1}{J}\sum_{i}\sum_{j}D_{\mathrm{pg}}(i,j) \tag{5.16}$$

$$D_{\mathrm{pg}}(i,j) = \sum_{t}P(i,j,t) - P_{st} \tag{5.17}$$

$$P(i,j,t) = g \cdot q(i,j,t) \cdot \eta \cdot h(i,j,t) \tag{5.18}$$

where $D_{\mathrm{pg}}(i, j)$ is loss in power generation over the period of simulation for long-term reservoir operation, $P(i, j, t)$ is electricity produced at time step t in simulation j for long-term reservoir operation for simulation i on short-term reservoir operation, P_{st} is averaged electricity that is produced over the period of simulation for long-term reservoir operation estimated from historical record, $q(i, j, t)$ and $h(i, j, t)$ are respectively averaged inflow and averaged effective head during time step t in simulation j for long-term reservoir operation for simulation i on short-term reservoir operation, g is gravity acceleration, and η is power generation efficiency that is assumed to be 0.85 in this case study.

5.5.3 *Generation of Inflow Prediction*

A synthetic generation model described in Sect. 5.3 was developed to generate a number of hourly inflow predictions for the coming 24 h for a Monte Carlo simulation. A linear function defined by Eq. (5.3) was assumed to represent error growth in the inflow prediction in this case study. The results of synthetic generation of inflow predictions employing observed values of inflow derived from a historical flood event as the true values are shown as an example here. A thousand of inflow prediction sequences for the coming 24 h were generated at every time step, assuming $\alpha = 0.02$ and $\beta = 0$ for the parameters of the error growth function (Eq. 5.3) in this analysis. The generation of pseudorandom values was conducted based on Box-Muller's method (Box and Muller 1958) in this study. The results of generation of values for random variable $w(t, i)$ in Eq. (5.2) are shown in Table 5.2. Mean, standard deviation, and rate of difference in standard deviation and skewness of the generated random values are shown here respectively for the generation of $w(0, 1)$, $w(28, 20)$, $w(38, 10)$, and $w(47, 1)$ as examples. The latter three were chosen as the examples because they were the random variables used to generate prediction errors for the peak of inflow, which was recorded at time step 48. On the other hand, skewness of random variable x_k can be calculated by the following equation:

Table 5.2 Generation results of values for random variable $w(t, l)$ (averaged values among 1000 generations)

Variables $w(t, l)$	Mean (Designed values)	Standard deviation (Designed values)	Skewness
$w(0, 1)$	−0.0007475 (0)	0.1628 (0.1686)	−0.008520
$w(28, 20)$	−0.8364 (0)	146.0 (143.8)	0.02377
$w(38, 10)$	2.112 (0)	69.56 (71.90)	−0.09297
$w(47, 1)$	−0.03195 (0)	7.315 (7.191)	0.07068

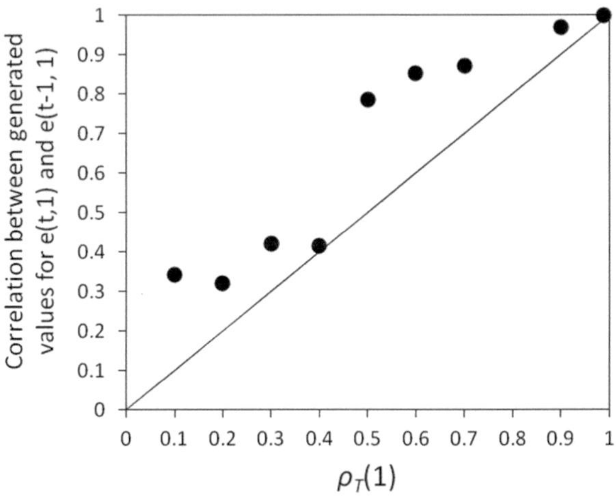

Fig. 5.5 Correspondence between designed serial correlation $\rho_{\mathrm{T}}(1)$ and correlation of generated prediction errors

$$C_s = \frac{\frac{1}{K}\sum_{k=1}^{K}(x_k - \bar{x})^3}{\left[\frac{1}{K}\sum_{k=1}^{K}(x_k - \bar{x})^2\right]^{\frac{3}{2}}} \tag{5.19}$$

where C_s is skewness, K is the number of samples, and $\bar{x}$ is mean value, respectively. It can be seen in Table 5.2 that values were basically generated as designed. Difference between generated and designed values for standard deviation was less than 3.5 % for each $w(t, l)$, and skewness was close to 0, which means the probabilistic distribution of generated values was not skewed. While some effect of sampling can be seen in the mean of values generated for $w(38, 10)$, the number of generations is considered much enough to diminish sampling effect from those results.

Correlation between errors in predictions for the next time step generated at successive time steps is shown as a scatter chart with designed serial correlation $\rho_{\mathrm{T}}(1)$ in Fig. 5.5. Values of serial correlation of values generated changing designed serial correlation ranged from 0.1 to 0.99 are shown in Fig. 5.5. It can be seen in

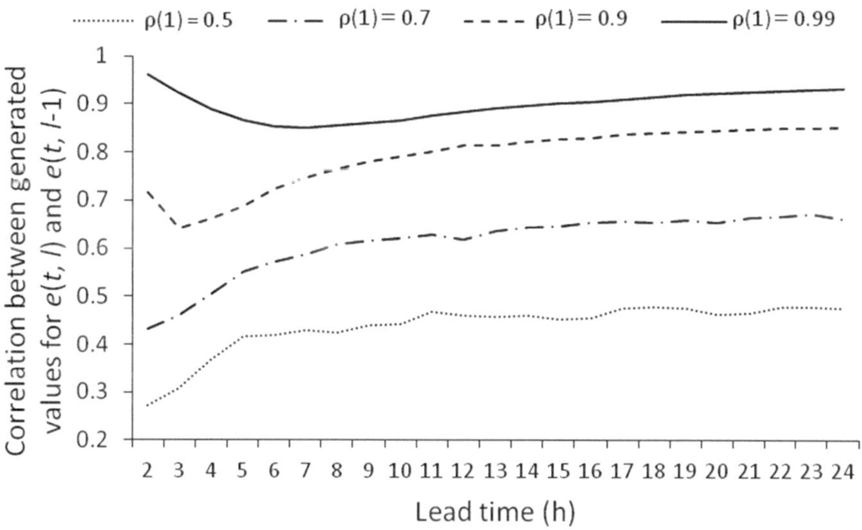

Fig. 5.6 Correlation between prediction errors e(*t*, *l*) and e(*t*, *l*–1) generated with each designed serial correlation over lead time of prediction

Fig. 5.5 that serial correlations of generated values were generally greater than the designed correlations. This is because serial correlation is not perfectly preserved in our model as the standard deviation of Gaussian white noise employed for $w(t, l)$ was not identical to the standard deviation of prediction errors as it is not constant over the whole lead time due to the assumption of $C_e(l)$ in this study. Nevertheless, it can be seen that correlation between the prediction errors generated at the successive time steps became greater as the designed serial correlation was greater.

On the other hand, correlation of generated prediction errors with those generated for the previous lead time is shown in Fig. 5.6. The results are shown for the cases where $\rho_L(1)$—designed serial correlation between errors predicted for two successive lead times—is set to be 0.5, 0.7, 0.9, and 0.99 respectively under the condition of $\rho_T(1) = 0.9$ in Fig. 5.6. It can be seen in Fig. 5.6 that correlation between prediction errors generated for the two successive lead times was close to the designed serial correlation especially when lead time of the prediction became long, while it was smaller than designed in early lead times of the prediction. The reason why correlation between prediction errors in early lead times was small can be considered because rate of difference in standard deviations of Gaussian white noises $w(t, l\text{-}1)$ and $w(t, l)$ was great, and the effect of random variable $w(t, l)$ was too strong compared to the effect of correlation in Eq. (5.2) in generation of errors with the error function of $C_e(l) = 0.02\,l$.

Effects of error growth function $C_e(l)$ defined by Eq. (5.3) on generation of prediction errors were also analyzed. Considering the definition of probabilistic distribution of prediction errors employed in the proposed generation method of prediction errors, $C_e(l)$ should be identical to the function of standardized root mean square error (RMSE) defined by the following equation:

$$\mathrm{RMSE}(l) = \sqrt{\frac{1}{I}\frac{1}{T}\sum_i \sum_t \left[\frac{e(i,t,l)}{I_o(t+l)}\right]^2} \tag{5.20}$$

where RMSE(l) is standardized RMSE of the prediction with lead time l and $e(i,t,l)$ is prediction error generated l hours ahead at time step t in simulation i on short-term reservoir operation. On the other hand, standardized mean error (ME) should be 0 if generation of prediction errors is conducted as designed. Standardized ME of the prediction with lead time l is defined by the following equation:

$$\mathrm{ME}(l) = \frac{1}{I}\frac{1}{T}\sum_i \sum_t \frac{e(i,t,l)}{I_o(t+l)} \tag{5.21}$$

In order to verify effects of each parameter for error growth function $C_e(l)$, effects of parameter α on standardized ME and RMSE of generated predictions were investigated changing values for parameter α from 0.01 to 0.03 with interval of 0.005. The results for ME(l) and RMSE(l) are respectively shown in Fig. 5.7a, b. It can be seen from Fig. 5.7a that ME(l) generally continued to be around zero, while

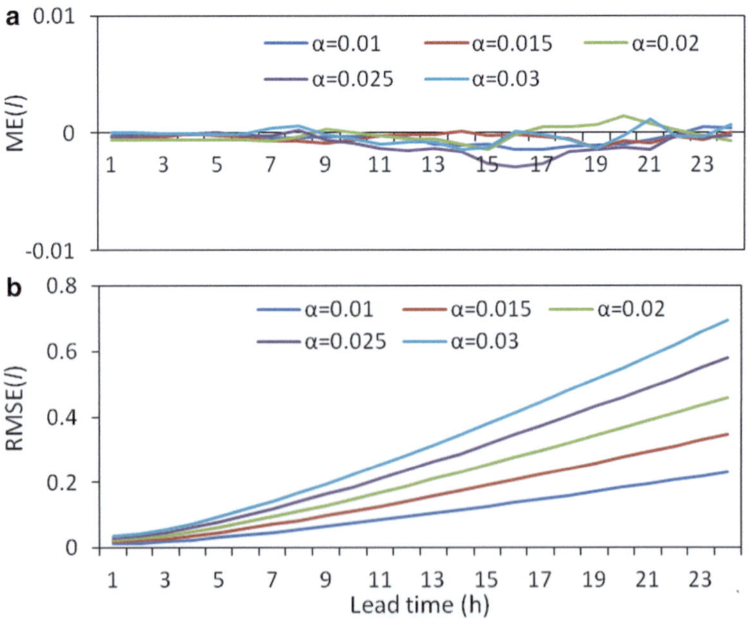

Fig. 5.7 Changes in statistics of generated prediction errors depending on the value for parameter α: (**a**) in ME(l) and (**b**) in RMSE(l)

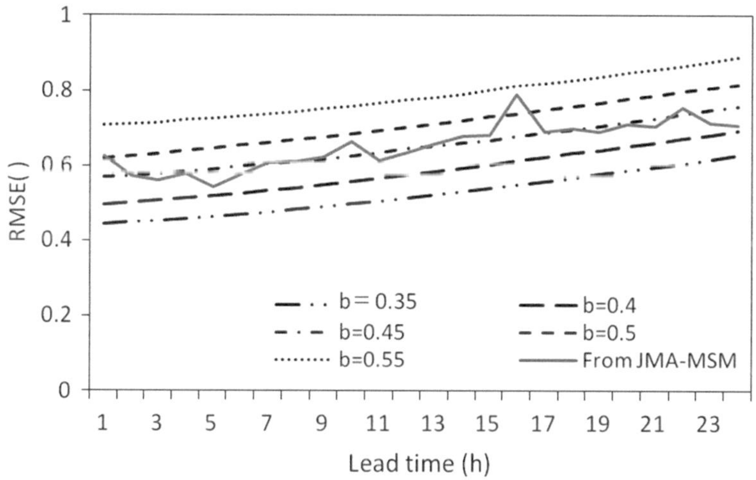

Fig. 5.8 Changes in RMSE(l) of generated prediction errors depending on the value for parameter β

it had relatively great absolute values where lead time is long and parameter α is great. On the other hand, it can be seen in Fig. 5.7b that RMSE(l) is growing as lead time becomes long following the error growth function employed.

Effects of parameter β of the error growth function were then verified. According to the report by Amai et al. (2014) on error characteristics of inflow prediction, which was estimated from rainfall prediction provided by the Meso-Scale Model of Japan Meteorological Agency—a short-term operational rainfall prediction available in Japan—the following function was employed for error growth function in this case study assuming α and β in Eq. (5.3) to be 0.0078 and 0.557, respectively:

$$C_e(l) = 0.0078l + 0.557 \quad (1 \leq l \leq 24) \tag{5.22}$$

Considering this report, parameter α of the error growth function for operational hydrological prediction was set to be 0.0078 in this analysis. On the other hand, four different values ranged from 0.35 to 0.50 with interval of 0.05 were considered in addition to 0.55 which was derived from the value for the operational hydrological prediction in Japan. The results of standard RMSE are shown in Fig. 5.8. The error growth function derived from the operational hydrological prediction linearly approximated by Eq. (5.22) is also shown in this figure for comparison. It can be seen in here that RMSE(l) of prediction errors were generated as designed generally. From these results, it can be considered that error structure of the prediction defined by the error growth function is preserved over synthetic generation.

5.5.4 Generation of Long-Term Inflow Regime

A thousand of long-term inflow regimes were synthetically generated by the use of the method described in Sect. 5.4 based on historical data of inflow at the Kamafusa Reservoir for 19 years from 1993 to 2011. Firstly, three parameters for the lognormal distribution were respectively estimated using historical inflow data for each 5-day long time step over the entire year (73 time steps a year). Serial correlation of inflow sequence was also estimated based on correlation analysis using historical data. Generation of inflow sequence was then conducted by the use of Eqs. (5.7), (5.8) and (5.9) with estimated parameters for each time step. A thousand of long-term inflow sequences over the entire year were generated in the same fashion. The results of the synthetic generation of inflow sequences over the entire year are shown in Fig. 5.9. The mean and standard deviation of a thousand values of inflow generated for each time step are shown in Figs. 5.9a, b, respectively, and correlation between inflow at time step t and that at time step t-1 estimated from the generated inflow sequences is shown in Fig. 5.9c. It can be seen in Fig. 5.9 that the generation model could simulate long-term inflow regimes of the Kamafusa Reservoir well. The inflow sequences after the time step when flood control operation of the reservoir ends were used in the Monte Carlo simulation on long-term reservoir operation for each target flood event in this case study.

5.5.5 Analysis of Prior Release Operation on Short-Term Reservoir Operation for Flood Management

Effects of conducting prior release operation of the target reservoir were analyzed through the Monte Carlo simulation using inflow prediction generated by the synthetic generation model developed in this study. In order to provide information on potential effectiveness of prior release operation for the target reservoir, parameters of the error growth function in the generation model of inflow prediction were decided based on the operational inflow prediction estimated from rainfall forecasts by JMA-MSM in this analysis. Parameter α in Eq. (5.3) was therefore set to be 0.0078, which is identical to that of the error growth function estimated for the operational prediction, and was not changed through the entire simulation. On the other hand, five values were considered for parameter β, namely, 0.35, 0.4, 0.45, and 0.5 in addition to 0.55, which was derived from the value for parameter β in the error growth function for the operational inflow prediction estimated from rainfall forecast by JMA-MSM. The first four values were employed in this analysis in order to investigate effectiveness of prior release operation by the target reservoir when it is conducted based on inflow prediction with improved accuracy compared to the current standard. Difference in effects of prior release operation considering inflow prediction with different degrees of error at the initial lead time can also be investigated through this analysis. On the other hand, the difference in effects

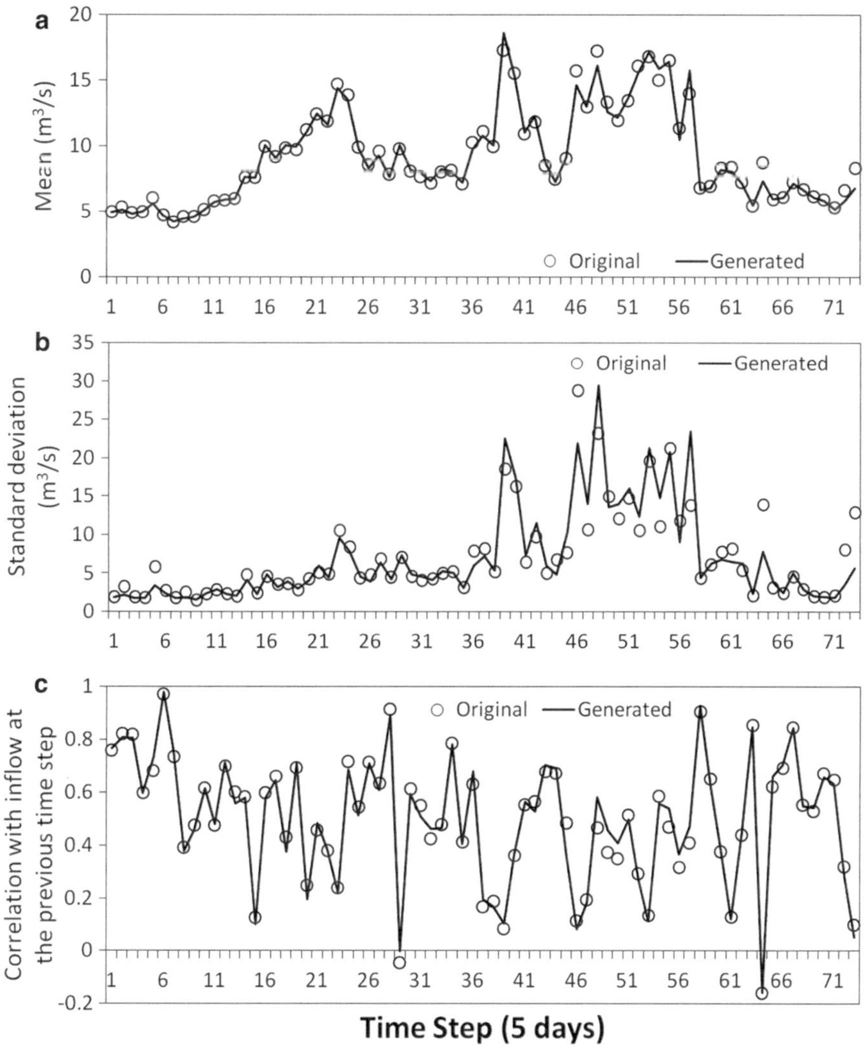

Fig. 5.9 Statistics of inflow generated for each time step: (**a**) mean, (**b**) standard deviation, and (**c**) correlation with inflow generated for the previous time step

of prior release operation with different values of parameter α was reported in Nohara and Hori (2015). The value for serial correlations of predicted inflow $\rho_L(1)$ and $\rho_T(1)$ was respectively set to be 0.9 in this analysis also based on the error characteristics of inflow prediction estimated from rainfall forecast of JMA-MSM. The Monte Carlo simulation of reservoir operation including prior release, flood control, and water utilization operations with consideration of inflow predictions generated with the settings described above for each flood event (from Events 1 to 15 with different scales), changing prior release operation policies (Operations

Table 5.3 The number of simulations where prior release operation was not completed to a designed volume before flood control operation starts out of 10,000 simulations for large or very large flood events

		Parameter β				
Operation policy	Storage water to be released	0.35	0.4	0.45	0.5	0.55
Operation I	2.4 MCM	0	0	0	0	0
	4.8 MCM	0	0	0	0	0
	7.2 MCM	0	0	0	0	0
Operation II (Wait and see)	2.4 MCM	9	9	7	15	10
	4.8 MCM	40	46	35	49	31
	7.2 MCM	453	364	341	291	277
Operation III (Interruption)	2.4 MCM	0	0	0	0	1
	4.8 MCM	4	6	1	3	2
	7.2 MCM	29	19	22	21	17

I to IV) and water amounts to be released by prior release operation (2.4 MCM, 4.8 MCM, and 7.2 MCM, only applicable to Operations I to III).

Results of the simulations for short-term reservoir operation including prior release and flood control are shown below. Feasibility of prior release operation was firstly analyzed for Operations I to III using 10,000 simulation results for 10 large or very large flood events from Event 6 to Event 15. The number of simulations where prior release was not completed to a designed volume before flood control operation starts ($q(t) > 300$ (m^3/s)) out of 10,000 simulations is shown in Table 5.3. It can be seen in Table 5.3 that Operation I performed no failure in completion of prior release to designed amounts of water regardless of the value for parameter β and designed amount of storage water to be released. Most simulations with Operation III with the interruption option also completed prior release operation before flood arrival regardless of the value for parameter β while there was failure to complete prior release operation in small number of simulations when storage water to be released was set to be 7.2 MCM. On the other hand, Operation II with the wait-and-see option failed completing prior release operation in many simulations especially when water amount to be released by prior release operation was set to be 7.2 MCM. This can be considered because the criterion to initiate prior release operation employed in Operation II was too conservative to release 7.2 MCM of water from the reservoir before the arrival of flood. Because the catchment of the target reservoir is steep and rainwater sometimes discharges to the reservoir within several hours, waiting for three hours to recognize flood occurrence might be too long for the target reservoir.

Looking into effects of parameter β on feasibility of prior release operation, no major difference in feasibility can be seen in the simulation results with different values for parameter β if water amount to be released by prior release operation is 2.4 MCM or 4.8 MCM. However, focusing on the results of the simulations with Operation II and 7.2 MCM of prior release, the rate of failure in completion of prior release operation decreased as parameter β became greater, which means that predictions included greater errors. The reason why this apparently contradictory

Table 5.4 The number of simulations where the exceptional operation to avoid from overflow from the reservoir was conducted during flood control, out of 1,000 simulations respectively for Events 11–14

		Parameter β				
Events	Peak rate of inflow [m^3/s]	0.35	0.4	0.45	0.5	0.55
Event 11	1708.5	0	0	0	0	0
Event 12	1413.5	0	0	0	0	0
Event 13	1964.2	0	0	1	0	1
Event 14	2852.2	764	847	875	902	927

result was observed can be considered because flood occurrence could be detected better considering predictions with greater error parameters as the chance to sample a positive great value for prediction error increases with greater values of error parameter β, which contributed to generate inflow prediction greater than flood inflow more frequently. It is therefore considered that prediction errors with a structure assumed in this study do not always degrade feasibility of prior release operation based on the inflow prediction.

Effects of prediction uncertainty on effectiveness of prior release operation for flood management are then investigated with the simulation results of reservoir operation with Operation IV. The number of simulations where exceptional operation to avoid from overflow from the reservoir was conducted during flood control operation is shown in Table 5.4. The results of the simulations with Operation IV for four very large flood events from Events 11 to 14 are shown in Table 5.4. Event 15 was excluded from this analysis because the event was not big enough to conduct the exceptional operation even with the actual flood control operation of the target reservoir without prior release. It can be seen in Table 5.4 that Operation IV could avoid from the exceptional operation in almost all of simulations for Events 11 to 13. This means flood control operation was successfully conducted without releasing water more than the designed maximal release rate (600 m^3/s for the target reservoir in this case study) thanks to prior release operation. On the other hand, the exceptional operation was conducted in many simulations for Event 14 where peak inflow was extremely large (2,852 m^3/s) compared to the designed flood inflow rate of the target reservoir (1,650 m^3/s). Moreover, the number increased as parameter β became greater.

On the other hand, maximum release rates during flood control operation averaged among 1,000 simulations for each flood event from Events 11 to 14 are shown in Fig. 5.10. It can be seen in Fig. 5.10 that maximum release rate was dramatically decreased with prior release operation regardless of the value for error parameter β. Even though Operation IV is an ideal flood control operation which is not yet feasible to introduce into the actual reservoir management based on the conventional flood management systems maintained in Japan for long years, the result shows a great potentiality of the advanced flood control operation with prior release operation based on real-time inflow prediction. It can also be seen in this figure that maximum release rate decreased as parameter β became greater. From

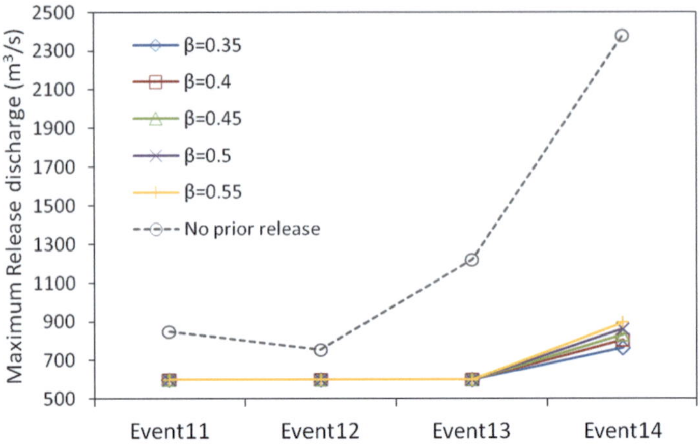

Fig. 5.10 Maximum release rates simulated in flood control operation over the flood event averaged among 1,000 simulations

the results shown in Figs. 5.9 and 5.10, it can be considered that smaller degree of prediction errors is preferred in prior release operation in terms of mitigating impacts of floods in the downstream.

5.5.6 Analysis of Prior Release Operation on Long-Term Reservoir Operation for Water Utilization

The Monte Carlo simulation on long-term reservoir operation was conducted to analyze impacts of prior release operation considering the inflow prediction on reservoir operation for water utilization. A thousand of long-term inflow scenarios were synthetically generated, and a thousand of simulations on long-term reservoir operation for water supply and power generation was respectively conducted with each inflow scenario considering the storage volume at the end of each simulation on flood control operation as the initial storage volume for long-term simulation for each flood event. For large or very large flood events (Events 6 to 15), no adverse impacts of prior release operation on water utilization were performed as water storage was recovered to the original storage level at the end of those flood events. Adverse impacts of prior release operation for five small flood events (Events 1 to 5) on water utilization were therefore analyzed here. Averaged impacts of prior release operation considering inflow prediction with each operation policy (Operations I to IV) on long-term reservoir operation for water utilization for those flood events are shown in Fig. 5.11. Averaged increase in drought damage over the simulation period for long-term reservoir operation compared to the standard value without conducting prior release operation is shown in Fig. 5.11a, while averaged loss in power generation from the standard value with no prior release is shown in

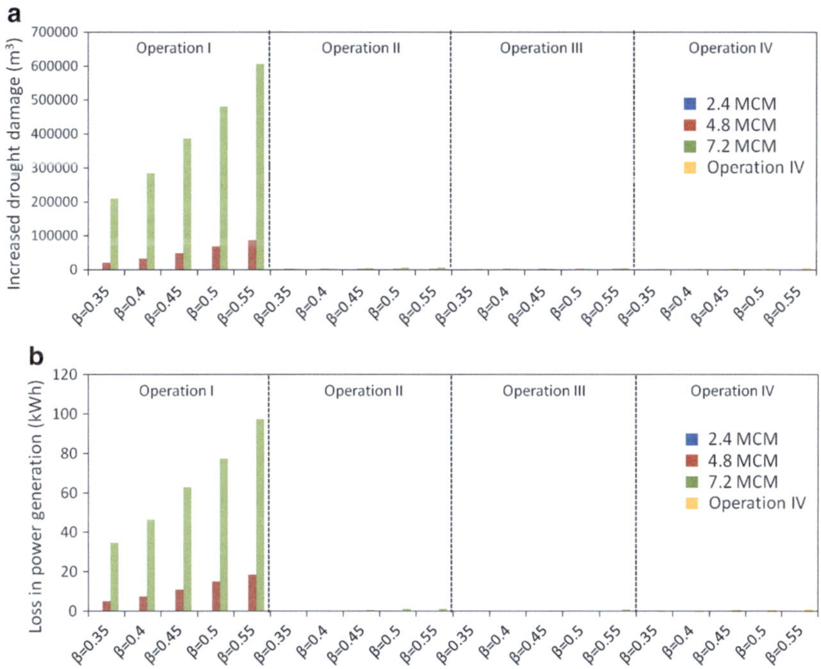

Fig. 5.11 Averaged impacts of prior release operation considering inflow prediction with each operation policy on long-term reservoir operation for five small flood events: (**a**) increase in drought damage for water supply and (**b**) loss in power generation

Fig. 5.11b. It can be seen in Fig. 5.11 that losses in both water supply and power generation were performed with Operation I when prior release was conducted to 4.8 MCM or 7.2 MCM. It can also be seen that greater loss was caused as error parameter β became greater with Operation I when the water amount to be released by prior release operation was 4.8 MCM or 7.2 MCM. It can be considered from these results that prior release operation considering inflow prediction with greater degree of error has more adverse impacts on water utilization. On the other hand, only a little damage or loss was performed in both water supply and power generation with Operations II to III, where some countermeasure against prediction uncertainty was considered. It can be therefore considered important to introduce measures which can correspond to prediction uncertainty or updates, like wait-and-see option in Operation II or interruption option in Operation III, in order to mitigate adverse impacts of prior release operation on water utilization. Considering the results in the simulations for short-term reservoir operation for flood management, Operation III with interruption option would be the most effective method for prior release operation with imperfect inflow prediction among three operation policies from Operation I to Operation III. On the other hand, there was also only little damages or losses calculated with Operation IV. Considering the ability of Operation IV to control large-scale floods which was shown in the simulations for short-term reservoir operation, the analysis suggests that prior release operation with Operation IV may be a very effective operation method for flood management with no major adverse impacts on water utilization.

Through the simulation, potential effects of prior release operation by a multipurpose reservoir on both flood management and water utilization were demonstrated. Effects of error structure of the inflow prediction on effectiveness of prior release operation were also demonstrated by the simulation. From the results of the simulation, it can be considered that a great degree of prediction errors, which was represented by error parameter β in this analysis, tends to degrade effectiveness of prior release operation on both flood management and water utilization. Only feasibility of prior release operation showed a negative correlation with feasibility of prior release operation if water amount to be released was 7.2 MCM. However, it was also demonstrated that adverse impacts on water utilization could be mitigated with prior release operation policies considering prediction uncertainty such as Operations II or III regardless of the degree of prediction errors. The potential effectiveness of Operation IV in flood management for very large flood events was also shown in this simulation with no major adverse impacts on water utilization. Even though it is an ideal flood control operation method, it shows the potential capability of multipurpose reservoirs. The simulation model developed in this study enables to assess effects of prior release operation considering inflow prediction with some error structure, which is considered helpful for reservoir managers to design those advanced reservoir operation. The results of this study are, however, demonstrated with certain assumptions in error structures of the inflow prediction. Because the hydrological values, which are to be predicted, often follow a skewed probabilistic distribution such as lognormal distribution, prediction errors can also follow a skewed distribution. Therefore, further simulations need to be conducted supposing other probabilistic distributions or serial correlations for prediction errors in order to comprehensively investigate effectiveness and impacts of prior release operation of multipurpose reservoirs.

5.6 Conclusion

A method to design and assess integrated reservoir operation for both the flood control and water utilization was developed with consideration of imperfect real-time hydrological predictions in order to enhance the capability of an existing multipurpose reservoir in this study. A Monte Carlo simulation model with synthetic generation model of hydrological predictions with a designed structure of prediction errors was developed in order to analyze effects of prior release operation considering imperfect hydrological predictions on both flood management and water utilization.

Through the simulation-based analysis, potential effectiveness of prior release operation by a multipurpose reservoir on flood management was demonstrated. On the other hand, potential adverse impact of prior release operation on water utilization can also be analyzed with the proposed simulation model. Effects of error structure of the inflow prediction on effectiveness of prior release operation were also demonstrated by the simulation. From the results of the simulation, it can be considered that a great degree of prediction errors, which was represented by error parameter β in this analysis, tends to degrade effectiveness of prior release operation on both flood management and water utilization. Only feasibility of prior release operation showed a negative correlation with feasibility of prior release operation if water amount to be released was great. However, it was also demonstrated that adverse impacts on water utilization could be mitigated with a certain operation policy corresponding to prediction uncertainty regardless of the degree of prediction errors. The potential effectiveness of advanced flood control operation method considering real-time inflow prediction, such as Operation IV in this study, in flood management for very large flood events was also shown in this simulation with no major adverse impacts on water utilization. Even though it is an ideal flood control operation method compared to the conventional flood management practice with reservoirs in Japan, it shows the potential capability of multipurpose reservoirs. These adaptive operation methods of a multipurpose reservoir considering real-time hydrological prediction can also be considered effective for more robust water resources management under climate change.

The simulation model developed in this study enables to assess effects of prior release operation considering inflow prediction with some error structure, which is considered helpful for reservoir managers to design those advanced reservoir operation. The results of this study are, however, demonstrated with certain assumptions in error structures of the inflow prediction. Because the hydrological values, which are to be predicted, often follow a skewed probabilistic distribution such as lognormal distribution, prediction errors can also follow a skewed distribution. Therefore, further simulations need to be conducted supposing other probabilistic distributions or serial correlations for prediction errors in order to comprehensively investigate effectiveness and impacts of prior release operation of multipurpose reservoirs.

References

Amai Y, Nohara D, Hori T (2014) Prior release operation of a multipurpose reservoir for flood control based on accuracy assessment of real-time hydrological prediction. In: Proceedings 19th IAHR-APD congress 2014, Hanoi

Box GEP, Muller ME (1958) A note on the generation of random normal deviates. Ann Math Statist 29(2):610–611

Chou FN-F, Wu C-W (2013) Expected shortage based pre-release strategy for reservoir flood control. J Hydrol 497:1–14

Datta B, Burges SJ (1984) Short-term, single, multi-purpose reservoir operation: importance of loss functions and forecast errors. Water Resour Res 20(9):1167–1176

Hoshi K (1997) 7.4.3 Generation of time series data. In: Japan Association of Hydrology and Water Resources (ed) Handbook of hydrology and water resources. Asakura Publishing, Tokyo, pp. 249–251 (In Japanese)

Hsu NS, Wei CC (2007) A multipurpose reservoir real-time operation model for flood control during typhoon invasion. J Hydrol 336(3):282–293

IPCC (2014) Summary for policy makers. In: Field CB, Barros VR, Dokken DJ, Mach KJ, Mastrandrea MD, Bilir TE, Chatterjee M, Ebi KL, Estrada YO, Genova RC, Grima B, Kissel ES, Levy AN, MacCracken S, Mastrandrea PR, White LL (eds) Climate change 2014: impacts, adaptation, and vulnerability, Part A: Global and sectoral aspects. Contribution of Working Group II to the Fifth Assessment Report of the Intergovernmental Panel on Climate Change. Cambridge University Press, Cambridge, UK/New York, pp. 1–32

Lattenmaier DP (1984) Synthetic streamflow forecast generation. J Hydraul Eng 110(3):277–289

Nohara D, Hori T (2014) Impact analysis of stochastic inflow prediction with reliability and discrimination indices on long-term reservoir operation. J Hydroinf 16(2):487–501

Nohara D, Hori T (2015) Planning support for prior release operation of a multi-purpose reservoir considering real-time hydrological prediction based on the Monte Carlo Simulation. In: Proceedings of the 36th IAHR World Congress, The Hague

Nohara D, Kojiri T, Smith PJ (2005) Heuristic DSS for reservoir operation considering global meteorological information, floods, from defense to management – Symposium papers. In: Van Alphen, van Beek, Taal (eds) Proceedings of 3rd international symposium on flood defense. Taylor & Francis Group, London, pp. 361–370

Sivaarthitkul V, Takeuchi K (1995) Assessment of efficiency increase of reservoir operation by the use of inflow forecasts: a case study of the Mae Klong River basin in Thailand. J Jpn Soc Hydrol Water Resour 8:590–601

Takeuchi (1990) Prediction accuracy of precipitation and practicability of anticipatory release operation policy. Annu J Hydraulic Eng JSCE 49:73–78 (In Japanese)

Thomas HA, Fiering MB (1962) Mathematical synthesis of streamflow sequences for the analysis of river basins by simulation. In: Maass A, Hufscmidt M, Dorman R, Thomas H, Marglin S, Fair G (eds) Design of water-resource systems. Harvard University Press, Cambridge, MA, pp. 459–493

Chapter 6
Optimization of Integrated Operation of Dams Using Ensemble Prediction

Satoru Oishi

Abstract Flood control is one of the most important issues of reservoir operation. Rivers in island countries like Japan, the Philippines, and Indonesia that have smaller reservoirs than continental countries need short-term reservoir operation for flood control. In Japan, typhoons give dominant amount of water to reservoirs. Prior releasing of water that makes effective use of the capacity of a reservoir requires the forecast of rainfall amount (hyetograph). Therefore, weather forecast of typhoons is indispensable for flood control. Oishi and Masuda (Study on optimization of the integrated dam operation using ensemble prediction in the upper reaches of the Nabari River. In: Proceedings of 35th IAHR world congress (IAHR), Chengdu, 2013) developed the reservoir control operation model using stochastic dynamic programming with one-week ensemble weather forecast. One-week ensemble forecast consists of 51 members, gives many kinds of weather variables including rainfall amount, and has a lead time of one week. In fact, the frequency of updating one-week ensemble forecast is a problem for using it. Therefore, a solution for the problem is proposed. For giving highly frequent updating, we propose to use typhoon ensemble forecast which issues four times a day, but it does not include rainfall amount. By using a similarity index with observed typhoon tracks and latest ensemble forecast result, a method to give reasonable typhoon ensemble forecasted rainfall amount has been developed. Showing the techniques and theories for managing water resources using advanced weather forecasting, we discuss about the possibility of adaptive countermeasure to manage the water resources by making the most of existing structure.

Keywords Reservoir operation • Prior releasing of water of dam • Flood control • Ensemble weather forecasting • Similarity index • Adaptive countermeasure • Stochastic dynamic programming

---

S. Oishi (✉)
The Research Center for Urban Safety and Security, Kobe University, Rokkodai, Nada, Kobe, Japan
e-mail: tetsu@phoenix.kobe-u.ac.jp

E. Kolokytha et al. (eds.), *Sustainable Water Resources Planning and Management Under Climate Change*, DOI 10.1007/978-981-10-2051-3_6

6.1 Introduction

Recently, torrential rain from localized severe storm, typhoon, and so on gives serious damage to river basin. Island countries like Japan, the Philippines, and Indonesia have very steep rivers in which flood water level increases very rapidly. It expands the damage of flood in these countries.

Reservoir operation is one of solutions to protect lower river basin from severe floods which generated from rapid increasing of water level. However, ad hoc operation may be conducted under very serious condition when second water increasing might be found after all capacity for flood control has been stored by previous flood. It means optimized flood control operation model should be designed by using prediction of inflow water with high accuracy in order to reduce the damage.

A lot of researches which deal with optimized dam operation have been published. For example, Takasao et al. (1982) have applied the dynamic programming (DP) with multi-reservoir and multi-evaluation-point constraints. Sayama et al. (2010) developed a system for multi-reservoir operation with DP and distributed hydrological model in large basin.

Recently, weather forecasting system gives accurate estimation of rainfall amount for several hours. However, it is still needed to improve the accuracy when it is applied to DP because small error of prediction propagates to lead wrong operation and error of prediction is inherent.

The ensemble weather forecast has been introduced for overcoming such problem led by error propagation. Ensemble weather forecasting gives stochastic concept in order to apply weather forecasting data to reservoir operation. Ensemble weather forecasting includes spread of estimated rainfall amount which shows the probability of forecasting and uncertainty of the estimation.

The first comprehensive study which dealed with ensemble weather forecast and dam operation has been conducted by Faber and Stedinger (2001) for improving the efficiency of hydraulic power plant. Nohara et al. (2009, 2011) estimated inflow discharge by taking each ensemble member into consideration. Eum and Kim (2010) studied the updating procedure of ensemble weather forecast for long-term reservoir operation.

In this chapter, the reservoir control operation model using stochastic dynamic programming with one-week ensemble weather forecast is explained by using a result of Oishi and Masuda (2013). One-week ensemble forecast consists of 51 members and gives many kinds of weather variables including rainfall amount. In fact, the frequency of updating one-week ensemble forecast is a problem for using it. Therefore, a solution for the problem is proposed. For giving highly frequent updating, we propose to use typhoon ensemble forecast which issues four times a day, but it does not include rainfall amount. By using a similarity index with observed typhoon tracks and latest ensemble forecast result, a method to give reasonable typhoon ensemble forecasted rainfall amount has been developed, and it will be explained in this chapter.

6.2 Capacity of Reservoir and Prior Releasing of Water

The total capacity of reservoir consists of flood control capacity, consumptive use capacity, inactive capacity, and dead capacity. Most of the time, there is an empty space without water in a reservoir. Volume of the empty space is defined as flood control capacity in most of reservoirs. The space will be filled by water during flood for decreasing water level at the lower part of river basin during flood. We call the activity to fill the reservoir by flood as "cutting a peak of discharge."

Flood control capacity and consumptive use capacity are clearly distinguished in order to avoid conflict. Therefore, a lot of multipurpose dams always look empty during rainy season because flood control capacity is without water all through the rainy season which is basically from May to October in Japan. However, in more than 99 % of days, the empty storage is useless. Moreover, recent numerical weather prediction gives better prediction of rainfall than decades before. Therefore, storing water in flood control capacity and releasing the same amount of water as flood control capacity before the rainfall comes can give benefit to water resources management. However, this kind of prior releasing of water in flood control capacity is not the target of the present chapter because it has more risk than the other type which is described in the following.

In the present chapter, the target is to release stored water in consumptive use capacity before the rain comes. There is not a risk of artificial flood but risk of artificial drought during releasing water in consumptive use capacity. Artificial drought happens when the amount of water released in prior to flood is more than the amount of water that rain gives. The risk of drought can be compensated by water given by rain forecasted. For example, we can release one million cubic meters when predicted total amount of discharge is more than one million cubic meters. Moreover the prior releasing of water gives safely controlled flood before the rain and cutting a peak of discharge.

The procedure of prior releasing of water with perfect prediction of rainfall having enough lead time consists of the following:

1. predicting rainfall,
2. calculating discharge,
3. calculating optimized amount of releasing water,
4. making a plan to release water under consideration of dam operational rule.

However, weather forecast is actually not perfect. There is uncertainty and reviews of the predictions are always necessary. The following section deals with the topics.

6.3 Numerical Weather Prediction System in Japan and Its Uncertainty

Numerical weather forecasting model has widely been used for predicting rainfall amount. However, incompleteness of observational data, not perfect understanding of atmospheric process, and chaotic behavior of nonlinearized numerical system lead to expansion of small error in a prediction. Therefore, recently ensemble numerical weather prediction system is developed as well as deterministic numerical weather prediction system.

Ensemble numerical weather prediction system starts from many initial conditions to analyze the spread of forecast. The spread is uncertainty in another word. Narrow spread gives relatively more certain forecast than wider spread. Usually, we assume the median of histogram of estimated rainfall amount calculated by ensemble weather prediction system. However, using each individual ensemble member for calculating the optimized reservoir control produces risk information generated by the optimization including probability of water level in lower river basin and confidential level of the operation.

The following sentences explain weather prediction system including deterministic one and ensemble one.

The operational weather prediction system in Japan Meteorological Agency (JMA) has two types, deterministic weather prediction model and ensemble weather prediction model. Deterministic weather prediction model consists of global spectral model (GSM), mesoscale model (MSM), and local forecast model (LFM). GSM uses hydrostatic assumption, whereas MSM and LFM are non-hydrostatic model. The difference between MSM and LFM is the resolution; MSM has 5 km interval grid, and LFM 2 km. Ensemble weather prediction model is based on GSM, and it consists of four types, weekly ensemble prediction system (WEPS), typhoon ensemble prediction system (TEPS), monthly ensemble prediction system, and seasonal ensemble prediction system.

JMA releases result of WEPS once a day from March 2001 by using a kind of general circulation model (GCM). JMA starts calculating WEPS from 21 Japan Standard Time (JST) which is Greenwich Mean Time (GMT) plus 9 h, and it issues WEPS at four JST after numerical calculation. WEPS contains 51 ensemble sets of prediction which has six hourly data, nine days of lead time, both longitudinal and latitudinal resolution of 1.25°, and four vertical layers (surface, 925, 850, 700, and 500 hPa). WEPS comprises physical variables as surface pressure, altitude, temperature, wind speed, wind direction, humidity, and surface rainfall amount.

JMA has also released results of TEPS since February 2008. JMA issues results of TEPS when typhoons exist or a tropical cyclone forecasted to be a typhoon exists. For making TEPS, JMA uses the same numerical model as WEPS. JMA calculates TEPS four times a day (03, 09, 15, 21 JST). TEPS contains 11 ensemble sets of prediction which has the position (longitude and latitude) and pressure of typhoon

Table 6.1 Numerical ensemble weather prediction system in JMA

Model	Region, resolution	Number of layers	Forecasting time initial time (UTC)	Number of members
Weekly ensemble prediction system (WEPS)	Horizontal resolution Entire earth About 40 km	60 layers	264 h 00, 12	27
Previous WEPS used until February 2014	Entire earth About 150 km		264 h 12	51
Typhoon ensemble prediction system (TEPS)	Entire earth About 40 km	60 layers	132 h 00, 06, 12, 18	25
Previous TEPS used until March 2014	Entire earth About 150 km		132 h 00, 06, 12, 18	11
Monthly ensemble prediction system	Entire earth About 55 km Abnormal weather detection	60 layers	18 days Saturday Sunday	25
Monthly ensemble prediction system	Entire earth About 55 km Monthly weather tendency	60 layers	35 days Tuesday Wednesday	25
Seasonal ensemble prediction system	Entire earth Atmosphere 180 km Ocean 100 km	Atmosphere 40 Ocean 50	7 months 5 days of interval	51

center and wind speed. By using TEPS, we can expect the following advantage: (i) high frequency of TEPS reduces error of prediction; (ii) prediction of typhoon track and its spread gives better reliability.

The detail specifications of JMA weather prediction systems are shown in Table 6.1.

6.4 Dam Discharge Optimization Model

6.4.1 Background

In this section, weekly ensemble weather forecast issued by Japan Meteorological Agency has been introduced into integrated operation of multi-reservoir system in order to develop a short-term flood control model which reduces the water level of the river. Stochastic dynamic programming is used for decision-making of releasing water from three reservoirs with the ensemble forecast. A flood that happened in

Nabari River basin where river authorities conducted their integrated operation of multi-reservoirs without the ensemble forecast was selected as a case study. Water level calculated by the proposed method has been compared with the result of actual decision made by the authority, one calculated under ideal condition of hundred-percent weather forecast and one brought by following an existing rule. Then the proposed method obtained the best solution.

6.4.2 Dynamic Programming for Deciding Dam Discharge

Dynamic programming (DP) developed by R. Bellman (1957) is a method of operational research. DP gives the optimized solution by dividing the problem into smaller subproblem. DP is based on the principle of optimality.

In the present chapter, the target of optimization is minimizing the discharge at evaluating point which is a certain point in the lower river basin. Then, the objective function up until time T is defined as follows:

$$\min_{O(n,t)} \sum_{t=1}^{T} \{D\left[Q(t)\right]\} \tag{6.1}$$

where $O(n, t)$ is the discharge from dam n at time $t(t = 1, \ldots, \mathrm{T})$, $Q(t)$ is the discharge at the evaluating point, and $D(Q(t))$ is the evaluation function at time t. The evaluation function in which we assume the damage of flood is proportional to the square of discharge when embankment is washed out is defined as follows:

$$D\left[Q(t)\right] = \left\{\frac{Q(t)}{Q_{id}}\right\}^2 \tag{6.2}$$

where Q_{id} is the design flood at the evaluating point. Please notice that the $D\left[Q(t)\right] > 1$ means exceeding the design flood (Takasao et al. 1982).

Constraints of dams which are decided by physical aspects of dams and their operational rule are formulated as follows:

$$\begin{aligned} S_{\min}(n) \le S(n, t) \le S_{\max}(n) \\ O_{\min}(n) \le O(n, t) \le O_{\max}(n) \end{aligned} \tag{6.3}$$

where $S(n, t)$ is the amount of stored water in dam n at time t, $S_{\min}(n)$ is the minimum amount of storage of dam n, $S_{\max}(n)$ is the maximum capacity of dam n, $O_{\min}(n)$ is the minimum discharge from dam n, and $O_{\max}(n)$ is the maximum discharge from dam n.

The continuous equation for stored water at each dam is formulated as follows:

$$S(n, t + 1) = S(n, t) + I(n, t) - O(n, t) \tag{6.4}$$

where $I(n,t)$ is the inflow to dam n at time t. The discharge at the evaluating point is calculated as summation of discharge from each dam, upstream, and remaining basin as follows:

$$Q(t) = \sum_i O(i,t). \tag{6.5}$$

Then, we define the problem as the minimization of cost function $f_t(s(1,t), s(2,t), s(3,t))$ from arbitral time t to the end T by optimizing the discharge from each dam.

By using the principle of optimality, the objective function has the following relationship:

$$\begin{aligned} &f_t(S(1,t), S(2,t), S(3,t)) \\ &= \min_{O_{\min}(n) \leq O^*(n,t) \leq O_{\max}(n)} \{D[Q(t)] + f_{t+1}(S(1,t+1), S(2,t+1), S(3,t+1))\}. \end{aligned} \tag{6.6}$$

Moreover, the future cost function $f_T(S(1,T), S(2,T), S(3,T))$ is uniquely defined as the following:

$$f_T(S(1,T), S(2,T), S(3,T)) = D[Q(T)]. \tag{6.7}$$

Therefore, we calculate the following equation step-by-step, $f_{T-1}, \cdots f_1$, and the initial condition of stored water in the dams was given as

$$\{S(1,1), S(2,1), S(3,1)\} = \{S^*(1,1), S^*(2,1), S^*(3,1)\};$$

then, the optimized solution is given as $f_1(S^*(1,1), S^*(2,1), S^*(3,1))$, and the optimized discharge from each dam is calculated as follows: $\{O^*(1,t), O^*(2,t), O^*(3,t)\}$

$$\begin{aligned} &\{O^*(1,t), O^*(2,t), O^*(3,t)\} \\ &= \min_{O_{\min}(n) \leq O^*(n,t) \leq O_{\max}(n)} \{D[Q(T)] + f_{t+1}(S(1,t+1), S(2,t+1), S(3,t+1))\}\cdot \end{aligned} \tag{6.8}$$

6.4.3 Stochastic Dynamic Programming for Deciding Dam Discharge

Stochastic dynamic programming (SDP) expresses the objective function as the expected value by considering discharge at each time with probabilistic distribution. Then, it is used for optimization using ensemble weather forecast.

The cost function is defined as follows by using stochastic process $P[I(t)]$ which is independent in terms of time series:

$$f_t(S(1,t),S(2,t),S(3,t)) = \min_{O_{\min}(n)\le O^*(n,t)\le O_{\max}(n)} \left\{ \sum_{I(t)} D\left[Q(t)\right] + f_{t+1}(S(1,t+1),S(2,t+1),S(3,t+1))P\left[I(t)\right] \right\}. \tag{6.9}$$

Generally, equivalent probability for each ensemble member can be assumed. Therefore, $P[I(t)]$ in Eq. 6.9 is defined as follows:

$$P[I(t)^m] = 1/M. \tag{6.10}$$

The objective function and cost function are defined as follows:

$$\min_{o(n,t)} \sum_{t=1}^{T} \left\{ \frac{1}{M} D\left[Q(t)^m\right] \right\} \tag{6.11}$$

$$f_t(S(1,t),S(2,t),S(3,t)) = \min_{O_{\min}(n)\le O^*(n,t)\le O_{\max}(n)} \left\{ \tfrac{1}{M} D\left[Q(t)^m\right] + f_{t+1}(S(1,t+1)^m, S(2,t+1)^m, S(3,t+1)^m) \right\}. \tag{6.12}$$

Then the optimized discharge from each dam, $\{O^*(1,I(1,t)^m,t), O^*(2,I(2,t)^m,t), O^*(3,I(3,t)^m,t)\}$, is calculated by using the following equation:

$$\begin{aligned} &\{O^*(1,I(1,t)^m,t), O^*(2,I(2,t)^m,t), O^*(3,I(3,t)^m,t)\} \\ &= \min\{\tfrac{1}{M} D\left[Q(t)^m\right] + f_{t+1}(S(1,t+1)^m, S(2,t+1)^m, S(3,t+1)^m)\}. \end{aligned} \tag{6.13}$$

6.4.4 Application Result

6.4.4.1 Target Dams

In the present chapter, the dam operation model described above has been applied to Nabari River basin where the cities in the lower basin were threatened by flood under typhoon no. 18 in 2009 (T0918, MELOR). Nabari River flows through Nabari City in Mie Prefecture and it is one of tributaries of Yodo river system. There is a confluence of Shorenji River and Uda River in the upper part, and Nabari River joins to Kizu River in the lowest end. Hinachi Dam locates in the upper part of Nabari River, Shorenji Dam in the upper part of Shorenji River, and Muro Dam in the upper part of Uda River. These dams are multipurpose dams which include purpose of flood control, irrigation, hydraulic power plant, and domestic water supply.

Typhoon T0918 made landfall in Chita Peninsula at 5AM JST of October 8, 2009, then it passed through Honshu Island to reach the southeast coast of Hokkaido Island. The typhoon killed 5 and injured 132 all over Japan, and it collapsed 4328 houses. The typhoon approached Nabari City in the dawn of October 8, 2009. The city was threatened by flood brought by the typhoon. Therefore, Kizukawa Dam integrated management office of Japan Water Agency (JWA) with Yodogawa

Dam integrated management office of Japanese Ministry of Land Infrastructure, Transport and Tourism (MLIT) took ad hoc operating procedure which has been examined in advance. In the procedure, they took the water level, rainfall amount, and storage capacities of three dams into account for integrating flood control action to store the water by reducing the discharge from dams. Then, the ad hoc procedure decreased the water level of Nabari point which is evaluating point for flood control by 1.5 m from calculated water level without taking action. The action contributed to avoid flood damage for 1,200 houses.

6.4.4.2 Dam Operational Rule and Dam Operation

The three dams in the upper part of the basin has their flood control rules which regulate flood control starting discharge. Before the inflow into each dam reached the flood control starting discharge, releasing discharge from each dam was equal to the inflow. Then, it reached the flood control starting discharge, releasing discharge was fixed to the flood control starting discharge, and the remaining amount of inflowing water should be stored by dams.

However, the dam integrated management offices took ad hoc operating procedure. The inflow into each dam reached the flood control starting discharge at the afternoon of October 7 and flood control has been started. The weather forecast at the moment predicted less rainfall amount than the one that actually happened. However, the typhoon route was changed at 2AM October 8; then rainfall amount increased to lead prediction of water level at Nabari point which exceeded the warning level. Therefore, dam integrated management offices took action for storing water in the reservoirs by reducing the releasing discharge from 3:15AM. This flood control action soon gave risk of shortage of capacity of Shorenji Dam which has the smallest capacity in the three dams at 3:40. Then, they investigated the situation of water level and remaining reservoir capacity carefully, then they increased releasing discharge from Shorenji Dam as well as decreased releasing discharge from Hinachi Dam which has the largest capacity in three dams. At 4:40, weather forecast estimated less amount of rain at Muro Dam basin than previous estimation, then they reduced the releasing water discharge from Muro Dam leading reduction of water level at Nabari point. Finally, they reconsidered the operation action 10 times within 2 h resulting to avoid flood in Nabari City. The inflow into and outflow from each dam have been recorded in hydrology and water quality database of MLIT (Database for Hydrology and Water Quality in Japan 2012) which has been used for the present chapter as data of comparison.

6.4.4.3 Modeling of Target Basin

The model of the target basin including Hinachi Dam, Shorenji Dam, and Muro Dam is shown in Fig. 6.1. The amount of releasing water discharge from each dam was optimized in the present chapter by calculating inflow of each subsection using forecasted rainfall amount.

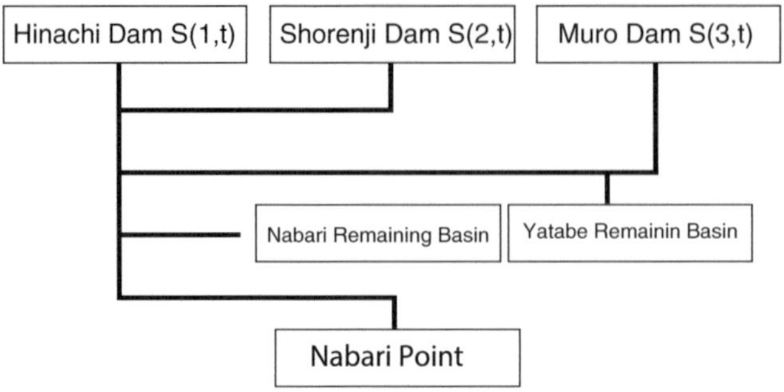

Fig. 6.1 Model of target basin

Table 6.2 Runoff coefficients and basin area

	Runoff coefficient	Basin area [km^2]
Hinachi Dam	0.39	75.8
Shorenji Dam	0.65	100.7
Muro Dam	0.40	134.9
Hinachi River	0.50	75.5
Shorenji River	0.50	100.0
Uda River	0.50	136.0
Yatabe remaining basin	0.50	70.8
Nabari remaining basin	0.50	51.4

6.4.5 *Flood Prediction by Using Ensemble Weather Prediction*

One-week ensemble weather forecast issued from JMA daily at 21:00 JST has been used in the present chapter as grid point value (GPV). The grid point of 33.75N and 136.25E was selected as the nearest point to Nabari City. The one-week ensemble weather forecast has 6 h interval for rainfall amount estimation; then, linear interpolation was applied to obtain the hourly rainfall amount.

AMeDAS rainfall amount at Nabari AMeDAS point (34.64N, 136.11E) was selected as data of perfect forecast.

Rational equation which is for calculating peak discharge in small river basin has been used for converting estimated rainfall amount into time series variation of discharge:

$$Q = \frac{1}{3.6} f r_i A_i \tag{6.14}$$

where $Q[m^3/s]$ is the discharge, f is the runoff coefficient, r_i[mm/h] is the rainfall intensity, and $A_i[km^2]$ is the basin area. Runoff coefficients and area were referred from Table 6.2 published by Kinki Regional Office of MLIT in November 2009.

H-Q equation for converting water level at Nabari point to the discharge is

$$\begin{cases} Q = 46.51(H - 2.2)^2 \ H \leq 5.24 \\ Q = 72.77(H - 2.8)^2 \ H \leq 5.25 \end{cases} \tag{6.15}$$

6.4.5.1 Objective Function and Constraints

The comparison among optimized dam operation with one-week ensemble weather forecasting issued from JMA, one with mean of ensemble weather forecasting and one with perfect forecasting using rain gauge observation data, operation under flood control rules regulated in dam operation manual, and ad hoc operation activity taken by the dam integrated management offices, has been conducted in the present chapter in order to evaluate the model. DP has been applied for mean of ensemble weather forecasting, perfect weather forecasting, and SDP for ensemble weather forecasting.

From H-Q equation at Nabari point, $Q_{id} = 1,893\,\mathrm{m}^3/\mathrm{s}$ was calculated from design flood water level of 7.99 m.

The constraints were decided as follows:

1. In case stored water amount is less than maximum normal level:
   Hinachi Dam $0 \leq S(1,t) \leq 1,430 \times 10^4\,\mathrm{m}^3$
   Shorenji Dam $0 \leq S(2,t) \leq 2,380 \times 10^4\,\mathrm{m}^3$
   Muro Dam $0 \leq S(3,t) \leq 1,840 \times 10^4\,\mathrm{m}^3$.
2. In case inflow into each dam exceeds flood control starting discharge, the amount exceeding flood control starting discharge should be stored in the reservoir. In the present chapter, prior releasing is allowed for optimized dam operation within the amount less than flood control starting discharge:
   Hinachi Dam $0 \leq O(1,t) \leq 300\,\mathrm{m}^3/\mathrm{s}$
   Shorenji Dam $0 \leq O(2,t) \leq 450\,\mathrm{m}^3/\mathrm{s}$
   Muro Dam $0 \leq O(3,t) \leq 300\,\mathrm{m}^3/\mathrm{s}$.

6.4.5.2 Optimization Results and Discussions for Applied Models

In the following sections, we defined terminology as follows for explaining the comparison among them:

SDP operation: stochastic dynamic programming optimization with ensemble weather forecast,

Ensemble mean DP operation: dynamic programming optimization with mean of ensemble weather forecast,

Perfect DP operation: DP optimization with actually collected rainfall data as a perfect weather forecasting,

Ad hoc operation: ad hoc operation activity taken by the dam integrated management offices conducted,

Regular operation: operation under flood control rules regulated in dam operation manual.

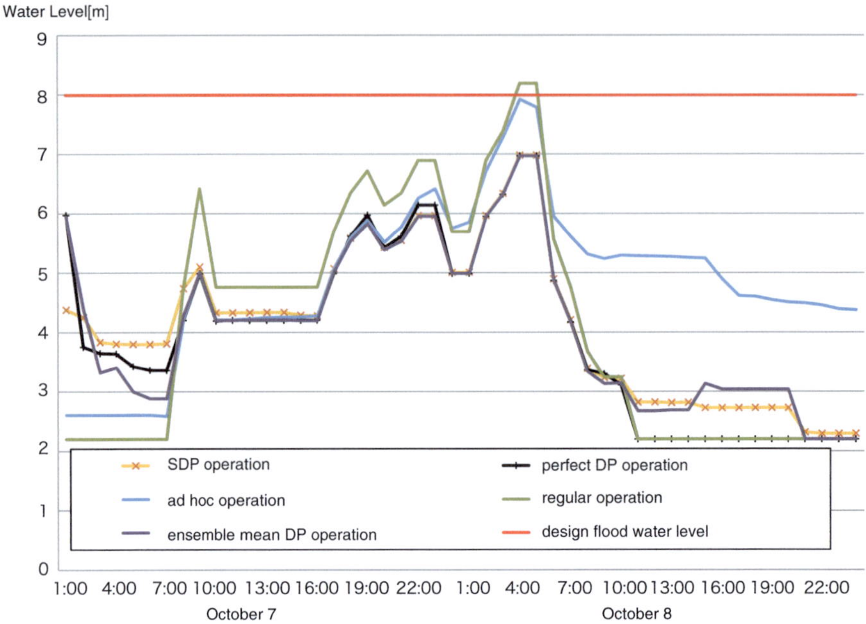

Fig. 6.2 Time series variation of water level at Nabari evaluating point by optimized flood control using SDP

The water level at Nabari flood evaluating point and stored water amount in each of Hinachi, Shorenji, and Muro Dam are shown in Fig. 6.2.

Figure 6.2 indicates that regular operation gives maximum water level (MWL) of Nabari point that was estimated as 8.2 m at 4:00 to 5:00AM JST of October 8. It means that regular operation could not prevent water level from exceeding design flood level. Ad hoc operation reduced MWL to 7.9 m which is less than design flood level; perfect DP operation and SDP operation give MWL of 7.0 m which is much less than ad hoc operation. It suggests that SDP using ensemble weather forecast gives better information for flood control action.

Deeper analysis inside each operation reveals that the SDP operation and perfect DP operation decide prior flood control operation by using Shorenji Dam and Muro Dam which was not conducted by ad hoc operation. Moreover, perfect DP operation stops releasing water from all dams after 17:00 of October 7, whereas SDP operation decides to reduce the releasing water from all dams to the lowest amount. The perfect DP operation decided to store water up to maximum flood control capacity, and SDP operation gave reserve storage as $3.2 \times 10^5\,m^3$ in Shorenji Dam and $5.1 \times 10^5\,m^3$ in Muro Dam. Hinachi Dam has so large capacity to store the flood discharge that SDP operation and perfect DP operation decided to store all of the inflow discharge into Hinachi Dam.

In ensemble mean DP operation did not give appropriate optimization because of the difference between mean rainfall amount of weather forecast and observed rainfall amount resulting that stored water in Shorenji Dam and Muro Dam exceeded their maximum storage capacity.

6.4.5.3 Discussion by Varying the Objective Function

The cause of ensemble mean DP operation exceeded the maximum storage capacities that were deeply investigated by assuming it as the formulation of cost function with square of discharge at Nabari point. Then, the formulation of cost function was changed into linear of the discharge and third power of it. Figure 6.3 shows the water level at Nabari point. In Fig. 6.3 ensemble mean DP operation was applied with cost function of square of discharge (shown as square of discharge), linear of discharge, and third powered discharge. Figure 6.3 contains the result of perfect DP operation and SDP operation both of which have cost function of square of discharge for comparison.

Figure 6.3 shows that higher water level at Nabari point was estimated by using cost function of linear and square of discharge leading to the decision of more amount of releasing water as prior flood control. Cost function using third powered discharge estimated lower water level at Nabari point leading to less amount of prior flood control releasing. The following decisions after 10:00AM JST of October 7 were almost the same. Therefore, the formulation of cost function affected prior flood control decision.

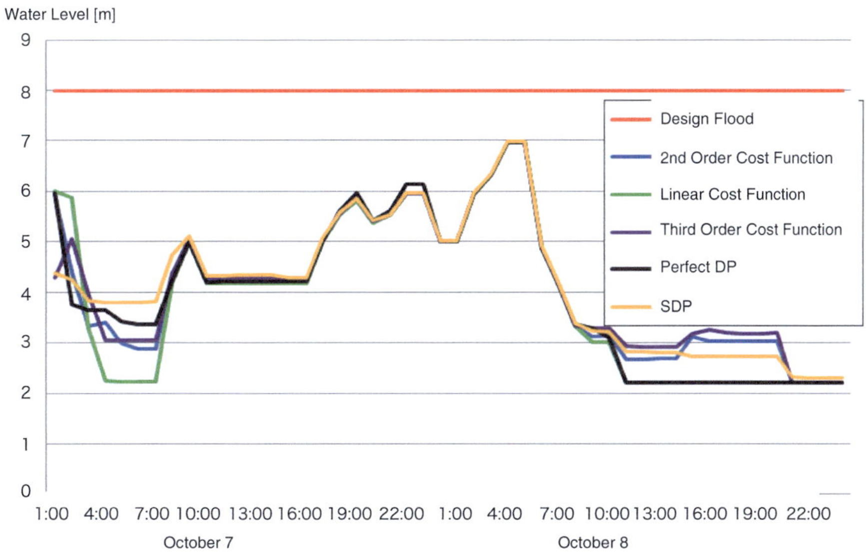

Fig. 6.3 Variation of water level by changing the order of the cost function

The difference of stored water amount at each dam was investigated in order to understand the effect of prior flood control decision to stored water amount. They decided that all inflow discharge into Hinachi Dam was stored and there was no difference. On the other hand, less prior flood releasing when they use cost function of third powered discharge resulted in the exceeding stored water amount of $3.2 \times 10^5\ m^3$ for Shorenji Dam and $3.3 \times 10^5\ m^3$ for Muro Dam from the maximum flood control capacity of each dam. The linear cost function increases the amount of prior flood releasing from dams resulting in that they remained reserve capacity from flood control capacity more than $3.5 \times 10^5\ m^3$ for Shorenji Dam and $3.0 \times 10^5\ m^3$ for Muro Dam.

In summary, higher-order cost function gave decision of less releasing water and reducing water level at Nabari point, whereas lower-order cost function gave more releasing water and reducing water level of each dam. Finally, the second order of discharge was appropriate to the cost function in the present case also.

6.4.5.4 Optimization of Dam Discharge by Taking Prediction with More Dangerous Rain

Instead of using mean of ensemble, larger and rapidly increasing rainfall forecasted in ensemble weather forecast was selected to optimize the flood control in order to consider the safer operation of dams. Safer operation was brought from dangerously forecasted rainfall which means an ensemble member with the largest amount of rainfall within ensemble members, a member with the shortest term from the starting rainfall to the peak of it, and members with multiple peaks. In the present chapter, the following ensemble members were used: (i) the member with total rainfall amount of 263 mm (E = 50), (ii) the member which predicted peak of rain earlier than the actual peak and strongest rainfall intensity of 17.3 mm/h (E = 43), and (iii) the member which has prediction closer to both (i) and (ii) with prediction of the total amount of rainfall as 251 mm and of maximum rainfall intensity as 16.7 mm/h. The actual total amount of rainfall was 188 mm.

Figure 6.4 shows water level optimized by using each ensemble member with dangerous prediction.

Figure 6.4 indicates that dangerous prediction increased prior flood releasing from dams in order to avoid exceeding amount of flood control capacity. Therefore, larger amount of reserve capacity remained as $4.0 \times 10^6\ m^3$ in Shorenji Dam and $3.0 \times 10^6\ m^3$ in Muro Dam. When E = 50 which predicted the largest amount of rainfall in the last half of rainfall period was used for optimization, it decided to release water from dams in the first half of rainfall period; then it decided to store water in dams. However, the stored water has exceeded total capacity for flood control at 15:00 JST of October 8; then it released water after the peak leading to rising water level at Nabari point by too much release from dams.

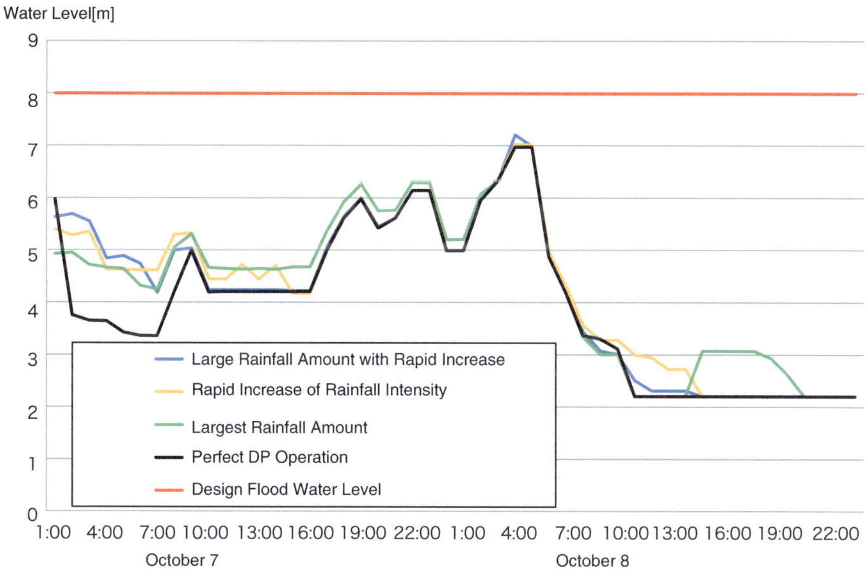

Fig. 6.4 Optimized water level by using ensemble members with dangerous prediction

6.5 Efficiency of Frequent Update of Ensemble Prediction

6.5.1 Background

Oishi and Masuda (2013) developed the reservoir control operation model using stochastic dynamic programming (SDP) with one-week ensemble weather forecast (WEP) made by weekly ensemble prediction system (WEPS) issued by Japan Meteorological Agency (JMA). WEP consists of 51 members and gives many kinds of weather variables including rainfall amount.

In fact, the frequency of issuing WEP is daily which is less frequent than the requirement and it raises a problem for using WEP. In the present chapter, a solution for the problem by using typhoon ensemble prediction (TEP) made by typhoon ensemble prediction system (TEPS) issued by JMA is proposed.

6.5.2 Uncertainty Analysis of TEPS and WEPS

TEPS has uncertainty because of their limitation of measurement, initial condition setting, and numerical specifications. Here, the problem that comes from numerical simulation is described.

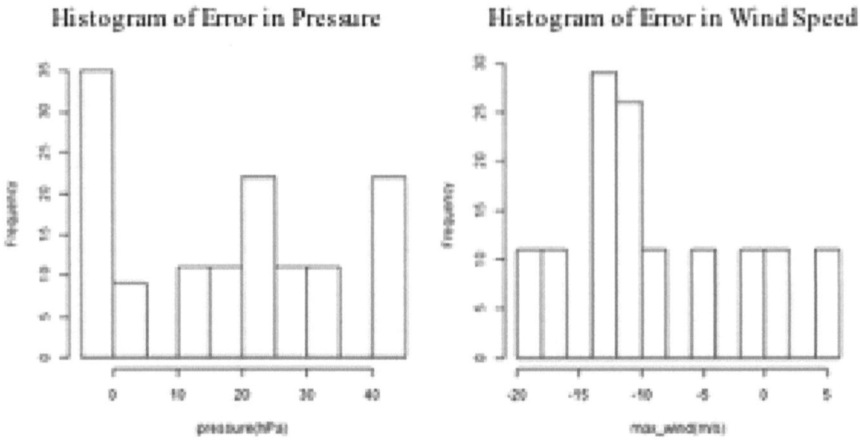

Fig. 6.5 Tendency of the prediction error in pressure (*left*) and wind speed (*right*)

The disadvantage of TEPS is their limitation of predicting pressure of typhoon center. For predicting the pressure of typhoon center accurately, it requires 5 km resolution of numerical simulation, whereas the GCM has 60 km resolution. The limitation of numerical resolution gives higher pressure as a prediction than actual situation. Then, the wind speed also is predicted weaker than actual. Figure 6.5 shows the tendency of the prediction error which is defined with observation level from the predicted value. The figure shows that pressure tends to be higher and the wind speed weaker in prediction.

As the purpose of ensemble numerical prediction, a WEPS has a range of predicted value which is called a “spread.” Figures 6.6 and 6.7 show a better example and a worse example of the spread of rainfall amount, respectively. Figure 6.6 shows the result of WEPS for typhoon no.17 in 2012 (T1217), JELAWAT, and Fig. 6.7 for typhoon no. 09 in 2011, MUIFA. Both Figs. 6.6 and 6.7 have horizontal axis showing time and three vertical axes showing the total amount of rainfall updating daily. A horizontal line shows the observed rainfall amount which is the correct value. Horizontal bars show the histogram of predicted amount of rainfall by WEPS. A set of horizontal bars starting from one vertical axis means a set of prediction spread obtained from a set of ensemble forecast issued daily. The bar graph in black color stretching downward from the horizontal arrow line in the lower part expresses hyetograph. Figure 6.6 shows the reducing of uncertainty of prediction where the spread of the left side of the figure is wide and one of the right side is sharp. Figure 6.7 shows the worst ensemble prediction among the present chapter which firstly gave lower amount of rainfall as forecast, and then prediction increased the amount of rainfall time by time and finally gave the overestimation.

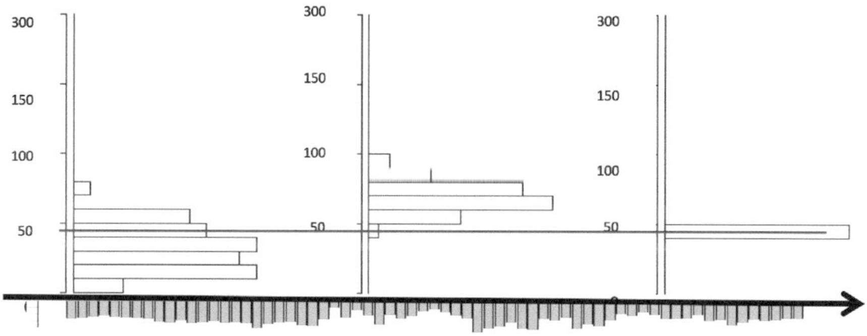

Fig. 6.6 Transition of spread of WEPS for T1217; *red line* shows the total amount of rainfall; three vertical axes show updating of WEPS and predicted rainfall amount; horizontal bars show histogram; vertical bar in *gray color* shows hyetograph

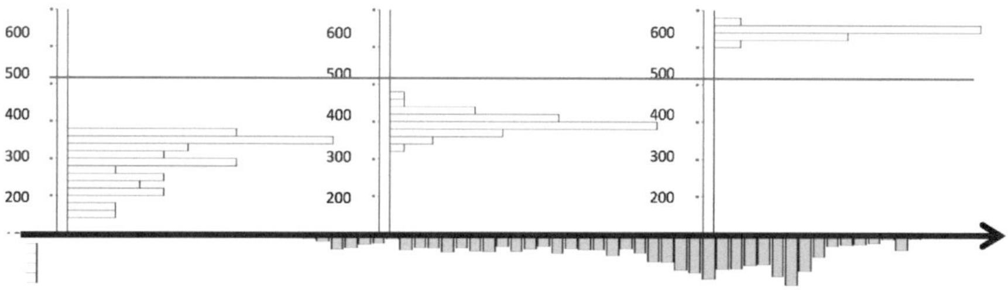

Fig. 6.7 Transition of spread of WEPS for T1109; *red line* shows the total amount of rainfall; three vertical axes show updating of WEPS and predicted rainfall amount; horizontal bars show histogram; vertical bar in *gray color* shows hyetograph

6.5.3 Combining of TEPS and WEPS

6.5.3.1 Development of the Combined Model

In order to obtain more frequent ensemble rainfall forecast by using TEPS (four times a day), a model that combines TEPS and WEPS (once a day) has been developed. The concept of the combined model uses similarity of physical variables between TEPS and WEPS. Moreover rainfall amount of similar WEPS ensemble member is assigned to corresponding member of TEPS.

In order to measure the similarity, first, the impact of each variable on forecasting rainfall amount has been analyzed by multiple regression analysis in which correlation among errors of each predicted variable in WEPS such as rainfall amount at a target point as an explained variable, longitude and latitude of typhoon center, distance between typhoon center and target point, and center pressure of the typhoon and wind speed as explaining variables. Table 6.3 shows results of multiple regressions. It shows that correlation coefficient R2 was less than 0.5 in 2012 and 2013. It means some revision of WEPS has been accomplished and numerical

Table 6.3 Result of multiple regressions among rainfall amount with the other physical variables

		t-value				
Date	R2	Distance	Latitude	Longitude	Pressure	Wind speed
918	0.2	0.71	−0.96	−1.31	6.00	2.61
1004	0.73	3.61	−1.4	3.35	−13.37	−0.17
1007	0.64	1.59	−7.46	6.65	9.63	6.06
1009	0.79	−5.88	−6.61	5.95	−14.65	−1.18
1109	0.68	−2.89	−4.98	0.75	5.74	−0.54
1115	0.74	−2.37	2.49	−4.62	1.06	−8.66
1204	0.03	1.03	−2.22	1.25	−0.85	−0.2
1210	0.24	3.91	0.25	−1.91	−4.3	−0.74
1215	0.47	1.22	−3.08	2.04	−10.28	0.88
1216	0.1	−3.83	3.36	−1.52	−2.74	−0.08
1217	0.2	1.52	1.77	−0.01	3.37	2.29
1304	0.38	1.23	5.67	−0.69	−6.19	−0.02

ensemble prediction has been improved to have less correlation in errors of physical variables including rainfall amount. Even correlation coefficient were low, t-value of the error of pressure toward the error of rainfall amount has significant value. Therefore, we are using pressure as an index of similarity. However, the track of the center of typhoons was thought to be an index of similarity; therefore, we are using distance as the other index of similarity. Finally, we proposed a couple of combined model of TEPS and WEPS by using similarity of distance which is called TEPS_WEPS(distance) model and using similarity of distance and pressure which is called TEPS_WEPS(distance, ps) model.

Figure 6.8 shows an example on T1217 where the forecast of WEPS was reasonable as shown in Fig. 6.6. It shows the time series variation of observed rainfall shown as amedas, mean of WEPS forecasts, TEPS_WEPS(distance) model forecasts, and TEPS_WEPS(distance, ps) model forecasts. According to Fig. 6.8, these models gave reasonable forecasts. Figure 6.9 shows the other example on T1109 where the forecast of WEPS was not good as shown in Fig. 6.7. When the WEPS was worse, the proposed model of TEPS_WEPS gave slightly better forecast.

Table 6.4 shows summary of the root mean square error (RMSE) of the forecast for all typhoon which the present chapter has dealt with. According to Table 6.4, TEPS_WEPS (distance, ps) model gave smaller RMSE than WEPS in 7 out of 14 typhoons. Deeper analysis for T1106 and T1112 has been conducted because TEPS_WEPS (distance, ps) gave much bigger RMSE in these typhoons. It was rain in the southern part of Shikoku Island where topography usually affects the rainfall distribution during passing of a typhoon. The model proposed in the present chapter did not take such a local topography into account. Then, the model gave worse forecast than WEPS model which took the topography into account. However, our target basin in the study was not in southern part of Shikoku Island. Then, we can use the result of model with reasonable accuracy.

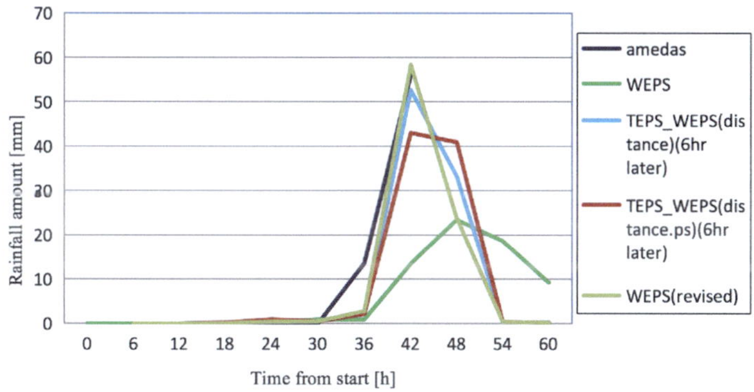

Fig. 6.8 Result of observed (amedas) and forecasted (the other) amount of rainfall in the case of T1217

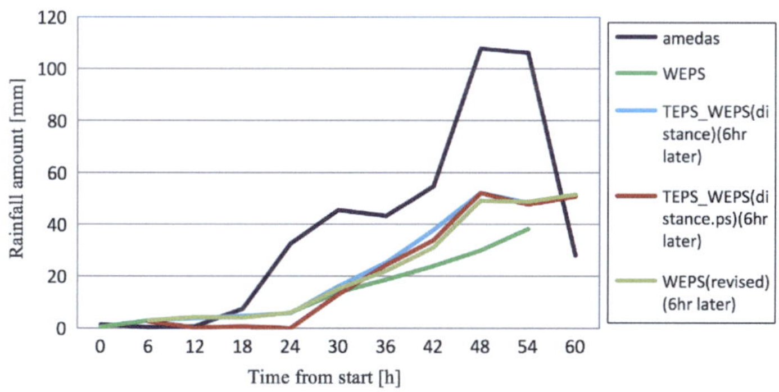

Fig. 6.9 Result of observed (amedas) and forecasted (the other) amount of rainfall in the case of T1109

6.5.4 Application of TEPS_WEPS Model to Reservoir Operation

The result is shown in Table 6.5 as damage function which gives smaller amount as better operation. Table 6.5 shows that the best operation has been done by using DP with perfect forecast and the second best was SDP by using all ensemble members of WEPS. TEPS_WEPS(distance) gave slightly worse result than DP by using ensemble mean.

Figure 6.10 shows the water level at a point at which the river authority evaluates the river water level for flood control. Figure 6.10 consists of various results in which red horizontal line was the design flood water level, gray is the result of operation under flood control rules regulated in dam operation manual, and light blue is one of ad hoc operations taken by the dam integrated management offices and ones

Table 6.4 Root mean square error of rainfall forecast made by WEPS and models proposed

RMSE	Available from 6 h later				Available from 12 h later		
(mm)	WEPS	TEPS_WEPS (distance)	TEPS_WEPS (distance,ps)	WEPS (revised)	TEPS_WEPS (distance)	TEPS_WEPS (distance,ps)	WEPS (revised)
918	18.9	9.1	9.6	8.7	9.9	8.3	8
1004	7.6	7.9	8.8	8.7	8.1	9	8.7
1007	13.6	15.4	10.4	10	7.4	10	10
1009	1.5	1.5	0.3	1.5	1.2	1.2	1.5
1106	34	33.1	32.8	33	33.1	33.1	33.2
1109	36.6	27.5	29.4	30.4	27.9	32.2	29.5
1112	45.3	46.2	45.8	45.7	45.8	45	45.6
1115	44.2	45.9	45.8	45.9	45.5	45.6	45.5
1204	33.1	23.3	24.4	24.6	23.2	20.1	24.6
1210	7.5	8.8	9.1	10.4	9.7	9.5	10.4
1215	9.5	13.3	14.6	13.7	13.8	13.6	13.6
1216	11.2	9.6	6	8.3	4.3	5.5	8.8
1217	16.8	4.5	7.1	4.2	4.7	6.7	4.2
1304	2.5	1.4	1.7	1.9	1.9	2.3	2.1
Total	16.9	14.8	11.5	14	14.3	11.3	13.9

Table 6.5 Result of DP and SDP using WEPS and TEPS_WEPS models with damage function

	Damage function using damage in DP		Damage function using damage in DP
Ad hoc operation actually conducted	5	TEPS_WEPS (distance) DP	2.39
Operation under flood control rules regulated in dam operation manual	6.13	TEPS_WEPS (distance)DP (revised)	2.47
DP with perfect forecast	2.24	TEPS_WEPS (distance, ps) DP	2.41
SDP with WEPS	2.31	TEPS_WEPS (distance, ps)DP(revised)	2.41
DP with ensemble mean of WEPS	2.34		

of SDP and DP by using WEPS and TEPS_WEPS model. The figure shows that cities might suffer flood when they conducted operation under flood control rules regulated in dam operation manual and that ad hoc operation which was actually conducted by the dam integrated management offices prevented cities from severe damage by flood. Moreover, the SDP and DP by using WEPS as well as the proposed method would give safe margin of 1.2 m of water level. Unfortunately, the present chapter did not improve flood control by proposing TEPS_WEPS models. However, TEPS_WEPS model gave more frequent updating of the forecast without reducing the accuracy.

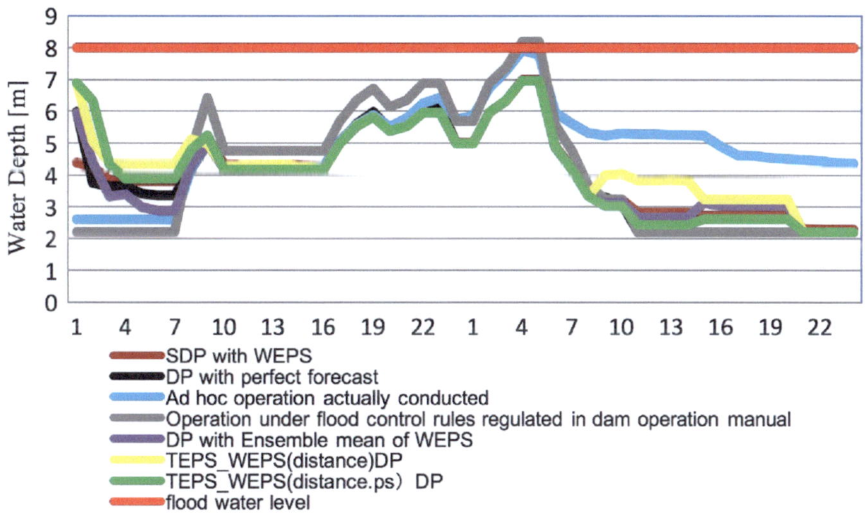

Fig. 6.10 Water level at the evaluating point in the river; *red horizontal line* shows design flood

These appropriate operations made by SDP and DP by using WEPS and TEPS_WEPS models come from the prior releasing of stored water from reservoir which depends on the forecasted amount of water exceeding the dam capacity. Then, the reservoirs had enough space to store flood.

6.5.5 *Conclusion*

The discussion in the present chapter is summarized into four following topics:

1. Stochastic dynamic programming (SDP) optimization by using ensemble weather forecast decreased the water level at Nabari evaluating point proofing effectiveness of ensemble weather forecast for optimization of dam operation.
2. DP optimization using mean of rainfall amount of ensemble weather forecast exceeded amount of stored water from capacity for flood control.
3. Evaluation of the formulation of cost function suggested that the second order of discharge proposed by Takasao et al. (1982) gave the best optimization.
4. When only dangerously predicted ensemble members were used for optimization, too much amount of water was released to cause the fear of flood at the lower part of the basin.
5. The authors proposed a method which gives highly frequent prediction of rainfall amount when typhoon comes. The method was named as TEPS_WEPS model, and they combined the results of weekly ensemble prediction system (WEPS) and typhoon ensemble prediction system (TEPS). The TEPS_WEPS models produced reasonable forecast of rainfall amount with 6 h updating frequency, whereas WEPS is updated every day.

6. The TEPS_WEPS models have been applied to the same typhoon event as Oishi and Masuda (2013) have dealt with. TEPS_WEPS models reduced water level by 1.2 m from operation under flood control rules regulated in dam operation manual and 0.9 m from ad hoc operation actually conducted by the dam integrated management offices.

These results show the efficiency of the models proposed in the present chapter. For better understanding of the advantage and disadvantage of the model proposed in the present chapter, application of the model into several typhoon events is needed.

References

Bellman R (1957) Dynamic programming. Princeton University Press, Princeton

Database for Hydrology and Water Quality in Japan. http://www.river.go.jp/. Last access on 1 Feb 2012

Eum H-I, Kim Y-O (2010) The value of updating ensemble streamflow prediction in reservoir operations. Hydrol Process 24:2888–2899

Faber BA, Stedinger JR (2001) Reservoir optimization using sampling SDP with ensemble streamflow prediction (ESP) forecasts. J Hydrol 249(1–4):113–133

Nohara D, Tsuboi A, Hori T (2009) A study on water release decision process of optimization models in the application of ensemble forecast to long-term reservoir operation. Annuals of Disaster Prevention Research Institute, Kyoto University, vol 54B, pp 753–764

Nohara D, Miki H, Hori T (2011) Applicability of multiple information sources of global meteorological condition to long-term reservoir operation. Annuals of Disaster Prevention Research Institute, Kyoto University, vol 52B, pp 699–709

Oishi S, Masuda H (2013) Study on optimization of the integrated dam operation using ensemble prediction in the upper reaches of the Nabari River. In: Proceedings of 35th IAHR world congress (IAHR), Chengdu

Sayama T, Tachikawa Y, Kanno H, Takara K (2010) Development of reservoir control optimization simulator by integrating a distributed rainfall runoff model and dynamic programming. Ann J Hydraul Eng 54:547–552

Takasao T, Ikebuchi S, Kojiri T (1982) A study of optimal flood control system considering the flood inundation probability in time and space. Annuals of Disaster Prevention Research Institute, Kyoto University, vol 25 B-2, pp 285–296

Part III
Adaptive Policy Under Climate Change: Case Studies

Chapter 7
Adaptation to Climate Change: Risk Management

Slobodan P. Simonovic

Friends are like stars,
you don't always see them
but, you know they are
always there. This is for
Toshi, my very close friend.

Abstract Adaptation to climate change is a challenge that is complex and involves increasing risk. Efforts to manage these risks involve many decision-makers, conflicting values, competing objectives and methodologies, multiple alternative options, uncertain outcomes, and debatable probabilities. Adaptation occurs at multiple levels in a complex decision environment and is generally evaluated as better–worse, not right–wrong, based on multiple criteria. Identifying the best adaptation response is difficult. Risk management techniques help to overcome these problems. Here, risk management is presented as a decision-making framework that assists in the selection of optimal strategies (according to various criteria) using a systems approach that has been well defined and generally accepted in public decision-making. In the context of adapting to climate change, the risk management process offers a framework for identifying, assessing, and prioritizing climate-related risks and developing appropriate adaptation responses. The theoretical discussion is illustrated with an example from Canada. It includes (a) the assessment of climate change-caused flood risk to the municipal infrastructure for the City of London, Ontario, Canada, and (b) analysis of adaptation options for management of the risk in one of the watersheds within the City of London – Dingman Creek.

Keywords Climate change • Flooding adaptation • Risk management

---

S.P. Simonovic (✉)
Department of Civil and Environmental Engineering, The University of Western Ontario, London, ON, Canada
e-mail: simonovic@uwo.ca

E. Kolokytha et al. (eds.), *Sustainable Water Resources Planning and Management Under Climate Change*, DOI 10.1007/978-981-10-2051-3_7

7.1 Introduction

Several definitions of adaptation are available in the climate change literature. The following are some of the examples as summarized by Simonovic (2012) following Smit et al. (2000):

The term adaptation means any adjustment, whether passive, reactive or anticipatory, that is proposed as a means for ameliorating the anticipated adverse consequences associated with climate change.

Adaptation to climate change includes all adjustments in behaviour or economic structure that reduce the vulnerability of society to changes in the climate system.

Adaptability refers to the degree to which adjustments are possible in practices, processes or structures of systems to projected or actual changes of climate. Adaptation can be spontaneous or planned, and can be carried out in response to or in anticipation of change in conditions.

According to some of the typologies considered, adaptation can be planned or spontaneous; passive, reactive, or anticipatory; etc. According to the IPCC (2013), adaptation "has the potential to reduce adverse impacts of climate change and to enhance beneficial impacts, but will incur costs and will not prevent all damages."

The tendency of systems (e.g., natural, social, and engineering) to adapt is influenced by certain system characteristics. These include terms such as "sensitivity," "vulnerability," "resilience," "susceptibility," and "adaptive capacity," among others. The occurrence and the nature of adaptations are influenced by these. Adaptation is often the result of interactions between climatic and other factors. It varies not only with respect to its climatic stimuli but also with respect to other non-climate conditions. It is important to highlight that the relationship between a changed climate system (e.g., higher temperatures, altered precipitation regime, etc.) and impacts on various systems is not necessarily linear. The role of adaptation (whether reactive or anticipatory, spontaneous or planned, etc.) is crucial for assessments of potential impacts of climate change.

Adaptation to climate change is a challenge that is complex and involves increasing risk. Efforts to manage these risks involve many decision-makers, conflicting values, competing objectives and methodologies, multiple alternative options, and uncertain outcomes (Noble et al. 2005). Risk management offers a decision-making framework that assists in the selection of optimal strategies (according to various criteria) using a systems approach that has been well defined and generally accepted in public decision-making. In the context of adapting to climate change, the risk management process offers a framework for identifying, assessing, and prioritizing climate-related risks and developing appropriate adaptation responses. Vulnerability assessment is a central element of risk management. Vulnerability assessment is increasingly useful for guiding adaptation, since it helps reveal local- and larger-scale system vulnerabilities for which adaptation measures may be necessary to prevent serious adverse consequences. Unlike "adapting," the concept "managing risks" seems, from many perspectives, much more clear. Risk management is a familiar concept, especially in disaster management, whereas the notion of “adapting” is still poorly understood by many. Risk management provides a means for addressing uncertainties explicitly. Without a risk management view, decision-makers often receive uncertain responses to their question “what are we adapting to?” Risk management is relatively easy to apply in practice.

Climate change policy, strategy, and implementation already use language and terminology of adaptation with increasing emphasis on the need for adaptation in the face of changing average climate and climate and weather extremes (Schipper and Burton 2009). Increasing demand exists for assessment and promotion of climate change-caused disaster risk management practice that can contribute to climate change adaptation. This requires increasing synergy, merging, and complementarity between these two currently and still largely differentiated practices. The major aim of this chapter is to present a risk management as a practical approach to climate change adaptation as it relates to risk of flooding.

7.2 Methodology for Assessment of Climate Change-Caused Flood Risk to Municipal Infrastructure

The methodology presented in this chapter is based on the work conducted under the City of London (Ontario, Canada) Climate Change Adaptation Strategy Phase 1, to conduct a general risk and consequence analysis of the City’s infrastructure to flooding caused by climate change (Peck et al. 2011; Bowering et al. 2014; Peck et al. 2014).

Risk is defined as the product of hazard and vulnerability. In the case of this assessment, the hazard is the climate change-caused flood event, and the vulnerability is the ability (or lack thereof) of the municipal infrastructure to withstand flooding. Risk includes the consequence of the flooding. Consequence is the economic impact of the flood event on the infrastructure (and study area as a whole). It is a measure of both physical damage to the infrastructure and economic impact caused by a loss of function. Practically, in this assessment the risk value is the consequence multiplied by the probability of the flood event occurring.

7.2.1 Overview of the Risk Assessment Methodology

The methodology includes both qualitative and quantitative information and applies it to the calculation of risk. This enables the inclusion of information from stakeholders, city engineers, and policy makers to provide a comprehensive risk assessment. The integrated risk assessment procedure developed in this work includes: (1) climate modeling, (2) hydrologic modeling, (3) hydraulic modeling, and (4) infrastructure risk assessment. The output from each step is used as input into the next step. Climate modeling approach based on the use of global climate model (GCM) data together with a weather generator (WG), a statistical downscaling tool, is used to provide precipitation data for a set of climate change scenarios. Precipitation data is transformed into flow data using the hydrologic model of the watershed. Flow information is processed through hydraulic analysis to obtain the extension and depth of flood inundation. Quantitative and qualitative risk calculation is performed in the next step to generate a detailed spatial distribution of flood risk to the municipal infrastructure due to climate change. The methodology in this study is specific to the flood hazard – identified as the most critical climate change impact for the City of London – but the general approach and methodology may also be applied to other hazards.

7.2.2 Climate Modeling

An original inverse impact modeling approach (Simonovic 2010) is used for assessing the vulnerability of river basin hydrologic processes to climate forcing. The approach consists of the following four steps:

Step 1. Identification of critical hydrologic exposures that may lead to local failures of water resource systems in a particular river basin. Critical exposures are analyzed together with existing guidelines and management practices. The vulnerable components of the river basin are identified together with the risk exposure. The water resource risk is assessed from three different viewpoints: risk and reliability (how often the system fails), resiliency (how quickly the system returns to a satisfactory state once a failure has occurred), and vulnerability (how significant the likely consequences of a failure may be). This step is accomplished in collaboration with local water authorities.

Step 2. In the next step, the identified critical hydrologic exposures (such as floods and droughts) are transformed into corresponding critical meteorological conditions (e.g., extreme precipitation events, sudden warming, prolonged dry spells). A hydrologic model is used to establish the inverse link between hydrologic and meteorological processes. Reservoir operation, floodplain management, and other anthropogenic interventions in the basin are also included in the model. In the City of London study, the US Army Corps of Engineers (USACE) Hydrologic Engineering Center Hydrologic Modeling System (HEC-HMS) is used to transform inversely extreme hydrologic events into corresponding meteorological conditions. HEC-HMS is a precipitation–runoff model that includes a large set of mix-and-match methods to simulate river basin, channel, and water control structures.

Step 3. A weather generator (WG) is used to simulate the critical meteorological conditions under present and future climatic scenarios. The WG produces synthetic weather data that are statistically similar to the observed data. Since the focus is mainly on extreme hydrologic events, the generator reflects not only the mean conditions but also the statistical properties of extreme meteorological events. The K-NN algorithm is used to perform strategic resampling to derive new daily weather data with altered mean or variability. In the strategic resampling, new weather sequences are generated from the historical record based on prescribed conditioning criteria. For a given climatic variable, regional periodical deviations are calculated for each year and for each period.

Step 4. In the final stage, the parameters of the WG are linked with GCMs, and an ensemble of simulations reflecting different future climatic conditions is generated. The frequency of critical meteorological events causing specific water resource risks is then assessed from the WG outputs.

The proposed methodology includes the assessment of climate change impacts for a range of climate scenarios (Simonovic 2010). A general suggestion is to consider the assessment of impacts for two extreme climate scenarios that will define the lower (CC_LB) and upper bound (CC_UB) of potential climate change. Two climate scenarios are derived by integrating a weather generator that perturbs and shuffles local historical data, with inputs from global climate models (GCMs). The lower-bound climate scenario (CC_LB) is obtained by perturbing and shuffling locally observed data with the assistance of a weather generator (WG) tool. Through the perturbation process, it allows the extreme (minimum and maximum) generated values to be outside of the historic range. In this way the character of the lower-bound scenario reflects the existing conditions (greenhouse gas emissions, land use, population, etc.) and their potential impact on the development of future climate. The upper-bound climate scenario (CC_UB) is derived by perturbing and shuffling historical data and combining them with the input from the global climate models. The choice of the GCM is made on the basis that the upper-bound climate scenario should represent the most critical impact of climate change for the location under consideration. The selection of the range of potential climate change through the use of two scenarios compensates for the existing level of uncertainty present in global modeling of climate change for a watershed. It is noted in the literature that the global models offer various predictions of future climate as a consequence of (1) the selected global model, (2) the selected global model simulation scenario, and (3) the spatial and temporal resolution of the selected global model. It is important to point out that both climate scenarios are equally likely as well as the range of climatic conditions between the two.

7.2.3 Hydrologic Modeling

The meteorological variables generated by the WG model are used as input data for a hydrologic model. The weather generator (WG) model generates daily precipitation and temperature variables at various stations within the river basin. However, the hydrologic model selected for the City of London study (HEC-HMS) requires extreme precipitation data with at least hourly resolution. In addition, spatial resolution of model input data has to be adjusted too. The temporal disaggregation and the spatial interpolation schemes are implemented to provide the necessary input data. The spatial interpolation, based on the inverse distance method and the location information for the measurement stations, is applied to obtain the meteorological data for each subbasin. The disaggregation procedure based on the method of fragments is implemented to convert daily data into hourly.

For the City of London study, daily data for 200 years is generated and any given year contains a number of events. The main objective of the hydrologic analysis is to perform the flow frequency analysis of extreme annual flood events. Therefore, 5-day annual extreme events that produce the largest annual events (200 events altogether) for the entire basin are selected.

A calibrated hydrologic model, HEC-HMS, is used to convert a climate input into flow data within the City of London. The annual extreme precipitation events for each of 200 years and both climate scenarios (CC_LB and CC_UB) – a total of 400 flood events – are selected and used as input into the HEC-HMS model. For each flood event, the streamflow values are calculated for each subbasin and each control point. Each hydrologic simulation run is done using a 5-day time horizon. The simulation results provide the essential hydrologic information for each subbasin and each control point for two climate scenarios and 200 years. Within the City of London, 171 locations of interest are identified – mostly representing input profiles for the hydraulic analysis (Eum and Simonovic 2010).

The frequency analysis is used to relate the magnitude of extreme events to their frequency of occurrence. The results of the hydrologic analyses are used as input into the hydraulic model that calculates the extension and depth of flood inundation for two regulatory flood return periods, 100 and 250 years. Flood frequency analysis of the hydrologic model output is conducted to provide the input for the hydraulic analysis. The method of L moments and three extreme event probability distributions are used, Gumbel, generalized extreme value (GEV), and Log–Pearson type III (Eum and Simonovic 2010). As expected, the results of flood frequency analysis for CC_UB scenario are showing an increase in both flood frequency and flood magnitude when compared with the CC_LB scenario. The difference between the two scenarios identifies the range of climate change flood impacts that may be expected at each location. Two return period flow values (100 and 250 years) for two climate scenarios (CC_LB and CC_UB) are then provided as input into the hydraulic analysis.

7.2.4 Hydraulic Analysis

The traditional process of floodplain mapping based on the hydraulic calculations of water surface elevations is adopted for the purpose of climate change food risk assessment to the municipal infrastructure for the City of London (Sredojevic and Simonovic 2010). A standard computer software, HEC-RAS, is used for hydraulic modeling and computation of water elevation in the basin. HEC-GeoRAS, an extension of ArcGIS, is used for the preparation of spatial data for input into a HEC-

RAS hydraulic model and the generation of GIS data from the output of HEC-RAS for the use in floodplain mapping.

The climate modeling provides meteorological data (precipitation) for hydrologic analysis. The hydrologic model is used to transform the climate data generated by the WG model into flow data that is required for hydraulic analysis. The methodology used in hydraulic analyses consists of three steps: (1) pre-processing of geometric data for HEC-RAS, using HEC-GeoRAS; (2) hydraulic analysis in HEC-RAS; and (3) post-processing of HEC-RAS results and floodplain mapping, using HEC-GeoRAS.

The first step in the pre-processing stage is the creation of a digital terrain model (DTM) of the river system in a triangulated irregular network (TIN) format. The TIN also serves for the delineation of floodplain boundaries and calculation of inundation depths.

After the completion of pre-processing stage, the hydraulic analysis is performed using the HEC-RAS modeling program for the computation of water surface profiles. The analysis starts by importing geometric data (GIS layers) generated in the previous stage. The hydraulic analysis is performed using flow data for two climate scenarios (CC_LB and CC_UB). For both climate scenarios, steady flow data is used for flow return periods of 100 and 250 years. Two water surface profiles (100 and 250) are generated within the boundaries of the City of London for both climate scenarios: 100 CC_LB, 100 CC_UB, 250 CC_LB, 250 CC_UB.

The post-processing of the water surface profiles is performed using the same maps that were used for the pre-processing of geometry data. Floodplain mapping is performed within the limits of the bounding polygon using the water surface elevations generated by the HEC-RAS.

7.2.5 *Risk Calculation*

A comprehensive risk assessment has been undertaken to better understand climate change-caused flood impacts on the municipal infrastructure and provide a measurement of risk as the basis for the development of climate change adaptation options. The main goals of the assessment are (1) to provide the level of risk to infrastructure that may be affected by flooding and (2) to prioritize areas of high infrastructure risk for future climate change adaptation planning decisions. The methodology is based on an integrated *risk index* for each infrastructure element considered. The risk index allows for the comparison among various locations that may be flooded. Each risk level for a particular location provides the source of risk (the type of infrastructure that may be affected) and relative contribution of each source to the overall risk.

7.2.5.1 Risk Index

Risk is defined as the product between a hazard and vulnerability in the context of the presented work (Peck et al. 2011). The vulnerability is defined here as “the shortfall in the ability of public infrastructure to absorb the negative effects, and benefit from the positive effects, of changes in the climate conditions used to design and operate infrastructure."

An original risk measure termed the risk index, R, is defined. This index is calculated for each infrastructure element and incorporates quantitative and qualitative data to address both objective and subjective types of uncertainty. The mathematical expression of the risk index is:

$$R_{ke} = P \times \sum_{i=1}^{3} (D_i \times IM_i) \tag{7.1}$$

where P is the probability of occurrence of the hazard event (dmnl); D_i is the economic loss for each impact category, i ($); IM_i is the impact multiplier (fraction of damage sustained for each impact); e is the infrastructure element; k is the infrastructure type from 1 to 6, (building, bridge, barrier, critical facility, pollution control plant, and road); and i is the impact category, from 1 to 3, representing function, equipment/contents, and structure. For a 100-year flood event, the probability, P, of occurrence in any given year is 1 in 100, or 1 %. Similarly, the probability of a 250-year event is 1 in 250 or 0.4 %.

The risk index is tabulated and normalized for each infrastructure element across each of the scenarios (100 CC_LB, 100 CC_UB, 250 CC_LB, 250 CC_UB). These values are then combined and displayed spatially using GIS in the form of risk maps. Risk is portrayed geographically by dissemination area (DA) classification, consistent with the Statistics Canada method of representing data. There are 527 within the City of London. Each DA is defined by Statistics Canada "a small, relatively stable geographic unit comprised of one or more adjacent dissemination blocks."

The risk index is used to aid in the prioritization of areas of infrastructure at risk. Equation (7.2) shows the calculation of risk to a dissemination area, q, for all infrastructure elements of interest (bridges, buildings, barriers, roads, critical facilities, and/or pollution control plants):

$$R_q = \sum_{e=1}^{m} R_{eq} \tag{7.2}$$

where q is the dissemination area and m is the number of infrastructure elements of interest.

7.2.5.2 Impact Multipliers (IM_i)

The second element of the risk equation represents the impact to the infrastructure as a result of its interaction with the flood hazard. The damages are both direct (such as a loss of structural integrity and components) and indirect (such as a loss of functionality). Damages resulting from flooding are extremely varied and include losses ranging from inconvenience to structural damage to death. This methodology considers three variables as a measure of these consequences: the loss of function (IM_1), loss of equipment (IM_2), and loss of structure (IM_3). Each of these factors (termed impact multipliers) is measured as a percent loss and calculated using both quantitative and qualitative information. They are incorporated into the risk index as demonstrated by expanding Eq. (7.1) as shown below:

$$R_e = \mathrm{P} \times (D_1 \times IM_1 + D_2 \times IM_2 + D_3 \times IM_3) \tag{7.3}$$

The quantitative data includes the ability of the infrastructure to withstand direct damages due to flooding in addition to actual inundation. The qualitative data includes information gathered through interviews relating to the decision-makers' expertise and experience. This includes the condition of the infrastructure and how that may affect its response to flooding. It is important to note that the measure of the impact multiplier may be different across the varying infrastructure types; however, they are consistent across any one particular infrastructure type.

Loss of function (IM_1) The loss of function impact multiplier, IM_1, measures the degree to which the infrastructure has lost its functionality. This is defined as the degree to which the infrastructure no longer functions at an acceptable level relative to which it was originally designed, as a result of flooding. The value of IM_1 is an integer belonging to [0,1] where 0 denotes no loss of function and 1 denotes the total loss of function.

Partial loss of function may occur in the case of critical infrastructure such as fire station, emergency management services (EMS), hospitals, and schools if some, but not all, of the access routes are blocked by floodwaters. The methodology assigns a fractional value of IM_1 depending on the number of incoming or outgoing major routes and the number of routes that are flooded. The relationship used to calculate IM_1 for fire stations and EMS buildings is:

$$IM_{1k} = \frac{(n - m)}{n} \tag{7.4}$$

where $k = 4$ (critical facility types); n is the total number of major access routes; and m is the number of routes obstructed by floodwaters. In the case of schools and hospitals, the loss of function multiplier is calculated based on the total number of access routes within one intersection from the building.

Loss of equipment (IM_2) The second impact multiplier, IM_2, estimates the percent of equipment lost as a direct result of inundation. Equipment is defined as contents or nonstructural components of the infrastructure. In the case of residential buildings, this would be the housing contents or anything that would be expected to be taken in a move. Transportation infrastructure (roads, bridges, culverts, and footbridges) and flood protection structures (dikes) do not have an IM_2 component.

Buildings and critical facilities have equipment values estimated using methods based on the building type and value and are estimated as 30 % of the total structure's value.

Loss of structure (IM_3) The final impact multiplier, IM_3, measures the percent structural loss of the infrastructure. This is the degree to which the structural integrity is compromised as a result of flooding. The flood depth was used in the calculation of IM_3 in addition to the infrastructure element's condition, age, capacity, and other knowledge gained during interviews with experts in each area. IM_3 is a measure of both quantitative and qualitative structural loss. The methodology uses an innovative approach in the incorporation of qualitative and subjective data with the quantitative measures. The qualitative portion uses fuzzy set theory to allow for subjectivity and differences of opinion with respect to the condition of the infrastructure, its failure mechanisms, and its response to flooding.

The deterministic element of IM_3 is calculated using stage-damage curves. These curves use the inundation depth as input to estimate the percent damage (LS) to the infrastructure (both structural and contents) as a result of flood inundation.

The qualitative element of IM_3 is used to quantify the subjective uncertainty associated with potential failure of the infrastructure system. Assessment of subjective uncertainty is conducted with the assistance of experts for various types of infrastructure. Qualitative component of IM_3 allows for the measure of partial failure as well as the impact of the structure's current conditions on its response to flooding as perceived by experts in the field. This measurement is termed the fuzzy reliability index (Simonovic 2009). The premise for the combination of the fuzzy reliability index with the quantitative structural loss measure is that the condition of the infrastructure will affect the amount of structural damage sustained by the infrastructure during a flood. The condition of the infrastructure is not quantified by the stage-damage curves, and therefore the input of those who are the most familiar with infrastructure may provide for the more accurate assessment of the risk.

Once combined with a flood event, the condition of the infrastructure will affect its structural loss measure. Therefore, to calculate IM_3 the fuzzy risk index and the deterministic measure must be combined. To represent this inverse relationship in the calculation of the loss of structure impact multiplier (IM_3), the following equation is used:

$$IM_3(CM) = \begin{cases} 1, & CM = 0 \\ \mathrm{Min}\left(1, LS \times \frac{1}{CM}\right), & CM > 0 \end{cases} \tag{7.5}$$

where IM_3 is the loss of structure impact multiplier used in Eq. (7.1), CM is the compatibility measure obtained through the fuzzy compatibility analysis (Peck et al. 2011), and LS is the percent loss of structure from the stage-damage curves ($LS \leq 1$).

Therefore in this study when CM is 0, the structure is deemed to be completely unsafe, or experiencing a total loss ($IM_3 = 1$). The stage-damage curves are assumed to represent the damage to a structure at a completely acceptable state. As such, for

CM less than 1, the risk to the infrastructure will increase proportionally. A *CM* value of 1 (completely safe) will yield $IM_3 = LS$.

7.2.5.3 Economic Loss

Economic loss refers to the potential monetary damage incurred by an infrastructure element as a result of a flood event. It is a value that is applied which provides a higher weight to those structures that are more expensive to repair or replace. This is in favor of the City's priority of protecting and investing in the infrastructure which could potentially cause the most interference as a result of a flood event. The economic loss factor is different for each piece of infrastructure. There is an associated economic loss value for each type of impact multiplier (IM_1, IM_2, IM_3) as shown in Eq. (7.3). These may be referred to as monetary losses due to loss of infrastructure's function (D_1), monetary losses associated with infrastructure's equipment (D_2), and, finally, the monetary loss incurred by damage to the infrastructure itself (D_3).

7.2.6 Results of the Analysis

The output of the analysis shows the risk to the municipal infrastructure both spatially and by itemized infrastructure. The spatial results are provided as risk maps, with the risk aggregated by dissemination area (DA). The risk values are also presented in tables accompanying the maps, which itemize the risk and consequence for each infrastructure element, by DA.

The results of the analysis (Fig. 7.1) for the City of London show that many areas of the city infrastructure are vulnerable to increased risk due to flooding caused by climate change. Under the upper-bound climate change scenario, the 1:100-year event (Fig. 7.1b) was found to be the most critical with respect to risk, while the 1:250-year event was found to be the most critical (Fig. 7.1d) with respect to damage (consequence).

7.3 Risk Management as Adaptation to Climate Change

The methodology presented in Sect. 7.2 is used to evaluate various alternative adaptation strategies through the comparison of results from the risk assessments performed for the *base scenario* flood risk areas with those from an *alternative strategy* that can be identified by the decision-makers. The *base scenario* flood footprints should be modeled both with no climate change and under the upper-bound climate change conditions. The *alternative strategy* flood footprints should be modeled solely using the upper-bound climate change conditions. The scenarios

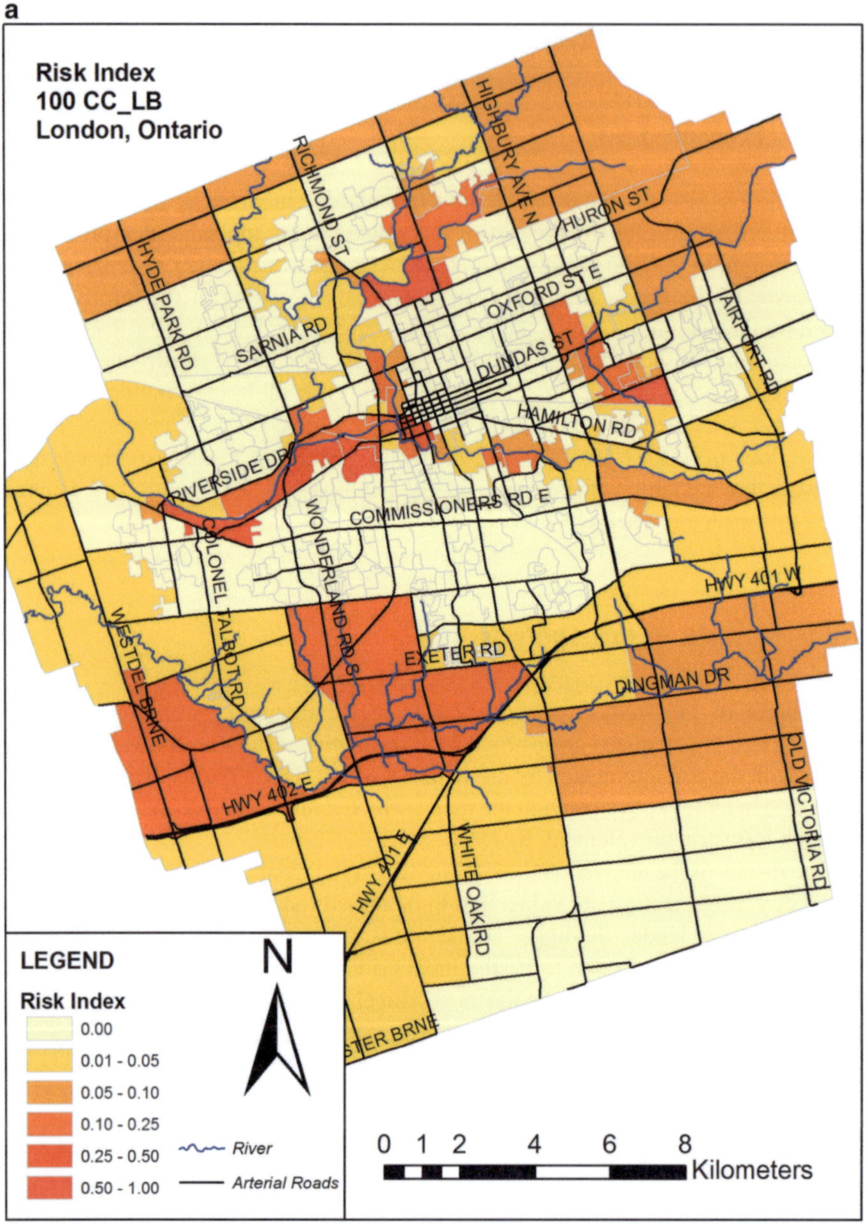

Fig. 7.1 Risk to all infrastructures for (**a**) 100 CC_LB, (**b**) 100 CC_UB, (**c**) 250 CC_LB, and (**d**) 250 CC_UB

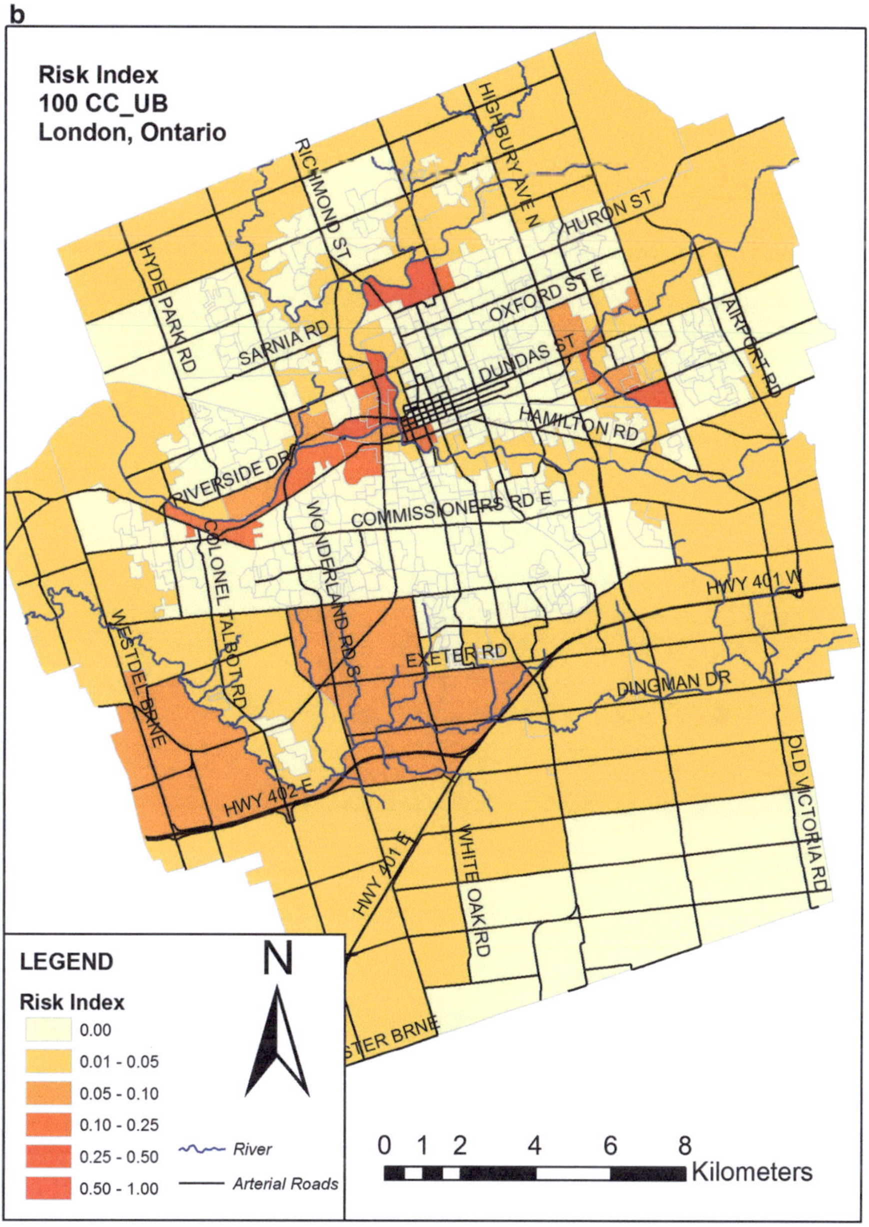

Fig. 7.1 (continued)

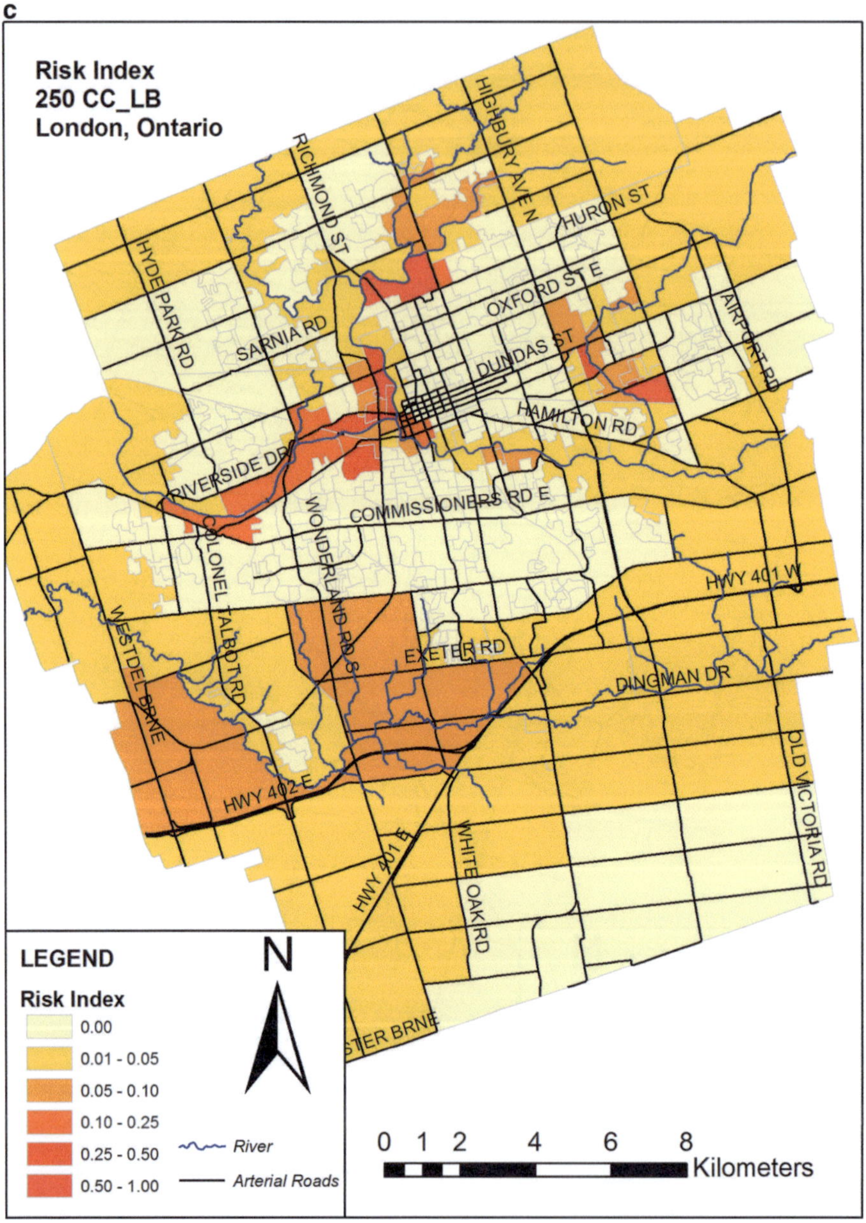

Fig. 7.1 (continued)

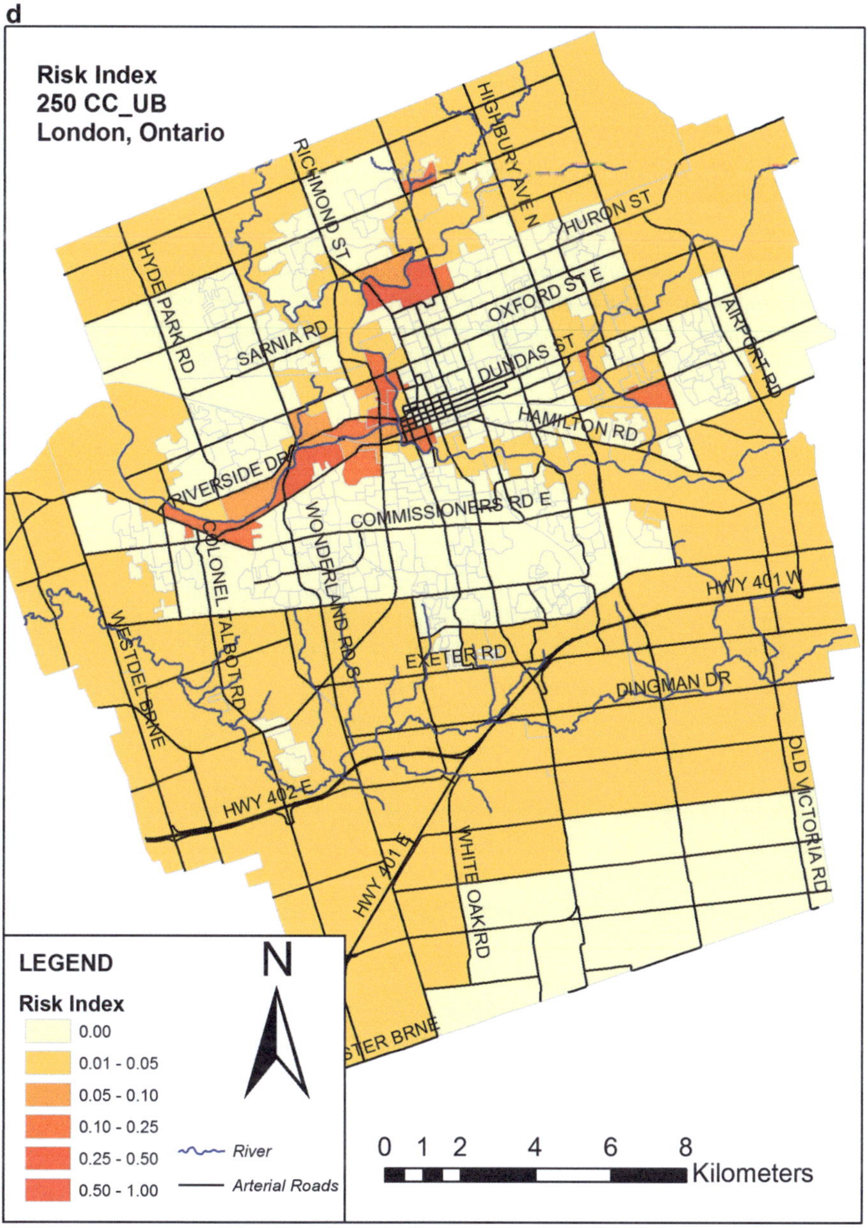

Fig. 7.1 (continued)

with no climate change represent the historical (local) precipitation conditions with perturbation of the data but no additional climate change data. Thus, the no climate change scenarios model the flood events based on locally observed data. The upper-bound climate change scenario is derived from a combination of historical data and the most critical results from various downscaled, global climate models (GCMs) as discussed in Sect. 7.2. Thus, this scenario represents the worst-case event. Either scenario is equally likely to occur; however, by modeling the worst-case event (upper-bound climate change) and the historical conditions (no climate change), a range is given which allows for better planning and risk management.

The presentation of the analysis of adaptation options for management of the risk is done using a case study of one of the watersheds within the City of London – Dingman Creek.

7.3.1 Study Area and Data

The Dingman Subwatershed spans multiple municipalities including the City of London, Middlesex Centre, and Thames Centre. The study area is the portion of the Dingman Subwatershed within the municipal boundaries of London, Ontario. The main watercourse is Dingman Creek which is a tributary of the Thames River. Smaller watercourses include Murray's Drain, Pincombe Drain, and Thornicroft Drain.

The area is mainly rural but it is continuing to develop. Highway 401 runs through the center of the area, and, as such, there is interest in the continued development of the area both commercially and residentially. Figure 7.2 (a) shows the study area and the municipal infrastructure within the study area including 16 bridges, 17 culverts, 17 schools, and 3 emergency service stations (fire, ambulance, and police). The risk analysis considered only the infrastructure impacted by the flood events including the aforementioned infrastructure as well as arterial roads, stormwater management facilities (SWMF) both existing and planned, pumping stations, manholes and outfalls, and noncritical buildings (commercial and residential).

The spatial units, dissemination areas (DAs), within the study area are shown in Fig. 7.2b. Since the DA is determined by population, it is evident that the northern portion of the study area is generally the most densely populated (many small DAs). Since the majority of Dingman Subwatershed is rural, the remaining DAs are quite large in area. This will show a generalized risk and consequence index over a large space, and as such, a detailed look into the risk tables for the results and discussion is required.

The risk analysis is performed on the base conditions with no climate change and with upper-bound climate change for two flood events (100 and 250 years) and two flood events under one alternative strategy with upper-bound climate change for a total of six different scenarios (Table 7.1). The first scenario is the future condition with the 2005 strategy in place (base conditions) for both the 1:100- and 1:250-year event with no climate change. The second scenario is the base conditions for both the

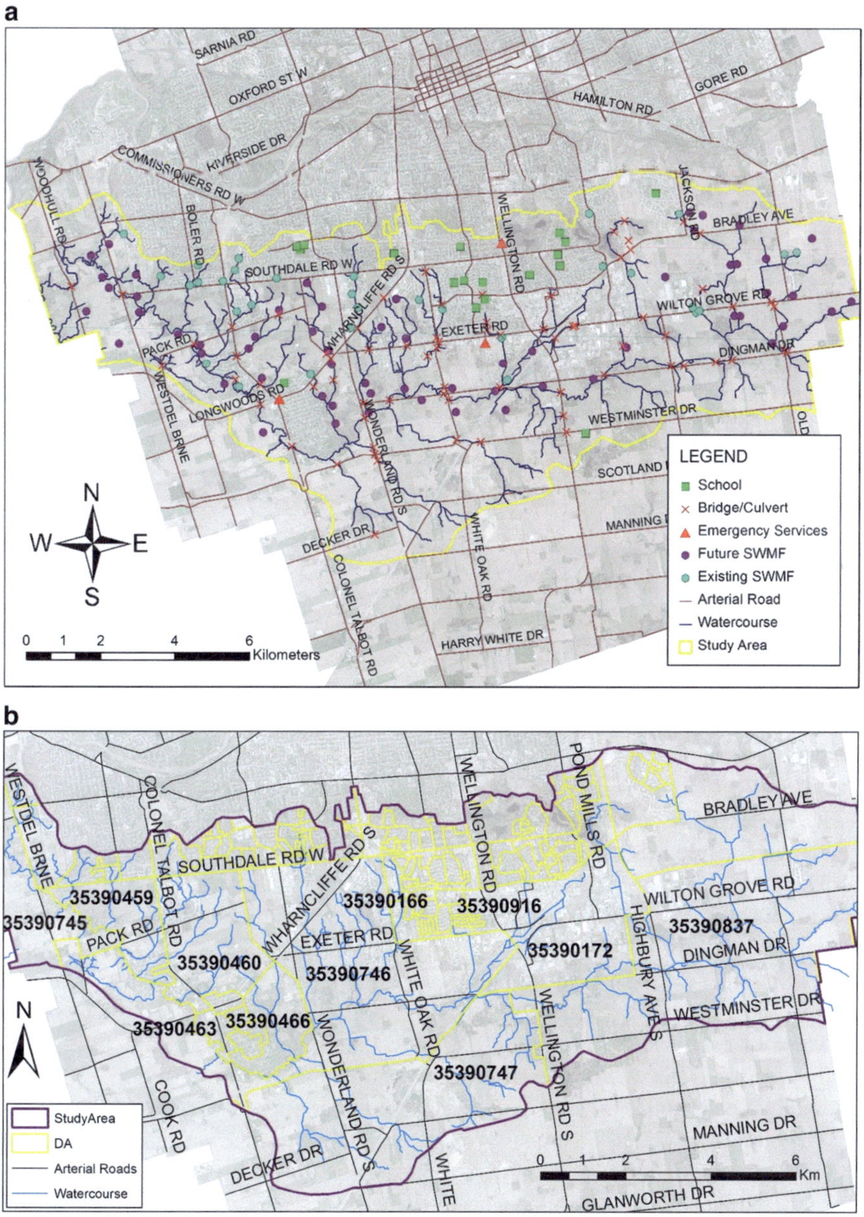

Fig. 7.2 Dingman Subwatershed: (**a**) study area infrastructure and (**b**) DAs in the study area

Table 7.1 Flood scenarios used in risk analysis

Scenario	Scenario name	Details
1	R36_100NC (base condition)	Future conditions with 2005 strategy
		1:100-year event, no climate change
		574.13 ha flood risk extent
2	R36_250NC (base condition)	Future conditions with 2005 strategy
		1:250-year event, no climate change
		670.59 ha flood risk extent
3	R36_100UB (base condition)	Future conditions with 2005 strategy
		1:100-year event, upper-bound climate change
		1072.47 ha flood risk extent
4	R36_250UB (base condition)	Future conditions with 2005 strategy
		1:250-year event, upper-bound climate change
		1192.87 ha flood risk extent
5	R39.4_100UB (alternative)	Changes in online storage
		1:100-year event, upper-bound climate change
		728.84 ha flood risk extent
6	R39.4_250UB (alternative)	Changes in online storage
		1:250-year event, upper-bound climate change
		1030.88 ha flood risk extent

1:100- and 1:250-year event with upper-bound climate change. The base condition (herein identified as R36) provides a basis from which the alternative strategy that includes climate change can be compared. The alternative strategy (herein identified as R39.4) is modeled for both the 1:100- and 1:250-year events using the climate change upper-bound scenario.

The alternative strategy was developed in the Dingman Subwatershed study as a component of the water resources, storm/drainage, and stormwater management strategy conducted by the City of London. Specifically, R39.4 addresses climate change by altering the proposed online facilities and increasing online storage to reduce flows in Dingman Creek.

Maps of each scenario showing the flood risk extent and infrastructure in the study area are available upon request. One example map is shown in Fig. 7.3 for the alternative scenario R39.4 for 1:250-year event under upper-bound climate change.

The purpose of the risk analysis is to measure the consequence and risk values of the infrastructure within the Dingman Subwatershed for each of the six scenarios. The alternative strategy can then be evaluated based on its contribution to the reduction of flood risk associated with climate change. Additionally, the impact of climate change will be shown by comparing the base condition both with and without climate change.

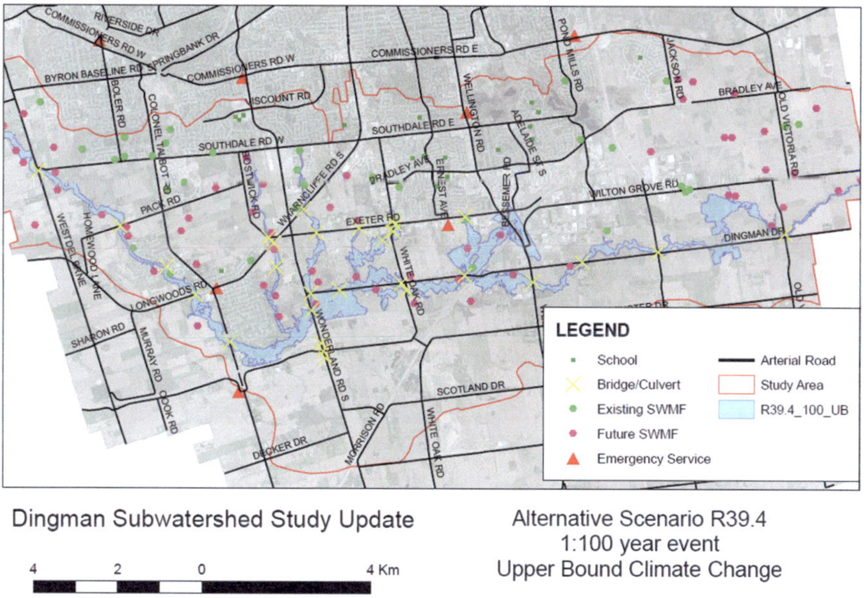

Fig. 7.3 Dingman Subwatershed: alternative scenario R39.4 inundation map

7.3.2 *Results and Discussion*

The results from the risk analysis for the six scenarios shown in Table 7.1 are explained here. The first section discusses the results from the comparison of the alternative strategy (scenarios 5 and 6) with the base case (scenarios 1, 2, 3, and 4) as well as the comparison of scenarios 1 against 3 and 2 against 4 to assess the impact of climate change. The second section examines the results for each of the scenarios separately. The results are available in the form of risk maps and risk and consequence tables. The tables (available upon request) can be studied along with the maps to provide complete understanding of risk. The consequence tables give damage values in CAD$ (2012). A risk map is presented for all six scenarios along with the corresponding tables organized by dissemination area. The risk is normalized for each scenario with the darker areas indicating a higher level of risk. The normalization of risk allows for a meaningful way to rank the risk in each area. The equation used for normalization of risk in each scenario is:

$$\overline{R_{DAj}} = \frac{R_{DAj}}{R_{j\max}} \tag{7.6}$$

where $\overline{R_{DAj}}$ = normalized risk index for dissemination area (DA), scenario j; R_{DAj} = risk index for dissemination area (DA), scenario j; and $R_{j\max}$ = maximum risk index in dissemination area (DA), scenario j.

The minimum risk is always 0 and the maximum normalized risk value is 1. Using the normalized risk, it is possible to draw conclusions about the relative risk in each area within a scenario. A word of caution when interpreting the normalized index is that large values tend to suppress the smaller values. Therefore, it is necessary to look at the accompanying risk tables to gain a clear description of the risk.

The overall risk to the Dingman Subwatershed is examined by comparing the base condition (with and without climate change) to the alternative option. Scenarios 1 through 4 are also compared to each other to show the contribution of climate change. This gives six case studies described below. Additionally, all maps and tables that are not presented here are available upon request.

Case Studies Six cases are used to enable comparison between the base condition and the alternative option both with and without climate change for each event: climate change (base condition) for the 1:100-year event and 1:250-year event and R39.4 (alternative option) for the 1:100-year event and 1:250-year event.

The comparison cases are:

Case 1: Contribution of Climate Change 100 – change in risk index and consequence index from scenario 1 to scenario 3

Case 2: Contribution of Climate Change 250 – change in risk index and consequence index from scenario 2 to scenario 4

Case 3: Comparison of R39.4_100UB with Climate Change – change in risk index and consequence index from scenario 3 to scenario 5

Case 4: Comparison of R39.4_250UB with Climate Change – change in risk index and consequence index from scenario 4 to scenario 6

Case 5: Comparison of R39.4_100UB Without Climate Change – change in risk index and consequence index from scenario 1 to scenario 5

Case 6: Comparison of R39.4_250UB Without Climate Change – change in risk index and consequence index from scenario 2 to scenario 6

The equation used to determine the change in risk (R) from the existing scenario to the alternative strategy is:

$$\text{Percent Change} = 100\left(\frac{R_{\text{alternative}} - R_{\text{existing}}}{R_{\text{existing}}}\right) \tag{7.7}$$

If there is no risk in the corresponding DA for the existing scenario ($R_{\text{existing}} = 0$), the area is given the maximum increase in risk and termed "infinite." To determine the actual consequence of this increase, the change in consequence is used. Change in consequence is presented as $CAD (2012) and is simply the difference in consequence between the alternative strategy and the base condition. In the first two cases, it is the difference in consequence between the base condition with climate change and without climate change. Thus, if the change is negative, the consequence has decreased relative to the base conditions. If the change is positive, the consequence has increased relative to the base conditions. The same

Table 7.2 Comparison case 1 and 2 results – contribution of climate change

		Case 1 Scenario 1 vs. scenario 3 (R36 100NC vs. R36 100UB)	Case 2 Scenario 2 vs. scenario 4 (R36 250NC vs. R36 250UB)
		Contribution of climate change	
Total	Risk increase (%)	**61.7**	**50.3**
	Consequence increase (CAD2012$)	**20,863,780**	**20,153,250**
Roads	Risk increase (%)	175.4	161.1
	Consequence increase (CAD2012$)	43,050	52,210
Bridges and culverts	Risk increase (%)	111.3	120.9
	Consequence increase (CAD2012$)	4,190,710	5,458,120
Buildings (noncritical)	Risk increase (%)	55.4	41.3
	Consequence increase (CAD2012$)	16,630,20	14,642,920

is true for change in risk. To determine the total change, the summation of risk and consequence is used in the formulae as opposed to the risk per DA.

The results from the comparison are shown in Tables 7.2 and 7.3. Table 7.2 shows the contribution of climate change by comparing base conditions with and without the upper-bound climate change (cases 1 and 2). Table 7.3 shows R39.4 compared against base conditions with and without climate change (cases 3, 4, 5, and 6). The total change considers the entire infrastructure within the study area that contributes to the risk and consequence index. For each scenario in the Dingman Subwatershed, this infrastructure consists of roads, bridges and culverts, and buildings. The change for each of these categories is listed below the total. From the results below, it is evident that climate change contributes to an increase in risk and consequence. However, this increase can be mitigated by using the proposed strategy.

7.3.3 *Discussion of the Results*

Cases 1 and 2 show that without any mitigation measures, under the upper-bound scenario, the risk to Dingman infrastructure will increase by an estimated 62 % in the 1:100-year event and by 50 % in the 1:250-year event with climate change. This corresponds to an increase in consequence of $20.9 M and $20.2 M, respectively. Spatially, the increase in risk is fairly evenly distributed along the Dingman study area with the areas south of Dingman Creek consistently showing a large increase in risk – see Fig. 7.4a and b. The greatest impact, with respect to risk, is on the roads with the risk increasing by 175 % in the 1:100-year event and 160 % in the

Table 7.3 Comparison case 3, 4, 5, and 6 results – R39.4

		Case 3	Case 4	Case 5	Case 6
		Scenario 3 vs. scenario 5 (R36 100UB vs R39.4 100UB)	Scenario 4 vs. scenario 6 (R36 250UB vs. R39.4 250UB)	Scenario 1 vs. scenario 5 (R36 100NC vs. R39.4 100UB)	Scenario 2 vs. scenario 6 (R36 250NC vs. R39.4 250UB)
		Online storage improvements			
Total	Change in risk (%)	**−19.2**	**−13.8**	**30.7**	**29.6**
	Change in consequence (CAD2012$)	**−10,496,490**	**−8,294,210**	**10,367,290**	**11,859,050**
Roads	Change in risk (%)	−44.9	−18.2	51.6	113.6
	Change in consequence (CAD2012$)	−30,390	−15,390	12,670	36,810
Bridges and =culverts	Change in risk (%)	−34.6	−22.9	38.1	70.4
	Change in Consequence (CAD2012$)	−2,755,920	−2,282,340	1,434,790	3,175,780
Buildings (noncritical)	Change in risk (%)	−16.5	−12.0	29.7	24.4
	Change in consequence (CAD2012$)	−7,710,180	−5,996,470	8,919,830	8,646,450

1:250-year event. However, the greatest impact with respect to consequence is on (noncritical) buildings with an increase of $16.6 M and $14.6 M, respectively.

Cases 3 and 4 explore the changes in risk and consequence under upper-bound (UB) climate change when flood risk mitigation measures are in place. In the scenario termed R39.4, online storage is increased, and proposed online facilities are modified in order to reduce the flows in Dingman Creek. The implementation of R39.4 would cause an overall reduction in risk (compared to R36) of nearly 20 and 14 % for the 1:100UB-year event and 1:250UB-year event, respectively. For both scenarios the spatial pattern of reduction is similar with the majority of the risk reduction occurring in the central and eastern portions of the study area as shown in Fig. 7.4c and d. The largest consequence reduction, in both events, is to the buildings at -$7.7 M (1:100UB) and -$6 M (1:250UB). For the 1:100UB-year event, the largest decrease in risk is to the roads at 45 % (−$30,400), while the bridges/culverts see a decrease in risk of 35 % (−$2.76 M). For the 1:250UB-year event, the largest risk reduction is to the bridges/culverts at 23 % (−$2.3 M), while the risk to roads is reduced by 18 % (−$15,400). R39.4 performs the best for the 1:100-year event (in both risk and consequence mitigation), although the results are similar decreases for both events. Thus, the strategy R39.4 successfully mitigates the risk due to climate change when compared with R36.

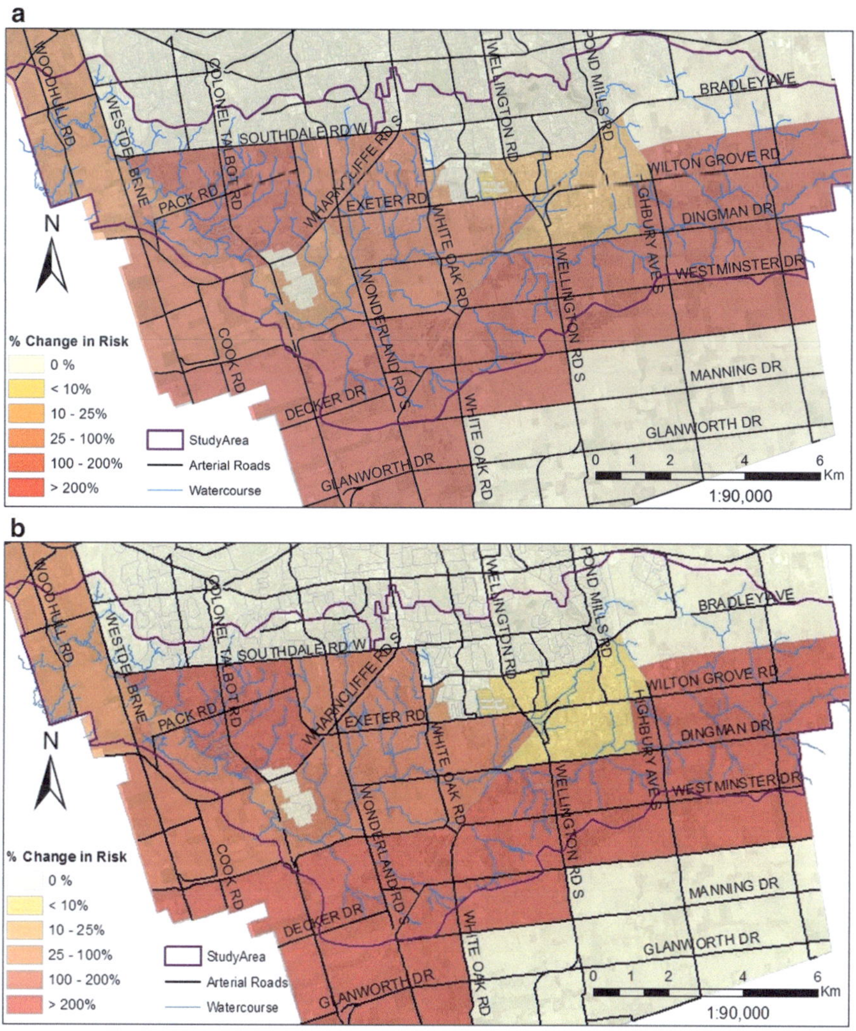

Fig. 7.4 Dingman Subwatershed: spatial distribution of change in risk (**a**) case 1, (**b**) case 2, (**c**) case 3, (**d**) case 4, (**e**) case 5, and (**f**) case 6

Cases 5 and 6 compare the results of the flood risk analysis for R39.4 (under UB climate change) against the results from R36 with no climate change. The overall risk increases by approximately 30 % for both the 1:100-year event and 1:250-year event, translating to a consequence increase of \$10.4 M and \$11.86 M, respectively. The patterns for risk mitigation are not similar for the two scenarios. In the 1:100-year event, Fig. 7.4e, the risk is unchanged for a large portion of the study area. The majority of the risk increase occurs in the eastern portion of the study area. The increase in risk to roads is 52 % (\$12,700), increase in risk to bridges/culverts is

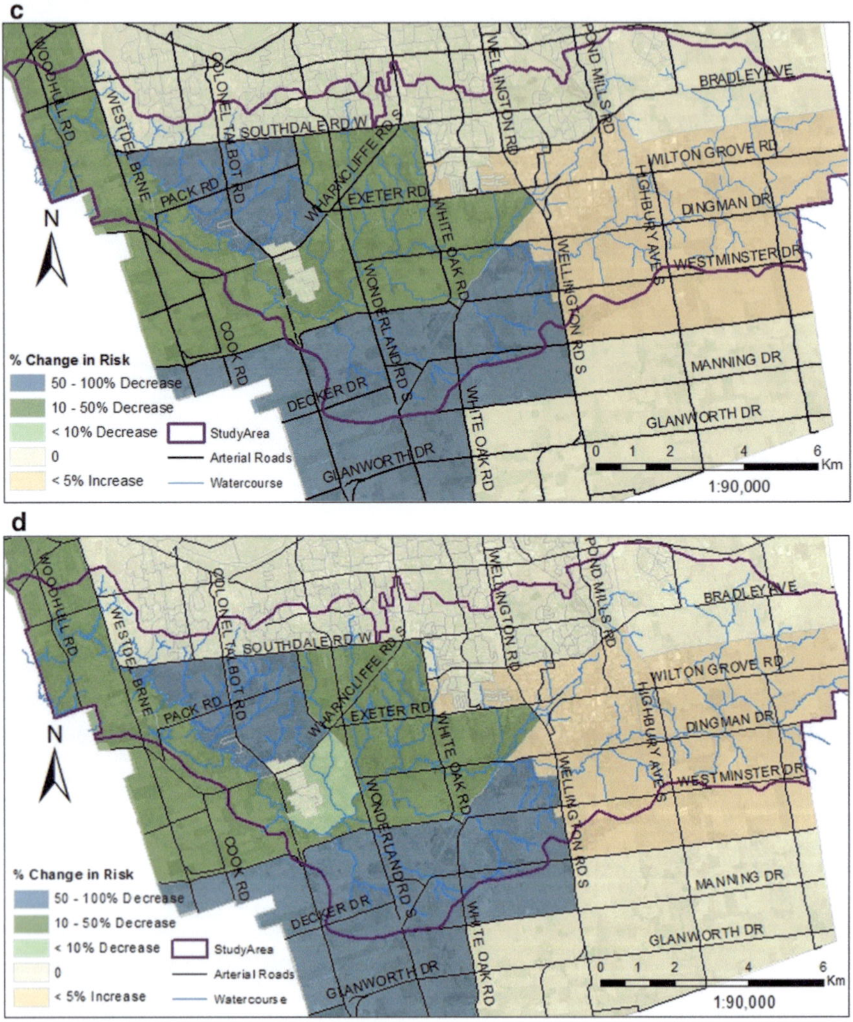

Fig. 7.4 (continued)

38 % ($1.4 M), and increase in risk to buildings is 30 % ($8.9 M). For the 1:250-year event, Fig. 7.4f, the distribution of risk increase is similar to that of case 2 as shown in Fig. 7.4b (base conditions compared to climate change only) but with lower overall risk increases. The majority of the increase occurs south of Dingman Creek, but the study area shows slight increases over most DAs. The risk to roads increases by 114 % ($36,800), to bridges/culverts the risk increases by 70 % ($3.2 M), and to buildings it increases by 24 % ($8.6 M).

Even though the mitigation efforts are in place, the R39.4 scenarios consider UB climate change, while the R36 scenarios that are in the comparison cases 5

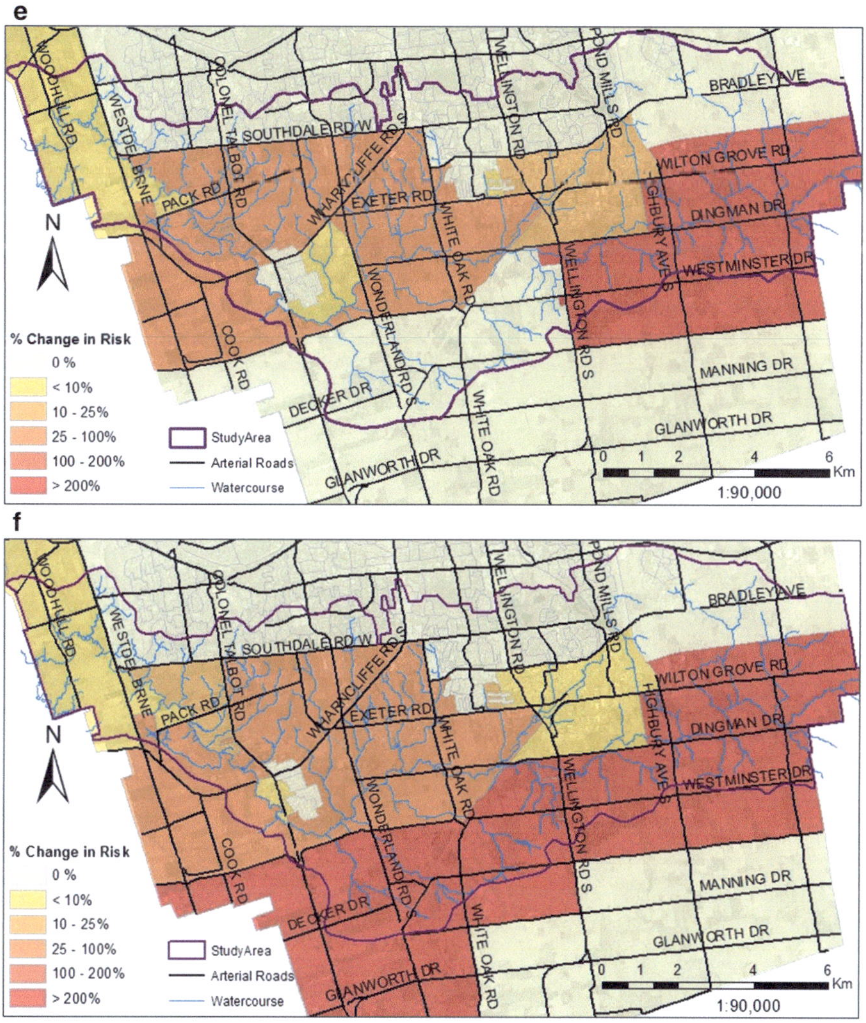

Fig. 7.4 (continued)

and 6 assume no climate change. Therefore, the risk will increase because the effects of climate change are larger than the effects of the mitigation. However, it is evident that the increases in risk and consequence are still less than those in the "do-nothing" scenarios (cases 1 and 2), and as such, it is concluded that the R39.4 alternative mitigates the negative impact of climate change on the municipal infrastructure due to flooding. Additionally, in the 1:100-year event, the distribution of risk increases changes (compared to the "do-nothing" approach alternative) such that the risk is completely mitigated in the southwest portion of the study area as shown in Fig. 7.4e.

7.3.3.1 Alternative Adaptation Option R39.4

Table 7.4 summarizes the impact of different rainfall flooding events on the infrastructure within the Dingman Subwatershed study area (the first two events are without climate change; the remaining events are under the upper-bound climate change scenario).

No schools, police stations, or hospitals are affected by any flood event. The only critical facility to be affected is an emergency medical service (EMS) building. The R36 1:100-year event (without climate change) is the best-case scenario or the scenario which demonstrates the smallest flood footprint and levels of inundation that are modeled in this study. Under the remaining scenarios, the inundation only increases in this area. Although the facility experiences some loss of access, the building itself does not become inundated in any scenario. A number of pumping stations are inundated. There are also numerous existing and planned SWMF that fall within the flood risk footprints. The outfalls and manholes within the flood

Table 7.4 Summary of infrastructure impacted by flooding

	R36 100NC	R36 250NC	R36 100UB	R36 250UB	R39.4 100UB	R39.4 250UB
Extent of flood risk area (flood risk footprint)						
Total flood risk extent (ha)	574.13	670.59	1072.47	1192.87	728.84	1030.88
Roads						
Length flooded (m)	3,100	3,756	8,279	10,404	4,269	8,282
Bridges/culverts						
# affected	18	21	28	29	21	27
Buildings						
# inundated	40	50	80	90	51	77
Critical facilities						
Restricted accessibility (EMS)	1	1	1	1	1	1
Pumping stations						
# inundated	3	3	4	4	3	4
Manholes						
# SAN submerged	60	68	88	100	65	90
# STM submerged	15	20	24	29	22	27
Outfalls						
# SAN submerged	1	1	1	1	1	1
# STM submerged	26	27	32	35	32	33

Table 7.5 Contribution of infrastructure to consequence value ($1,000 CAD2012)

Alternative strategy	**R36**	**R36**	**R36**	**R36**	**R39.4**	**R39.4**
Flood event	**1:100NC**	**1:250NC**	**1:100UB**	**1:250UB**	**1:100UB**	**1:250UB**
Roads	25	32	68	85	37	69
Bridges and culverts	3 767	4 514	7 958	9 972	5 202	7 690
Buildings (noncritical)	30 014	35 488	46 644	50 131	38 934	44 134
Total	**33 806**	**40 034**	**54 670**	**60 187**	**44 173**	**51 893**

risk area (flood risk footprint) are included for completeness. The only manholes of concern are those without flood-proofing.

Table 7.5 shows the contribution of the infrastructure to the total consequence value in each scenario. Note that the contribution of roads to the total consequence value is very low compared to the other categories. From these results it is seen that the highest consequence contributors in all scenarios are the buildings, while the lowest are the roads. Under the 1:100-year event, with upper-bound climate change, the alternative R39.4 has the lowest total consequence of $44.2 M. Under the 1:250UB-year event, in the same scenario, R39.4 has the lowest total consequence of $51.9 M. With no mitigation measures in place, the 1:100UB-year event experiences an estimated $54.7 M in damages, while the 1:250UB-year event has an estimated $60.2 M in damages (the highest of all scenarios).

Figure 7.5 shows the risk index for each scenario. The 1:100-year events show a higher risk index than the 1:250-year events under the same conditions. This is because the probability of the 1:100-year event occurring is higher than that of the 1:250-year event which counteracts the fact that the 1:250-year event has a higher consequence value. Of the scenarios under the upper-bound climate change, R39.4 has the lowest risk values for both the 1:100-year event and 1:250-year event. The scenario with the highest risk is the 1:100UB-year event under base conditions (R36). The scenario with the lowest risk is the 1:250-year event (no climate change) under base conditions (R36).

The normalized risk index and the consequence value for each scenario are obtained by DA. Figure 7.6a and b shows the relative (normalized) risk index, for base conditions (R36) and alternative option R39.4, for the 1:100-year event with upper-bound climate change. The risk has been normalized for each scenario individually, so the maps depict the relative ranking of spatial risk, independent for each scenario. Under R36 (Fig. 7.6a), the areas at highest risk are north of Dingman Creek. It is mainly due to the large number of bridges and buildings that are inundated. There is also a commercial area at the corner of Wellington and Exeter which contributes to a high-risk value in this DA.

In the R39.4 scenario (Fig. 7.6b), the areas of high risk are the same. The reasons for high risk are also due to the large number of bridges and culverts inundated as well as the commercial area at Exeter and Wellington. However, in this scenario, there is no risk south of Westminster Rd. and west of Wellington Rd S. This is

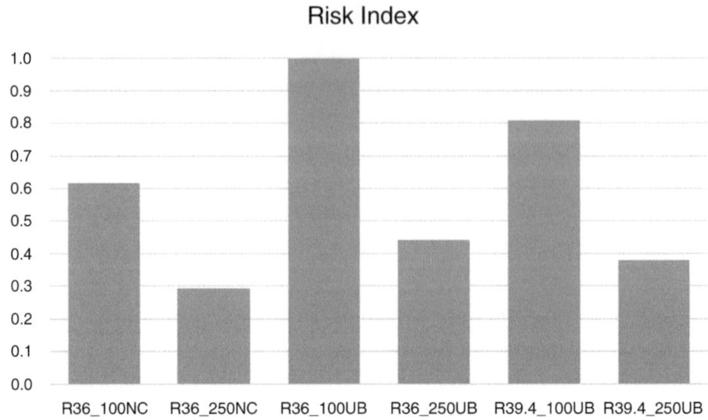

Fig. 7.5 Risk index for all scenarios

due to the fact that the mitigation measures in R39.4 prevent the flood waters from reaching this area.

Figure 7.6c and d shows the distribution of normalized risk under the 1:250-year event with upper-bound climate change. Figure 7.6c shows the distribution with base conditions (R36). The distribution is similar to that of the R36 1:100-year event. The bridges and buildings north of Dingman as well as the commercial center at Exeter and Wellington and Exeter and White Oak Rd. comprise the high-risk areas. Under R39.4 scenario, the risk distribution remains very similar. Thus, the spatial distribution of risk for all scenarios (except the R39.4 1:100-year event) is very similar due to the inundation of certain areas in all scenarios. The high-risk areas are those with a higher population density and more bridges and culverts and commercial centers.

7.4 Conclusions

A methodology is developed to quantify the risk to the municipal infrastructure from climate change-related flooding. The risk is measured using a combination of flow/frequency, stage-damage, and damage/frequency curves. The measure of risk is termed the *risk index* and calculated for each infrastructure element within a municipality. The risk is aggregated and summed by spatial unit and presented in the form of risk tables and maps. The risk index takes into account both quantitative and qualitative information obtained from research and interviews with technical experts.

The Dingman Creek case study results clearly demonstrate that climate change causes an increase in both risk and consequence with respect to the municipal infrastructure under flood conditions. The developed methodology can be effectively used

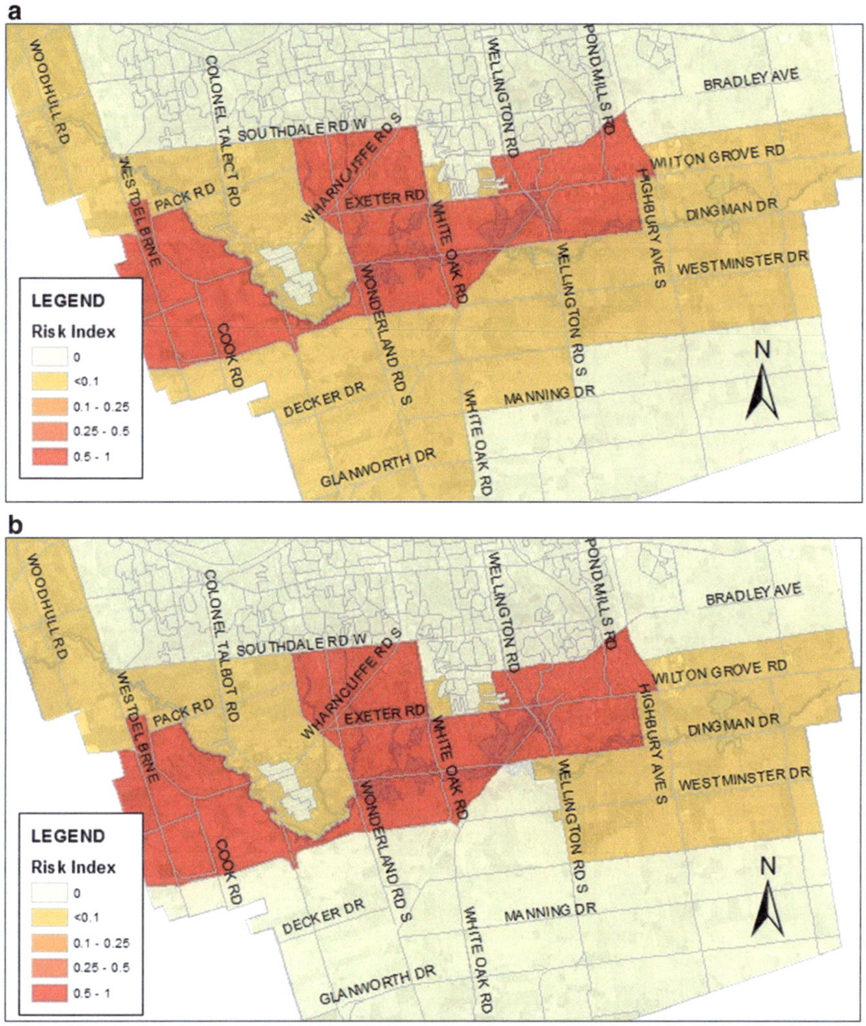

Fig. 7.6 Dingman Subwatershed: normalized risk index – (**a**) base conditions R36 1:100-year event UB, (**b**) alternative R39.4 1:100-year event UB, (**c**) base conditions R36 1:250-year event UB, and (**d**) alternative R39.4 1:250-year event UB

to assess the risk associated with various options and therefore provides support for climate change adaptation through risk management. The results from the application of the methodology to a municipality will lead to a better policy and informed decision-making.

Fig. 7.6 (continued)

Acknowledgments The author would like to acknowledge the assistance provided in the preparation of this Chapter. The numerical analyses are completed by Mrs. Elizabeth Bowering-Taylor. Dingman Creek study results are obtained from the *Slobodan P. Simonovic Consulting Ltd*. The City of London provided the infrastructure data. Thank you.

References

Bowering E, Peck A, Simonovic SP (2014) A flood risk assessment to municipal infrastructure due to changing climate part I: methodology. Urban Water J 11(1):20–30

Eum H-I, Simonovic SP (2010) City of London: vulnerability of infrastructure to climate change, background report 1 – climate and hydrologic modelling. Water Resources Report No. 068, pp 103. Department of Civil and Environmental Engineering, The University of Western Ontario. Available on line.

IPCC (2013) Summary for policymakers. In: Stocker TF, Qin D, Plattner G-K, Tignor M, Allen SK, Boschung J, Nauels A, Xia Y, Bex V, Midgley PM (eds) Climate change 2013: the physical science basis. Contribution of Working Group I to the Fifth Assessment Report of the Intergovernmental Panel on Climate Change. Cambridge University Press, Cambridge, UK/New York

Noble D, Bruce J, Egener M (2005) An overview of the risk management approach to adaptation to climate change in Canada. Global Change Strategies International, Ottawa, Canada. Report prepared for Natural Resources Canada – Climate Change Impacts and Adaptation Directorate, pp. 29

Peck A, Bowering E, Simonovic SP (2014) A flood risk assessment to municipal infrastructure due to changing climate part II: case study. Urban Water J 11(7):519–531

Peck A, Bowering E, Simonovic SP (2011) City of London: vulnerability of infrastructure to climate change final report. Water Resources Research Report. No. 36, pp. 66. Department of Civil and Environmental Engineering, The University of Western Ontario. Available on line.

Schipper L, Burton I (eds) (2009) The Earthscan reader on adaptation to climate change. Earthscan, London

Simonovic SP (2012) Floods in a changing climate-risk management. Cambridge University Press, Cambridge, UK/New York, 179 pp

Simonovic SP (2010) A new methodology for the assessment of climate change impacts on the watershed scale. Curr Sci 98(8):1047–1055

Simonovic SP (2009) Managing water resources: methods and tools for a systems approach. UNESCO/Earthscan James & James, Paris/London, 640 pp

Smit B, Burton B, Klein RJT, Wandel J (2000) An anatomy of adaptation to climate change and variability. Clim Chang 45:223–251

Sredojevic D, Simonovic SP (2010) City of London: vulnerability of infrastructure to climate change – background Report 2 – hydraulic modelling and floodplain mapping. Water Resources Research Report. No. 069, 147 pp. Department of Civil and Environmental Engineering, The University of Western Ontario. Available on line.

Chapter 8
Adaptation to Climate Change: Decision Making

Young-Oh Kim and Eun Sung Chung

Abstract In spite of the recent global effort in mitigating greenhouse gases, the temperature of the earth continues to increase because we are already committed past emissions. Therefore, adapting to the changing climate is an immediate challenge that requires choosing a successful strategy. This chapter reviews classical decision-making methods and discusses their limitations when applied to climate change adaptation planning. Three novel decision-making methods, robust decision making (RDM), real option analysis (ROA), and dynamic adaptive policy pathways (DAPP), are discussed, and their applications are then introduced in water resources planning under different climate change scenarios. Such methods should be either "robust" or "adaptive" for decision makers to capture the nonstationary and uncertain characteristics of climate change. As its name indicates, the RDM method focuses on the "robust" perspective and chooses an alternative that performs satisfactorily over a wide range of scenarios. In contrast, both the ROA and DAPP methods focus on the "adaptive" perspective and have a decision tree framework where risk is spread over time. ROA allows decision makers to delay or abandon the chosen alternative rather than just implementing it without modification, while DAPP introduces the tipping point concept that offers a systematic way of when to switch between alternatives. However, these advanced decision-making methods are resource intensive; thus, a continuous administrative effort and institutional as well as technical supports are required for their success in the climate change era.

Keywords Adaptation • Climate change • Decision making • Dynamic adaptive policy pathways • Robust decision making • Real options

---

Y.-O. Kim (✉)
Department of Civil & Environmental Engineering, Seoul National University,
1 Gwanak-ro, Gwanak-gu, Seoul, Republic of Korea 08826
e-mail: yokim05@snu.ac.kr

E.S. Chung
Department of Civil Engineering, Seoul National University of Science and Technology,
54-325 Gongneung-ro 232, Nowon-gu, Seoul, Republic of Korea 01811
e-mail: eschung@seoultech.ac.kr

E. Kolokytha et al. (eds.), *Sustainable Water Resources Planning and Management Under Climate Change*, DOI 10.1007/978-981-10-2051-3_8

8.1 Introduction

Global warming is an inevitable reality that has been proved with observed data, which many believe is an anthropogenic climate change caused by greenhouse gas emissions. In spite of global efforts on mitigation of these emissions during the last two decades, climate change and its impacts show no signs of abating. Mitigation is a fundamental action, but a few decades may be required for it to be effective. Therefore, at present, "adapting" to the changing climate is urgent. This alternative action is rather natural because evidences of the impacts of climate change on the world can be seen almost every day and will continue to be seen in the uncertain climate change era.

The word "adaptation" appeared in the English literature in the early seventeenth century (Orlove 2009); however, its concept in climate change began receiving full attention after the Intergovernmental Panel on Climate Change (IPCC) officially defined it as "the adjustment in natural or human systems in response to actual or anticipated climate stimuli or their effects which moderates harm or exploits beneficial opportunities" (IPCC 2001), and United Nations Framework Convention on Climate Change (UNFCCC) emphasized "adaptation requires urgent attention and action on part of all countries" (UNFCCC 2002). More recently, several studies on climate change adaptation (CCA) have been carried out, ranging from concept (e.g., Smit et al. 2000; Adger et al. 2005; Hallegatte 2009; Orlove 2009; Birkmann 2013) to practice (e.g., Wheeler 2008; Tompkins et al. 2010; Preston et al. 2011; Lesnikovski et al. 2015) and from theory-oriented books to comprehensive national reports (e.g., NRC 2010; Ranger et al. 2010; WUCA 2010).

The scope of CCA nearly covers all the activities of our lives and environments. Resizing suit production, resetting vaccination timing, and changing the crops that we grow are considered as typical CCA examples. CCA can be categorized as autonomous or planned, proactive or reactive, short range or long range, and localized or widespread (Smit et al. 2000; Pelling 2011, Chapter 2). "Water" is one of the most important "climate-sensitive" sectors urgently requiring adaptation as demonstrated by Tompkins et al. (2010), who reported that water supply and flood sectors occupy the largest portion (34 %) of 300 CCA activities in the United Kingdom. Policies, plans, and projects that are created or improved to cope with the anticipated problems of water resources due to climate change can be key forms on how CCA activities are carried out. Agents in charge of CCA include national and local governments, private sectors, community organizations, and individuals.

The success of CCA depends not only on which option is selected but also on the time, location, and the method of implementing the selected option. Therefore, in the CCA process, we are often at a crossroad, and a rational "decision-making (DM)" theory is required. Although conventional decision-making theories have been developed in the past, adapting to climate change seems to be a new challenge that is beyond the realm of classical decision-making theories. This is mainly because climate change is characterized as "nonstationary" and a large degree of, sometimes deep, uncertainty. It should be noted that a time series is nonstationary when its key statistics, such as mean and variance, are functions of time, while deep uncertainty implies that a future event and its likelihood of occurrence are unknown. For example, the "average" temperature of the earth has been rising (i.e., indicating nonstationary characteristics), but its future projections relevant to DM are highly heterogeneous in terms of space and time (i.e., indicating deep uncertainty). Therefore, a successful decision making for CCA (DMCCA) should consider these two attributes effectively in its process.

Recent studies demonstrated that DMCCA should be based on "robust" and "adaptive" concepts to capture the nonstationary and deep uncertainty characteristics of climate change. A DM method that aims at favoring "robustness" performs satisfactorily over a wide range of plausible future scenarios compared to those that focus on optimality, which performs best only for a few scenarios. DMCCA alternatives that dominate other alternatives for a few scenarios but does not function properly for other scenarios should be avoided when the objective is to achieve robustness. On the other hand, an "adaptive" DM method refines its decision iteratively by continuously monitoring and learning from its previous outcomes and new information. In normal decision-making situations, such an iterative and continual process is seldom necessary because it requires considerable money, time, and effort. Therefore, robust and adaptive decision-making methods must adapt to changing climates and absorb a wide range of climate conditions. The word "flexible," which leaves options open (Walkers et al. 2013), has been widely used as a substitute for both words, "robust" and "adaptive," as robust decisions should be flexible over a wide range of scenarios and adaptive decisions should be flexible over the time horizon. The "flexible adaptation pathways" slogan of New York City (NPCC 2010) is a good example adopted in practice. Other expressions can also be used similarly for the above three words; for instance, Hallegatte (2009) recommended no-regret, reversible, soft, safety-margin, or faster-rotating strategies for CCA.

As DMCCA implies an entire process rather than referring only to an outcome, a well-designed framework embedding the robust and adaptive concepts is necessary to effectively support decision makers. The report published by United Kingdom Climate Impact Programme (UKCIP) entitled "Climate adaptation: risk, uncertainty and decision-making" (called RUD hereafter; Willows and Connell 2003) first proposed a comprehensive and systematic framework for DMCCA, which consists of eight stages: (1) identify problems and objectives, (2) establish decision-making criteria, (3) assess risk, (4) identify options, (5) appraise options, (6) make decisions, (7) implement decisions, and (8) monitor, evaluate, and review. This staged framework can be aided with a web-based tool called "the UKCIP adaptation wizard" (http://www.ukcip.org.uk/wizard/#.VehXyU3ot1M). More recently, IISD (2009) and NRC (2010) also proposed similar frameworks; the former focused on a policy making process, while the latter focused on an adaptation planning process. The contents of this chapter cover the entire DMCCA framework but focuses primarily on stage 5 of the RUD report, i.e., more specifically on how to quantitatively generate a good decision for a given problem. In particular, three DMCCA theories that have been developed during the last decade, the robust decision making (RDM), the real option analysis (ROA), and the dynamic adaptive policy pathway (DAPP) approaches, are introduced, and their applications are discussed in the field of water resources.

8.2 Classical Decision-Making Theory

8.2.1 Concept and Basic Structure

Water resources planning and management problems have been favorite examples in DM theories. Concrete mathematical descriptions and various practical applications can be easily found even in textbooks of water resources (systems) (e.g., Loucks et al. 2005). This section revisits the classical DM theory and methodology not only because they offer the fundamentals for advanced DM theories and methods introduced in the following sections but also because they have been often revised or combined with advanced DM methods for CCA.

In general, decision problems consist of a "decision" and "state variables," which imply a decision is made as the function of a certain state. For instance, calculating the quantity of water to be supplied to a city for a future period depends on the quantity of water that will be available during the planning period and the quantity of water currently secured for the city. In this case, the quantity of water to be supplied is the decision variable, while the current and future estimates of water availability is the state variable. In general, these variables can be either continuous or discrete; however, in almost all examples of this chapter, the decision variable is represented by "alternatives," and the state variable is expressed as "scenarios." Table 8.1 shows an example of a DM problem where three alternatives and three scenarios are involved. For example, we can consider a hypothetical case where three water supply facilities (such as expanding the capacity of an existing dam, construction of a new desalination plant, and reinforcement of water conservation programs) are reviewed as alternatives under three climate scenarios (such as wet, medium, and dry futures).

Table 8.1 Example of decision alternative, state scenario, and performance criterion of a DM problem

	Scenario 1	Scenario 2	Scenario 3	Average	Max
Alternative 1	50(0)	46(6)	48(6)	144/3	50(6)
Alternative 2	42(8)	52(0)	52(2)	146/3	52(8)
Alternative 3	34(16)	44(8)	54(0)	132/3	54(16)
				Max = 146/3	Min = 50(6)
Choice				Alternative 2	Alternative 1 (Alternative 1)

The value in each cell indicates the economic benefit of each alternative if each scenario occurs, and the value in parenthesis represents "regret"

A combination of each decision alternative and each state scenario produces a unique "outcome" (i.e., each cell in Table 8.1), which should be measured with a "performance criterion." The ultimate purpose of a decision analysis is to choose a preferred alternative by comparing the performance criterion values of the alternative candidates. The common choice for the performance criterion is monetary valuation, and typical examples are "net present value" (or "benefit-cost ratio") of a well-known cost-benefit analysis (CBA). Table 8.1 also shows a monetary valuation example (called the payoff matrix), where the value of each cell indicates the economic benefit of each water supply alternative if each scenario occurs. However, real decision problems often face various types of nonmonetary and intangible measures (for instance, the amount, running length, or sometimes, frequency of water deficit; and the aesthetic value of water). Contrary to CBA, a "cost-effectiveness analysis" (CEA) can handle a few nonmonetary measures because it does not evaluate the benefits of alternatives in monetary terms, but it chooses the least-cost alternative that meets a prespecified benefit target. The performance criterion is sometimes converted to an "objective function" in order to solve it mathematically.

The methodology to choose a preferred alternative depends on the "goal" a decision maker pursues. Most decision problems aim at minimizing the loss or maximizing the benefit. The cells in the fifth column of Table 8.1 show the "expected" benefits of the alternatives over three scenarios, and Alternative 2 is chosen because it has the maximum average value. Such a DM is traditionally well structured with the "optimization" theory, which basically consists of an objective function, constraints, and decision variables. Both CBA and CEA can be categorized as (constrained) optimizations. The real optimization problems usually suffer from nonlinear, non-concave functions with huge sets of decision and state variables and uncertainties; however, various solution search algorithms are also available, which include linear programming, dynamic programming, nonlinear programming, and genetic programming that have been developed over the last 60 years.

However, one may also pursue a "safer" goal to optimization, especially when the stakes are very high when she/he loses. The last column in Table 8.1 lists the maximum values from individual rows, and the minimum value of these maximums is given by Alternative 1. This "minimax" approach based on benefit values may be a "pessimistic" (or "conservative") strategy, but it has the disadvantage of considering only the extreme payoff. An approach that can consider all payoffs is the minimax "regret" (Savage 1954), which is illustrated with values in parenthesis in Table 8.1. The regret implies opportunity loss and is defined as the difference between each payoff and the best possible payoff in a column. In Table 8.1, minimizing the maximum regret results in the "best of the worst" alternative, i.e., Alternative 1. The regret concept has been favorably adopted for the RDM theory (Sect. 8.3) where an alternative that performs reasonably well over all scenarios is preferred to an optimal alternative that performs best for a certain group of scenarios, but its performance is very poor for other groups. Stevenson and Ozgur (2007) compared three alternative choice strategies:

- The maximin strategy (Wald 1949) is a conservative approach. It consists of identifying the worst (minimum) payoff for each alternative and selecting the alternative with the best (maximum) of the worst payoffs. In effect, this DM theory sets a limit on the potential payoff, and the actual payoff cannot be less than this amount.
- The maximax approach is the opposite of the maximin approach. The best payoff for each alternative is identified, and the alternative with the best maximum is selected.
- The minimax regret approach (Savage 1954) minimizes the worst-case regret. In this approach, the regret is defined as the difference between the actual payoff and the payoff that would have been obtained if a different course of action had been chosen. The worst-case regret is the regret in which the best option was chosen (Loomes and Sugden 1982).

In several real problems, there exist multiple performance criteria, some of which often conflict with other criteria. In order to handle multiple criteria in decision problems, multi-criteria analysis (MCA) or multi-criteria decision analysis (MCDA) has been widely used because it considers both quantitative and qualitative data in prioritizing adaptation strategies. All MCA approaches make options and their contribution to different criteria explicit, and all require the exercise of judgment. However, they differ in the manner they combine tangible and intangible data. Formal MCA techniques usually provide an explicit relative weighting system for different criteria. The main role of MCA is to resolve the difficulties faced by human decision makers in handling large amounts of complex information in a consistent way. Therefore, MCA can identify a single most preferred strategy, rank or outrank all strategies, shortlist a limited number of strategies for subsequent detailed appraisal, or distinguish acceptable from unacceptable possibilities (Dodgson et al. 2009). Therefore, MCA provides a systematic framework for assessing and scoring CCA strategies for a wide range of climate change scenarios and decision criteria, which are expressed in physical, monetary, or qualitative ways. The weights on criteria and scenarios can be derived through stakeholder participation or expert involvement (Werners et al. 2013; Kim and Chung 2015).

Ranger et al. (2010), Watkiss and Hunt (2013), and Kunreuther et al. (2014) concisely summarized the classical DM theory and methodology and discussed their implications for CCA, focusing on three traditional methods, CBA, CEA, and MCA. Please refer to these references for more details.

8.2.2 Handling Uncertainty and Scenario-Based Approach

In general, uncertainty makes DM problems difficult and complex. Most classical DM methods have been developed assuming that probabilities of random variables are well identified. Uncertainty is often captured using multiple scenarios, which runs a specific DM method for individual scenarios independently and combines the calculated outcomes with the given probabilities. It should be noted that the "implicit" optimization approach utilizes the same concept as a scenario-based DM approach, while in an explicit optimization approach (e.g., stochastic dynamic programming), uncertainty is expressed with an explicit formula with respect to probability rather than scenarios.

Table 8.1 illustrates a very simple example of a scenario-based DM approach, whereas problems that involve a sequence of decisions are best manipulated using decision trees (such as Figs. 8.3, 8.4, and 8.5 in Sect. 8.4). The fourth column in Table 8.1 assumes that the occurrence of all scenarios is equally likely, and scenario probabilities should be updated as new information becomes available, which consequently changes the choice of alternatives with a new weighted average. Stevenson and Ozgur (2007) also compared equal and weighted averages by using the maximin, maximax, and minimax approaches:

- The Hurwicz criterion, also referred to as weighted average, offers a compromise between the maximax and maximin criteria for the DM theory. This approach requires the DM theory to specify a degree of optimism in the form of a weighting coefficient. Possible values range from zero to one, where the DM theory is more optimistic for values closer to one (Hurwicz 1945).
- The equal likelihood criterion offers a method that incorporates more information. This approach treats the equivalent plausibility as if each scenario was equally likely to occur and focuses on the average payoff for each row by selecting the alternative that has the highest row average in a decision matrix.

As mentioned in the previous section, the minimax (as well as the maximax and maximin approaches) approach does not require any probability information and thus could be suitable for decision problems relevant to CCA where probabilities are not well defined.

For the past 30 years, the scenario-based approach has received much attention because a large number of climate scenarios (e.g., those based on SRES of AR4 and RCP of AR5) have been developed using relatively precise monitoring data and high-performance computer simulation results. In general, a climate change impact assessment is initiated by inputting these climate scenarios to a chain of general circulation models downscaling methods, hydrological models, and flood or drought vulnerability models. Most previous studies on CCA that aimed at reducing the projected vulnerability have often used this "predict-then-act" approach to predict the consequences of adaptation strategies and recommend an optimal response, that is, an action that is better than the rest. This "top-down" approach seems natural and conceptually straightforward, but it should address the uncertainty inherent in the process of selecting CCA strategies by estimating the likelihood that different future scenarios will occur (Lempert and Schlesinger 2000). Therefore, the probability-based approach has been useful for understanding comprehensive future climate change patterns and guiding specified adaptation policies, although identification of reasonable scenario probabilities has been a key issue (Schneider 2001). However, Lempert et al. (2003) argued that probabilities should be used with scenarios only if they contain solid information, while Moss and Schneider (2000) emphasized on probabilities based on subjective judgment. A few IPCC contributors have initiated a process to guide its assessment report toward characterizing uncertainties with probability distributions that represent the consensus of the scientific community (Giles 2002), while IPCC (2005) warned that the community may converge to an expressed value and become overconfident. These controversies over estimating and using scenario probabilities resulted in the development of the RDM theory, which will be addressed in Section 8.3 where limitations of the predict-then-act approach will also be discussed.

8.3 Robust Decision Making

8.3.1 Description

As mentioned in Sect. 8.1, RDM focuses on the robust perspective, as its name represents. The alternative that RDM chooses for a decision problem should perform reasonably well over a wide range of uncertainties. In general, RDM relies on three key components: (1) multiple scenarios of plausible future climates (Lempert et al. 2003), (2) a robustness criterion, and (3) one of the following approaches: (a) MCA or MCDA under complete uncertainty (Ranger et al. 2010; Chung and Kim 2014; Kim and Chung 2015), (b) assessed risk of policy (Lempert et al. 2004; Dessai 2005; Dessai and Hulme 2007), (c) vulnerability and robust response (Lempert et al. 2013; Bloom 2014), and (d) decision scaling. Regarding the first key component, similar to traditional multi-scenario methods, RDM characterizes uncertainty by considering multiple future scenarios. In some cases, these multiple scenarios will be represented by multiple future scenarios of the world such as scenarios of SRES of IPCC AR4 and RCP2.6, 4.5, 6.5, and 8.5 of IPCC AR5. RDM can also incorporate probabilistic information in a quantitative manner, but it rejects the view that a single joint probability distribution represents the best description for a deeply uncertain future. Rather, RDM uses ranges or, more formally sets of plausible probability distributions to describe deep uncertainty. For the second component, RDM uses robustness rather than an optimality criterion to assess an adaptation strategy, whereas the traditional subjective utility framework ranks alternative decision options based on the best estimated probability distributions. There exist several definitions of robustness, but all incorporate some types of satisficing criterion. For instance, a robust strategy can be defined as one that performs reasonably well compared to alternatives across a wide range of plausible future scenarios (Lempert et al. 2006). Lastly, RDM employs various analysis frameworks to characterize uncertainty and to help identify and evaluate robust strategies. They are summarized in the following sections. For strengths and weaknesses of these approaches, refer to the studies of Hall et al. (2012), Hallegatte et al. (2012), and Weaver et al. (2013).

8.3.2 *Fuzzy MCDA-Based RDM Approach for Sustainability*

MCA has high relevance for CCA (Werners et al. 2013). However, the general disadvantage of MCA is that scoring and weighting can be quite subjective, influenced by stakeholders and decision makers involved in the DM process. Therefore, recent MCA studies have applied a fuzzy set theory to consider the uncertainty of scoring and weighting. Furthermore, various variants of MCA methods have been proposed to derive robust priorities from several uncertain scenarios such as TOPSIS coupled with the minimax regret approach (Kim and Chung 2014). In addition, several methods, such as entropy, principal component analysis, Delphi technique, and analytical hierarchy process (AHP), have been introduced to derive subjective and objective weights (Porthin et al. 2013; Kim and Chung 2015), and sensitivity and uncertainty analyses were performed for obtaining the final decision (Hyde et al. 2005; Kang et al. 2013; Song and Chung 2016). Now, several combinations have emerged to solve the small part of deep uncertainty for robust prioritization such as the minimax regret approach with TOPSIS (Kim et al. 2015), fuzzy AHP (Sanneh et al. 2014), fuzzy TOPSIS (Chung and Kim 2014; Kim et al. 2015), and fuzzy VIKOR with entropy (Kim and Chung 2015).

In classical decision-making problems, the ratings and weights of criteria are known precisely, whereas in the real world, it is unrealistic to assume that the knowledge and representation of a decision maker or expert are precise. A few criteria are difficult to be defined or measured because of their inherent vagueness and complexity (Ducey and Larson 1999). The effectiveness of adaptation strategies varies based on the used climate change scenarios, as shown in Fig. 8.1. Because the evaluation of alternatives must handle the imprecision of established criteria and estimated performances, the development of a fuzzy DM model is necessary to address either qualitative (unquantifiable or linguistic) or incomplete information (Philles and Andriantiatsaholiniaina 2011). Therefore, most widely used MCA methods have been extended to solve problems in a fuzzy environment and to apply to practical decision problems (Carlsson and Fullér 1996; Greco et al. 1999; Liu 2007; Vadrevu et al. 2010; Turskis and Zavadskas 2011). Therefore, Chung and Kim (2014) developed a new analytical robust prioritization framework for CCA strategies by using a fuzzy set theory, multiple climate change scenarios, various MCA techniques, sustainability concepts, and strategies for DM under complete uncertainty.

Kim and Chung (2015) used the VIKOR method for sustainable compromise solutions to suggest a set of adaptation solutions based on the regret aversion of decision makers such that the unknown preferences of decision makers can be considered and introduced into the fuzzy concept to represent the uncertainties of future climate change scenarios and expert valuations of the relative importance of decision criteria, which consists of social, economic, and environmental factors for sustainability. Furthermore, it employed the individual importance of each climate change scenario to account for the difference in information significance of various climate change scenarios by using an objective weight, specifically Shannon entropy-based weight.

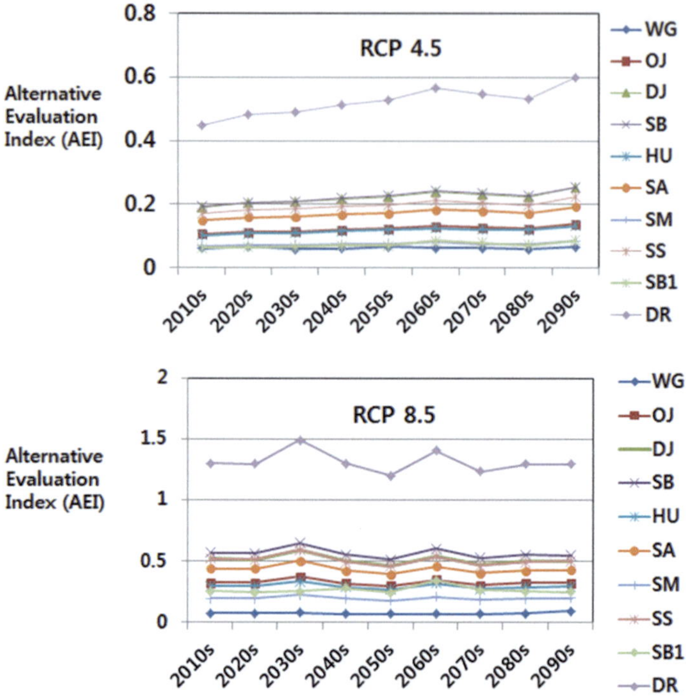

Fig. 8.1 Decadal variations of alternative evaluation index for RCP4.5 (Jeon 2013)

In order to reflect the marginal utility of allocation capacity, Kim et al. (2015) developed an iterative framework for robust reclaimed wastewater (RWW) reuse allocation that considers several climate change scenarios. Based on alternative evaluation index (AEI), which includes various social, economic, and environmental criteria for sustainability (Chung and Lee 2009), the concept of incremental AEI (IAEI) was introduced to rank the best site by incrementally allocating facility capacity for RWW reuse to sites under two climate change scenarios. The minimax regret strategy was employed to consider the uncertainty inherent in climate change scenarios. The consequent robust ranking from the IAEIs was used to determine the final allocation of unit water quantity in a given iteration. This iteration continued until the total allocated water quantity satisfied the maximum available facility capacity for RWW reuse.

In addition, the RDM approach coupled with recent MCA tools can provide support to develop several real robust solutions for a highly changing climate environment with the consideration of sustainability, vulnerability, and stakeholders or decision makers. Several MCA techniques have been developed to consider various types of uncertainties on the selection of criteria, quantification of weights, derivation of performances, and appraisal of options. Thus, the MCDA method can be selected by criteria such as the type of data (quantitative, qualitative, or both), use of weights, use of subjective and objective weights, derivation of outranking, incomplete (or complete) information, and additional analyses such as sensitivity and uncertainty analyses. Critically, MCDA can reasonably consider the uncertainty of huge data, which is prevalent in the field of CCA strategy determination.

8.3.3 Assess-Risk-of-Policy Approach

Several researchers have proposed robust strategies for a wide range of plausible climate change scenarios. Additionally, Dessai and Hulme (2004) reviewed three critical questions within the context of the predict-then-act approach: (1) Why might we not need probabilities of climate change? (2) What are the problems in estimating probabilities? (3) How are researchers estimating probabilities? The CCA strategies can be determined using the following criteria: goals and motivation of the policy analysis, the unit of analysis, the timescale, and training of the analyst. That is, in the context of climate change, the conditional and provisional approach has various problems in deriving the probability density functions of future climate variables. Furthermore, Lempert et al. (2004) noted that the predict-then-act approach has often been used successfully for CBA, but that climate change violates its assumptions as follows: First, climate change is associated with radically diverse decision contexts and geography and timescales. Second, climate change is closely related to the conditions of deep uncertainty, where decision makers do not know or cannot agree on: (1) the appropriate models to describe interactions among the variables of a system, (2) the probability distributions to represent uncertainty about key parameters in the models, or (3) how to value the desirability of alternative outcomes.

Therefore, the assess-risk-of-policy framework has been proposed to develop CCA strategies that would perform satisfactorily with maximum insensitivity to uncertainty. The performance of the selected options would not be unstable in cases where the negative impacts resulting from a large number of plausible climate change scenarios of various GCMs and regional circulation models (RCMs) need to be reduced. These can be called well-hedged strategies in terms of risk assessment. In spite of these strengths, this framework requires several types of subjective decision components, which require the proper choices of the various simulation models and the values of input parameters for the selected models, because the selected models and their parameters can critically affect the responses of adaptation options in all scenarios.

In some simplified cases, where plausible limited scenarios are accepted or significant features do not vary largely, the predict-then-act approach will be more effective and reassuring for decision makers. However, as climate change scenarios of AR5 can even predict future climatic variables, such as rainfalls and temperatures, with consistent characteristics, the limited scenarios and simple patterns might not be realistic anywhere and anytime. Therefore, this approach should carefully consider the following question: "What plausible models, parameter values, and alternative strategies have been neglected that might make a proposed robust strategy vulnerable?"

While the predict-then-act approach identifies uncertainties in plausible forecasted scenarios and strong candidate strategies are agreed upon in general, the assess-risk-of-policy framework can suggest implicit criteria that can help analysts, engineers, and scientists in reducing the uncertainty as much as possible. The application of these criteria might cause the analysis to identify the uncertainties or focus on those parts of the problem where the uncertainties are most precisely characterized (Metlay 2000). That is, this approach suggests all probable vulnerabilities of the selected options and helps decision makers select a strategy with the most acceptable vulnerabilities. In the end, the assess-risk-of-policy is much more subjective because it forces decision makers to choose strategies having relatively low uncertainties and high performances in the vulnerable area and period. The predict-then-act approach that takes less time and uses a small amount of human and physical resources can be more useful if the approach is improved to produce high precise forecasts for a longer period. However, in situations where the future is almost completely uncertain and the negative impacts are perfectly unknown or difficult to articulate hedging strategies for a wide range of climate change scenarios and vulnerabilities, the assess-risk-of-policy framework can produce more realistic and reasonable decisions.

8.3.4 Vulnerability-and-Response-Option Approach

Recently, a vulnerability-and-response-option analysis framework was developed to characterize uncertainty and to help identify and appraise robust adaptation strategies with the constructive participation of stakeholders, following the procedure: (1) structure problem, (2) choose candidate strategy, (3) evaluate strategy against large ensemble of scenarios, (4) characterize vulnerabilities, and (5) identify and assess options for ameliorating vulnerabilities (Lempert and Grove 2010). In the first step, the engineers set up the problem structure and then analysts collect the relevant data and select simulation models, which can predict the performance and status of strategies at an acceptable level. This approach is totally different from the predict-then-act approach and has a critical distinction in the uncertainty of probabilistic forecasts or multiple scenarios. This approach considers a few candidate strategies and includes current government policies or several new policies proposed by key stakeholders through public communication. Or the optimum strategy from a traditional subjective utility analysis can be selected as the initial candidate.

The third step characterizes the vulnerabilities of the candidate strategy. These vulnerabilities identify regions and scenarios where the candidate strategy fails to meet the target performance of decision makers. In order to identify these vulnerabilities, the candidate is appraised against several plausible climate change scenarios, each representing different combinations of key uncertain climatic factors. In this framework, statistical techniques based on the concept of robustness are frequently used to determine scenarios having a poor performance for decision makers. If the performance of a strategy frequently deviates from the critical thresholds of all criteria, which are determined by stakeholders and experts, the strategy will be removed from the candidate list. In addition, cost will be the most important criteria if decision makers have budget limits. If the cost is larger than the threshold, the strategy will not be considered. In this case, a combination of strategies, which are also accepted by stakeholders and decision makers, can be leading candidates considering the total cost and synergy performance.

In the final step, this framework identifies the strong candidate strategies that might improve the vulnerabilities of the candidate strategy. Because these new strategies often exploit more or less adaptivity, an analyst should provide the trade-off. Therefore, decision makers could use this relationship to determine whether or not to adopt the new strategy. If a particular strategy is selected, analysts repeat the process from the second step by using this new candidate strategy. Alternatively, if decision makers or stakeholders do not select original and new candidate strategies using the trade-off among vulnerability, adaptivity, and performance, we return to the first step to restructure this decision problem in a way that might yield more desirable robust options for stakeholders or decision makers (Lempert and Grove 2010).

Other studies (Lempert et al. 2013; Grove et al. 2013; Bloom 2014) applied this vulnerability-and-response-option approach by customizing it to the characteristics of the study area. This process, as shown in Fig. 8.2, begins with a decision structuring step in which planners define goals, uncertainties, and policy choices under consideration. Analysts then use computer models to generate a

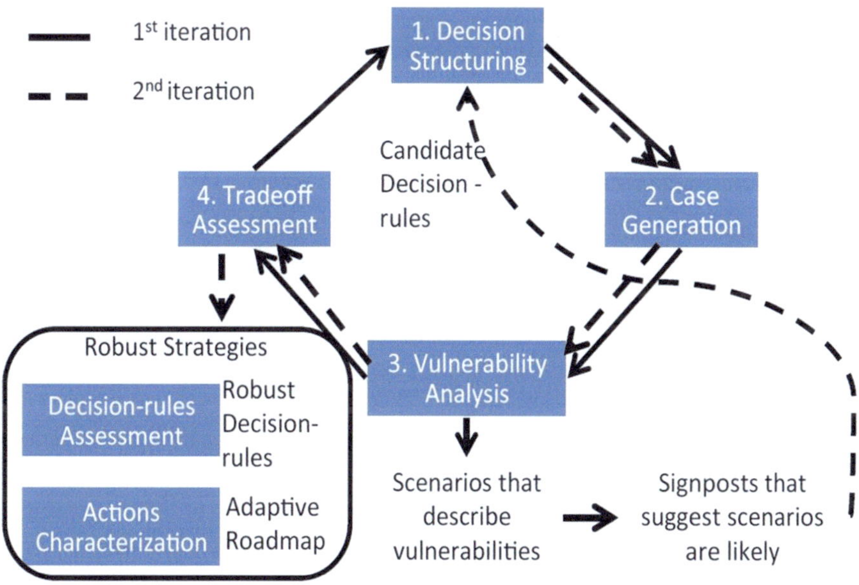

Fig. 8.2 Expanded procedure for vulnerability-and-response-option approach to generate adaptive strategies with multiple iterations (Bloom 2014)

large database of simulation runs in which each case represents the performance of a proposed policy in one plausible future scenario. Computer visualization and statistical analysis of this database help the planner identify clusters of scenarios that illuminate the vulnerabilities of the policies. These scenarios can then help planners identify potential new ways to address those vulnerabilities or evaluate whether these choices are worth adopting through a trade-off analysis. The process continues until planners settle on a robust strategy (Lempert et al. 2013).

8.3.5 Decision Scaling

Traditional approaches, such as the predict-then-act approach, can be characterized as top-down assessments because they start producing climate change projections from a GCM and then downscale them at coarse resolutions. In general, relatively few GCMs are used to estimate future climate scenarios because of the high computational efforts and several hours required. Further, socioeconomic impacts and hydrological components should be analyzed with additional simulations, which are also time-consuming. Nevertheless, the top-down approach has several limitations including the uncertainty of greenhouse gas emissions, errors of simulation models, differences between several GCMs, and consideration of the full range of plausible future climate scenarios.

In contrast to the top-down approach, McMahon (2007, 2011) linked reservoir reliability with climate conditions in regional studies, providing a basis for a climate scenario assessment independent of complex procedures in order to introduce the influences of climate change into a streamflow time series. It is called a bottom-up approach. It assesses the socioeconomic condition of the study system and identifies vulnerabilities or risks related to climate change, thereby focusing the analysis on the system or decision without considering GCM projections. A few studies (Johnson and Weaver 2008; Lempert et al. 2006) used GCMs as scenario generators. Other studies (Dessai et al. 2009; Sarewitz and Pielke 2000) did not consider any GCM projections at all because they believed that the information used for climate change scenarios is uncertain. However, it has been criticized by the reason of using no GCM in the bottom-up approach. Through decision scaling, the process of GCM projections is incorporated into the identification of critical climate conditions by using bottom-up analysis.

The decision scaling method finds the missing link between the insight into vulnerabilities obtained from the bottom-up process and all the information available from the climate change projections of GCMs or RCMs. Therefore, it can identify the type of climate changes that would cause critical problems. Because GCM-based climate projections offer a constricted view of possible future climate scenarios (Stainforth et al. 2007), decision scaling avoids using these data to identify vulnerabilities. Instead, an analyst conducts a more extensive vulnerability analysis to understand the sensitivity of the system to changes in climate variables. A climate response function is then derived to define the performance of the system and relevant planning thresholds in terms of the underlying climate statistics. By describing the system performance thresholds in climatic units, the process empowers the planner to assess climate risks by obtaining and comparing various sources of climate information. This may include, but is not limited to, GCM-based projections. It has been argued that such an approach would enable planners either to discount the need for GCMs in cases where the performance of a system linked to climate variables is found to be poor or to tailor specific decision-critical thresholds to fit with the available climate information (Brown and Baroang 2011; Brown and Wilby 2012; Turner et al. 2013).

8.3.6 Discussion

RDM has recently become one of the most effective tools in finding strategies for CCA. Other variants of RDM such as ROA, DAPP, and the others originated from the concept of RDM. The latter uses multiple views of the future climate scenarios and robustness criteria as the basic concepts for analysis. The four typical approaches presented in this article have been developed and improved with many real applications in the world for the past 10 years. However, more innovative approaches that cannot be characterized by the four classifications or their improved versions have appeared and will be developed to determine the robust strategies using various DM techniques.

From the recent studies, RDM shows three representative features. First, stakeholder participation and public involvement are important in the RDM process. The vulnerability-and-response-option approach and decision scaling have been improved to include the opinions from various stakeholders including residents and decision makers in a quantitative and qualitative manner. MCDA-based RDM has been extended to consider the preferences of stakeholders and enhance the importance of robustness criteria. Public hearings and surveys for collecting opinions, workshops for capacity building, and public meetings for conflict resolution and consensus building are therefore required. In severe conflict situations, litigation can be conducted in the courtroom. Second, uncertainty or sensitivity analyses have been incorporated in the general RDM process to reduce the uncertainties in the climate change scenarios and DM components. MCDA-based RDM used fuzzy set theory to include the uncertainties of the weighting values of decision criteria, performance values of strategies, and the importance of climate change scenarios. Decision scaling also helps derive the division map of the climate space through optimal decisions based on the climate response functions such as temperature and precipitation. Third, the anthropogenic factors such as social and economic components are included in the RDM process using vulnerability. The IPCC (2001) conceptualized vulnerability from a system's perspective. According to this, a system is judged to be vulnerable if it is exposed to climate change impacts, if it is sensitive to these impacts, and if it has a low capacity to cope with these impacts. A general conceptual model for vulnerability including social, economic, and environmental components has emerged in the climate change literature, similar to the wider use of the concept (Kelly and Adger 2000; Adger 2006). MCDA-based RDM, vulnerability-and-the-response-option, and decision scaling approaches have been extended to combine the vulnerability concept with the previous RDM process. These studies follow the bottom-up analysis which begin with the assessment of the socioeconomic system of interest followed by the identification of climate states that impact a decision and finally identification of vulnerabilities or risks related to climate and decision criteria (Brown 2013).

One of the main aims of this study is for the RDM approach to be used in the field. For this, the three mentioned features are necessary. Application of various outstanding approaches such as public involvement, uncertainty sensitivity analyses, and vulnerability is therefore expected to be prevalent in the national and regional water resource plans in the near future.

8.4 Real Options Analysis

8.4.1 Description

ROA has also been often referred to as a DM methodology for CCA (e.g., Watkiss et al. 2013). RDM described in Section 8.3 spreads risk over scenarios, but ROA spreads risk over "time" by using "options." The word "real" is derived from real assets to which the option pricing theory is often applied for the valuation of investment. Although initially coined by Myers (1977) for the financial market, ROA recently received attentions from engineering decision makers since de Neufville (2003) revisited its concept and merits with an emphasis on flexibility under uncertainty in the field of design and planning of various engineered projects. With ROA, one can operate each alternative differently; for example, one can abandon, delay, or alter the alternative as time proceeds (i.e., as more information is available). Such different operations of an alternative are called "options," which, however, are not the obligation but the right. It should be noted that several literatures mix both the terminologies, "alternative" and "option." However, this is not the case in this chapter.

Simply, ROA is an extension of the classical CBA called the "discounted cash flow" approach, where a stream of in and out cash flows over the lifetime of each alternative is predicted and the most economical (often called the optimal) alternative is chosen. Once the decision is made, the chosen alternative is implemented without any modification until it is completed. In other words, the classical discounted cash flow approach has only the "now and never" option without any flexibility over time. In the ROA procedure, however, the chosen alternative can be modified using a few options. For example, one can delay the construction timing of a project (alternative) until justification of the project proves more concrete by comparing the cost and benefit of waiting for such a delay with more information. Moreover, the project can be constructed step-by-step by watching how the future evolves rather than constructing it all at once. Using an option with ROA, the alternative that has been appraised as economically infeasible with the traditional economic analysis may be considered as a feasible solution. This difference is called the "option value." The crucial term in both the discounted cash flow and ROA approaches is net present value (NPV), which is defined as the difference between the present value of benefits (e.g., the flood damage reduced by a flood mitigation structure) and the present value of costs (e.g., the installation and the maintenance and operation costs of the flood mitigation structure).

8.4.2 Typical Steps

Similar to other DM methods, ROA identifies problem objectives and constraints, alternatives and their options available, and scenarios of key uncertainties (WUCA 2010). A ROA model that best depicts the decision problem being considered is then chosen from several techniques, such as the Black-Scholes analytical approach (e.g., Lau et al. 2006), the binomial lattice approach (e.g., Michailidis and Mattas 2007), the Monte Carlo simulation approach (e.g., Jeuland and Whittington 2013), and the decision tree approach (Borison and Hamm 2008). The chosen ROA model calculates the financial values of individual combinations of alternatives and options as a function of time with given monetary information such as price change. This procedure is repeated for each scenario combination of key uncertainties, and the expected performance for each option is assessed using the given scenario probability. The following paragraphs introduce examples of two different approaches of ROA to illustrate the steps of the binomial lattice and the decision tree approaches, respectively.

The first ROA example tested four flood mitigation planning alternatives for a Korean basin (called the Korean flood mitigation planning (KFP) example hereafter), each of which indicates a unique set of various structural and nonstructural flood mitigation measures such as raising existing dams; constructions of new dams, wetlands, and retention ponds; and improvement of reservoir operations (Ryu 2014). One of the final products of this example was the option matrix for each alternative. Figure 8.3 shows examples of the "option matrix" where three options, such as invest (I), delay (D), and abandon (A), for each alternative, are employed during the so-called investment opportunity period of 30 years from 2016 to 2045. Figure 8.3 presents an obvious contrast between the two different alternatives. The abandon option first appears in the year 2029 for Alternative 1 (Fig. 8.3a), while, for Alternative 4, the abandon option appears in the year 2024 (Fig. 8.3b), 5 years earlier than Alternative 1. On the contrary, the invest option becomes available from the beginning of the investment period (i.e., the year 2016) for Alternative 1 (Fig. 8.3a), while, for Alternative 4, the invest option is delayed until the year 2029 (Fig. 8.3b).

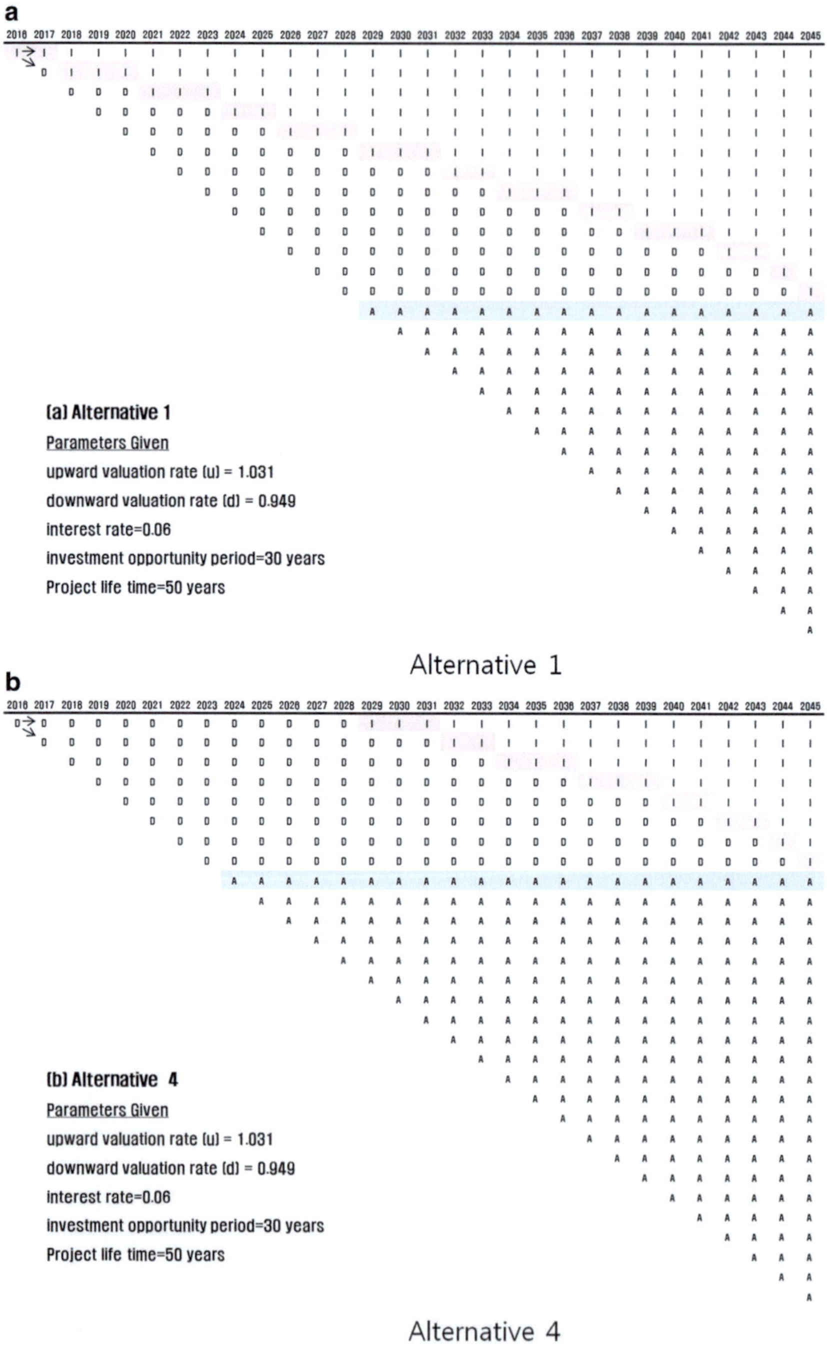

Fig. 8.3 Examples of the option matrix in ROA (*I* invest, *D* delay, *A* abandon options)

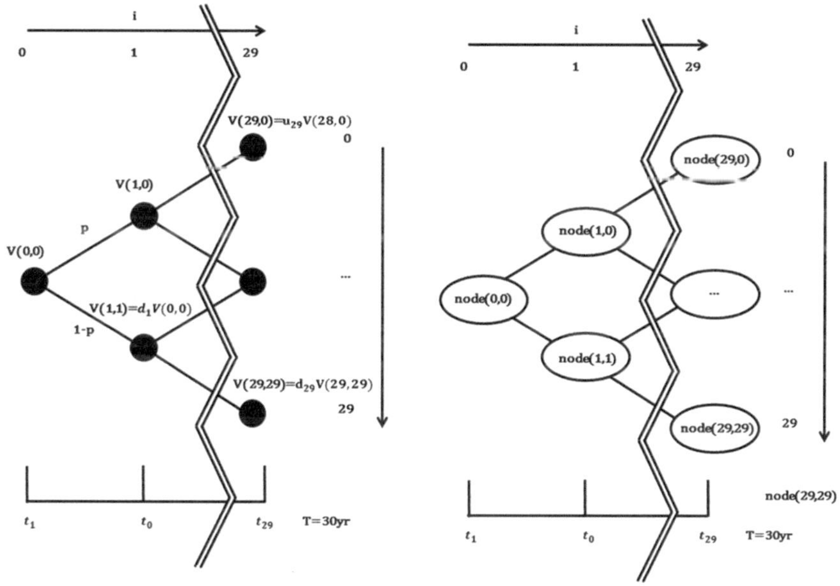

Fig. 8.4 Example of the binominal tree in ROA

At each time step of ROA, one can include possible scenarios to reflect uncertainty, and thus, ROA can be visualized with a set of decision trees. In the KFP example, ROA was characterized with a binomial decision tree where only two scenarios are considered as shown in Fig. 8.4. The initial value $V(0,0) = V_0$ at t_0 can either go up to $V(1,0) = u_1 \cdot V_0$ with a probability p_0 or down to $V(1,1) = d_1 \cdot V_0$ with a probability $1\text{-}p_0$ at the next period t_1, and thus, in general, $V(i,j) = u_{i-j} d_j V_0$ for node (i,j) at t_i, $i = 0, 1, 2, 3, \ldots T$ and for $j = 0, 1, 2, 3, \ldots T$ where T is the investment opportunity period. The parameters u_i, d_i, and p_i should be given or estimated to initiate the ROA procedure, which is seldom feasible in most real applications; thus, reasonable approximations are often required. With these parameters, the first task of the ROA procedure is to create the value tree at all nodes assuming the alternative being considered is implemented (i.e., invested).

For other options, such as the abandon and delay options, the values at all nodes are also calculated by a similar methodology, but their cost should cover the loss (e.g., the flood damage in this example) when the project alternative being considered is not implemented (i.e., delayed or abandoned). The maximum of these values over all the options at each node is called the "expanded NPV (ENPV)," which is the addition of NPV and the "option premium." The difference between ENPV and NPV at t_0 for each alternative is called the "option premium," which indicates the worth of using options throughout the entire investment opportunity period. In the KFP example, the NPV obtained using the discounted cash flow approach was positive for Alternative 1, while it was negative for Alternative 4.

However, in ROA, the ENPV was positive for Alternative 4 because of a high value of its option premium, which results from the utilization of the options shown in Fig. 8.3, although the ENPV for Alternative 1 is still larger than that for Alternative 4. In other words, both the two alternatives are now economically feasible with adequate options. Therefore, one may choose Alternative 4 instead of Alternative 1 if some other benefits of Alternative 4, other than the economic feasibility, dominate those of Alternative 1. A typical example for those "other" benefits is one that comes from an eco-friendly alternative.

The next ROA example described in this section is a hypothetical case study (called the Australian water utility planning (AWP) example hereafter) by Borison and Hamm (2008) to illustrate the decision tree approach, which provides more generality and flexibility than the binomial lattice approach; however, it is difficult to implement it in practice. Their final option analysis is summarized in Fig. 8.5, where three alternatives ("Dam," "Desal," "Mod Desal") and three uncertainty factors ("Dam Feas," "'Desal Cost," "'Inflows") are considered for each of the four time periods, such as 2008–2009, 2010–2011, 2012–2013, and 2014–2027. Here, the acronyms of Dam, Desal, and Mod Desal represent three construction alternatives of new assets, such as a dam, a large desalination plant, and a modular desalination plant, respectively, while the acronyms of Dam Feas, Desal Cost, and Inflows represent uncertainties related to dam feasibility, desalination plant construction cost, and catchment inflows, respectively. The objective of the AWP example is to maximize the net value of water supply provided by portfolios of the above three

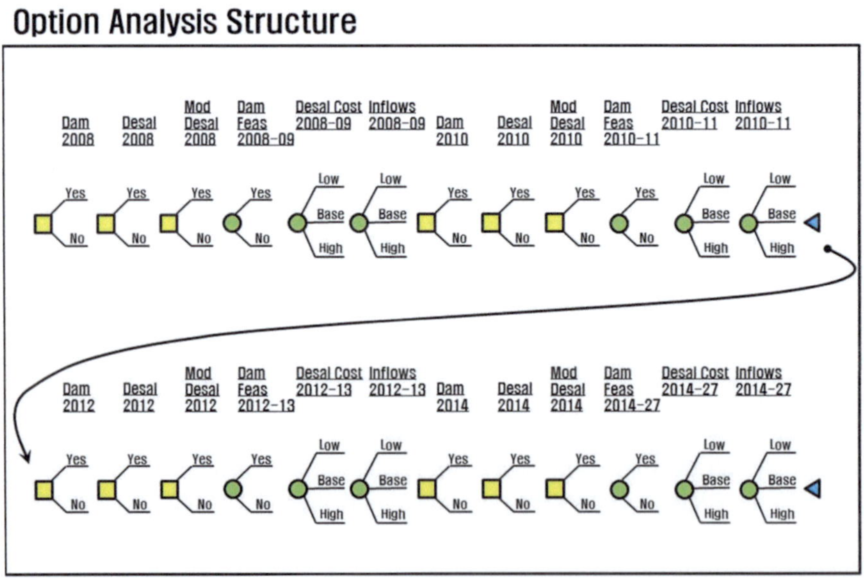

Fig. 8.5 Option analysis structure of the Australian water utility planning example (Borison and Hamm 2008)

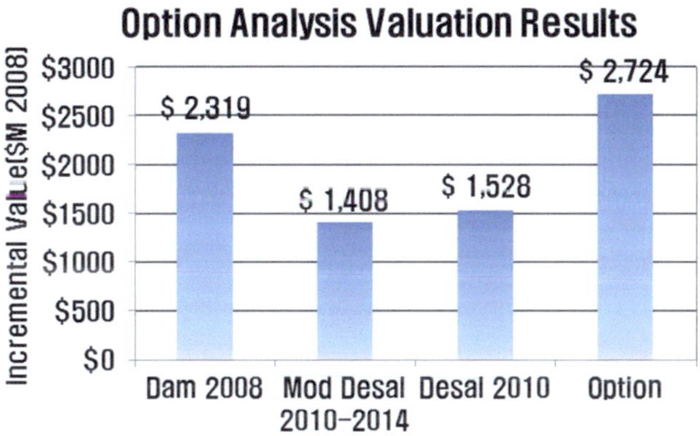

Fig. 8.6 Option analysis valuation structure of the Australian water utility planning example (Borison and Hamm 2008)

assets (Borison and Hamm 2008). The AWP example considers eight combinations of the above three alternatives by using the "advance" option, which consists of advancing nothing, Dam only, Desal only, Mod Desal only, Dam and Desal, Dam and Mod Desal, Desal and Mod Desal, and a combination of all three. The report of Borison and Hamm (2008) also showed the estimation of the probability of each uncertainty source through rigorous sensitivity and uncertainty analyses.

It should be noted that there are two differences between the KFP and the AWP examples. First, the AWP example considers more than a single source of uncertainty; furthermore, some of them are not binomial. Secondly, mainly because of such non-binomial complexities, the AWP example handles the possibilities of alternatives and uncertainties simultaneously, which produces approximately 400 million different paths (Borison and Hamm 2008). The resulting valuation of this option analysis (as illustrated in the last bar in Fig. 8.6) is a weighted average using the probabilities of all the paths. The first three bars in Fig. 8.6 indicate the results of a conventional economic analysis with the consideration of uncertainties, but which does not consider options. Figure 8.6 reports that the (incremental) option value for the AWP example is roughly estimated as $400 million when "option" is compared with "Dam 2008," which is the best fixed alternative (Borison and Hamm 2008). It should be noted that the metric of the y-axis denotes the incremental value of an alternative or optional portfolio over "As Is."

8.4.3 *Discussion*

As ROA has been tested and evolved by various sectors in the field of the financial markets since several decades, its potential for application to CCA also seems high if scenario probabilities are well defined and economic data are provided.

Therefore, although few, the number of ROA applications for CCA DM in the field of water resources planning is increasing. Lau et al. (2006) explored the potential use of the Black-Scholes model with ROA for urban drainage risk management to ultimately evaluate the funding strategies of sewerage companies. More recently, Steinschneider and Brown (2012) analytically linked the real option concept to the derivation of reservoir operation strategies as a function of reservoir inflow forecast scenarios and concluded that the real option significantly improves water supply performance of the case study reservoir system in America. The analytical ROA approach is easy to implement but hard to understand and express its output effectively.

Similar to the KFP example described in this section, Michailidis and Mattas (2007) applied the binomial ROA approach to an irrigation dam in Greece to test three options that delay, double, and abandon the investment and concluded that the classical discounted cash flow approach could underestimate the ability of the management under random future. Jeuland and Whittington (2013) conducted a highly computational ROA study based on the Monte Carlo simulation and incorporated option strategies into the design and operation of new dams along the Blue Nile in Ethiopia under climate change uncertainty. They concluded that investment planning based on the real options framework can be robust for bad situations and flexible enough to take advantage of good situations. The simulation-based ROA has been rarely applied to water resources planning; however, a few general-purpose simulation softwares can facilitate its applicability with the use of more flexible options (Borison and Hamm 2008).

As noticed in the AWP example described in this section, the ROA procedure can be the most general and powerful approach within decision tree structures, which assist decision makers to conceptualize and visualize adaptation planning (Borison and Hamm 2008). The most recent example of the decision tree approach is Woodward et al. (2014), who applied ROA to the flood mitigation planning for the Thames Estuary of the United Kingdom and concluded that the optimal adaptive strategies based on ROA and the multi-objective genetic algorithm dominate deterministic and rigid strategies. The backward-moving dynamic programming is often combined with this decision tree ROA approach to facilitate its heavy computations. Pol et al. (2014) introduced a probability distribution of water level increase in a dike investment decision problem, solved it within the dynamic programming framework, and concluded that the uncertainty information provides more impact on dike heightening strategy than future learning.

In summary, ROA is a well-proven economic analysis theory, which is flexible enough to adjust investment timing as well as strategy by using options. The decision trees typically generated by ROA are informative for decision makers to conceptualize their adaptive planning procedure. Past ROA studies for water resources planning indicates that ROA is most relevant to large, capital-intensive infrastructure investments, often comparing traditional investing-now versus waiting options (Borison and Hamm 2008). In spite of these advantages, ROA is resource intensive and requires well-identified probabilities and economic data, which is often not feasible in practice.

8.5 Dynamic Adaptive Policy Pathways

8.5.1 Description

Coined by Haasnoot et al. (2013), DAPP offers another DM methodology for different climate change scenarios. In particular, DAPP is based on an adaptive approach to capture flexibility over time, which implies one can choose near-term action keeping options open for the future (Haasnoot et al. 2013). The long terminology results from a combination of two precedent methodologies called "adaptive policy making (Walker et al. 2001)" and "adaptation pathway" (Haasnoot et al. 2011, 2012), but the meanings of both "dynamic" and "adaptive" seem identical as Walker et al. (2013) classified DAPP as a "dynamic" robust methodology. Haasnoot et al. (2013) and Walker et al. (2013) further clarified that adaptive policy making was rooted in the "assumption-based planning" (Dewar et al. 1993) methodology, and the "sell-by date" concept in DAPP was adopted from the "adaptation tipping point" (Kwadijk et al. 2010) methodology.

In spite of the combined history, the essence of DAPP is to create a so-called adaptation pathways map that enables decision makers to transfer to a different planning pathway (i.e., a different action or alternative) in order to reflect changing conditions. As shown in Fig. 8.7, the DAPP adaptation pathways map consists of solid and dashed lines, circles, and vertical bars that represent "action effective in all scenarios," "action not effective in scenario U," "transfer station to new action," and "adaptation tipping point of an action," respectively. It should be noted that scenarios here imply different projections of climate and hydrology that are usually formed as an ensemble. As x- and y-axes denote time and actions, respectively, each pathway represents a unique decision action over time. It is possible to transfer from one action to another action only at a tipping point. This pathway concept of DAPP

Fig. 8.7 Example of DAPP adaptation pathways map (Haasnoot et al. 2013)

has been adopted from the adaptation pathways methodology (Haasnoot et al. 2013; Walker et al. 2013). It should be noted that a tree created with ROA described in Section 8.4 provides a few "option" pathways for a given alternative, but the DAPP map provides various "action (i.e., alternative)" pathways. Therefore, each different option pathway in the ROA tree should be considered as a unique action pathway in the DAPP map.

8.5.2 Key Steps

DAPP provides a systematic procedure that consists of ten steps, including the step for developing the adaptation pathways map, as shown in Fig. 8.8. In general, the first half of the procedure follows the adaptation pathway procedure, while most of the second half follows adaptive policy making. In particular, the concepts of contingency actions and triggers as well as iterative approaches employed in Steps 3, 4, and 5 and also between Steps 9 and 10 have been adopted from adaptive policy making, while the use of transient scenarios and the concept of pathways map have been adopted from adaptation pathway.

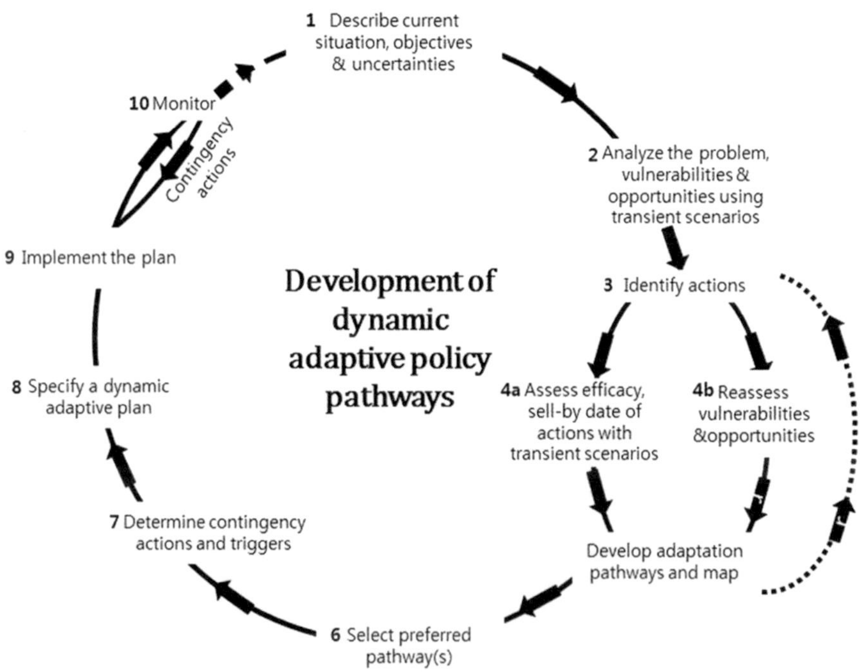

Fig. 8.8 DAPP procedure (Haasnoot et al. 2013)

Detailed tasks at each step are well described in Haasnoot et al. (2013) that originally proposed DAPP. This section summarizes the key outputs from individual steps and then aims at describing how these outputs are related and produced in the following paragraphs. Among the various outputs, the key outputs of DAPP are the definition of "success" (Step 1), the action to be listed in the y-axis of the DAPP map as well as the pathway candidates (Step 3), the sell-by date of each action (Step 4), the "promising" pathways and their scorecard (Step 5), the "preferred" pathways (Step 6), the contingency actions and triggers (Step 7), the initial adaptive plan (Step 8), and the continuous monitoring result (Steps 9 and 10).

The success defined in Step 1 is usually expressed with quantitative thresholds such as the allowable number of failures (e.g., how many times of water shortage can be allowed?), the allowable length of failure mode (e.g., how long can the water shortage be allowed to last?), and the magnitude of allowable failure (e.g., how much of water shortage can be allowed as a maximum?). The success is used for determining the sell-by date in Step 4, which is defined as the year in which a particular action no longer meets the objective (i.e., does exceed the predefined success threshold). As the sell-by date depends on the choice of a particular scenario, it is expressed with a box plot as shown in Fig. 8.9. Most DAPP studies employed the median or the average values from box-whisker plots. However, the application of these values may be problematic when the range of a box is large because the median or the average sell-by dates may not represent the reality.

Various action candidates corresponding to predefined objectives can be developed in Step 3, ranging from structural (e.g., raising dam heights, building new facilities, and dredging channel bottoms), semi-structural (e.g., rainwater harvest-

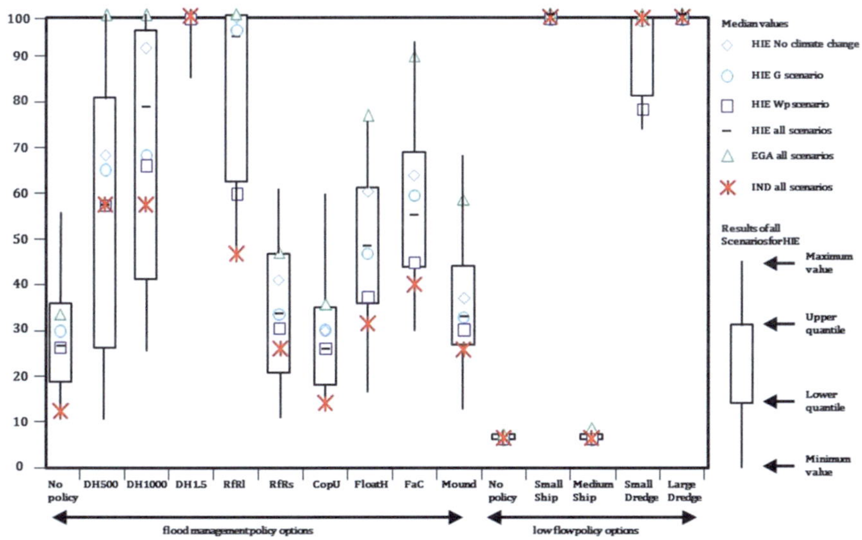

Fig. 8.9 Example of sell-by dates (Haasnoot et al. 2012)

ing, expanding wetland areas, and changing to porous pavement), and nonstructural measures (e.g., improving dam operations, reducing water demand through education, and installing an early warning system). On the contrary, the pathway candidates in Step 3 represent all the combinations of action sequences evolving over time. They should be narrowed down to "promising" pathways (Haasnoot et al. 2013) through an iterative reassessment procedure between Steps 3, 4, and 5. Some actions cannot be combined into a promising pathway because they may be mutually exclusive or their orders may be illogical. One can carry out this screening and refinement procedure together with stakeholders (Haasnoot et al. 2013). At the end of Step 5, a scorecard, which evaluates individual pathways with a few criteria that represent benefit and cost, is generated through quantitative and qualitative vulnerability assessments. As multiple scenarios exist, the score for each pathway should be calculated as an expected value.

As mentioned before, from Step 6, DAPP follows the "adaptive policy making" procedure, which is somewhat principle oriented rather than providing a specific guideline for each step. Based on the scorecard created in Step 5, a couple of "preferred" pathways are recommended in Step 6. One can apply the robustness concept in this step to reflect the deep uncertainty of climate change. In other words, a pathway may be preferred, which is expected to perform satisfactorily overall performance criteria and over the entire range of scenarios. Haasnoot et al. (2012, 2013) emphasized that the definition of robustness and the resulting preference depends on societal perspectives and tested DAPP with a hypothetical example with three perspectives such as hierarchist, egalitarian, and individualist. Haasnoot et al. (2013) and Walker et al. (2013) pointed out that the selected preferred pathways should be further refined by designing anticipated contingency plans with signposts and triggers (i.e., thresholds of signpost in Step 7). The initial adaptive plan is then finally determined, implemented, and monitored after this step, but a specific guideline, especially for choosing an initial adaptive plan from the preferred pathways, has not been provided in the current version of DAPP.

8.5.3 Discussion

The applications of DAPP are currently very limited because DAPP has been structured very recently. Although the adaptive policy making and adaptation pathway approaches have been applied to various cases (Walker et al. 2013), DAPP has been tested only for a few hypothetical examples of water management for the Rhine delta of the Netherlands. Therefore, for the water sector at least, exploring more real-world case studies with DAPP is expected to validate its efficacy in practice and improve its detailed procedures and schemes. As pointed out by Haasnoot et al. (2012), determining robust pathways for real cases may not be straightforward because of interrelated pathways and multiple targets that were not dealt with in simplified, hypothetical case studies.

Haasnoot et al. (2012) and Walker et al. (2013) also emphasized that computational supporting tools are important for DAPP because a large number of runs are generally required to handle an ensemble of scenarios, several combinations of actions and pathways with different performance criteria, and iterative loops. This is usually true for other DMCCA approaches, especially for RDM described in Section 8.3. For DAPP applications, exploratory modeling and analysis (EMA), a methodology that uses computational experiments to explore various types of uncertainties related to climate and hydrology scenarios (i.e., natural uncertainty), model structures and parameters (i.e., modeling uncertainty), and perspectives (i.e., societal uncertainty) has been used as a computational supporting tool (Walker et al. 2013), although the details are not covered in this section. See an online support for EMA (http://simulation.tbm.tudelft.nl/ema-workbench/contents.html) in addition to Haasnoot et al. (2012) and Walker et al. (2013) for details.

It should be noted that concepts of adaptive pathways and sell-by date (or the adaptive tipping point) are not completely new in DAPP, For example, Reeder and Ranger (2011) applied the "route-map approach" to the Thames Estuary 2100 project, which was a pioneering real case study for CCA planning. Figure 8.10 shows the route map developed for the Thames Estuary 2100 project that is very similar to Fig. 8.7 of DAPP, but actions and pathways are depicted as functions of threshold level (i.e., maximum water level rise) on the x-axis. The role of these threshold levels is the same as that of the sell-by dates or the adaptive tipping points of DAPP, sometimes called "adaptive turning points" (Werners et al. 2013) as well, at which the actions can be switched. Setting an appropriate threshold level is

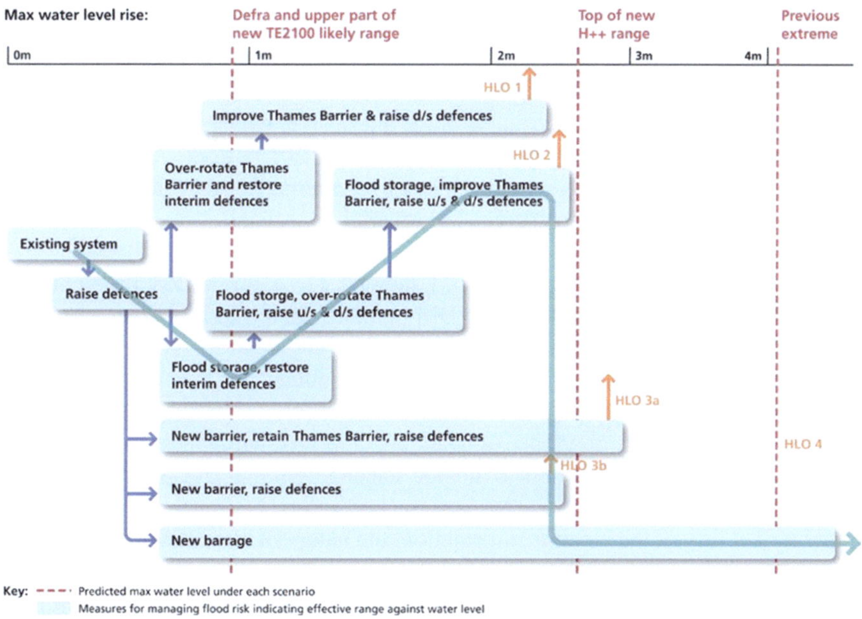

Fig. 8.10 Example of the route-map approach: the Thames Estuary 2100 project (Reeder and Ranger 2011)

based not only on an engineering analysis but also on social consensus building between stakeholders. Therefore, the DAPP and the route-map approaches have another advantage of promoting public involvement in their procedures.

In summary, DAPP provides the most general DM tool, which covers various alternative (or action) adjustments, switches, and combinations, especially when compared to RDM and ROA. A large degree of generality in ROA makes its applicability higher while its implementation difficult. When incorporated with the policy-first and decision scaling approaches (Sect. 8.3), the DAPP and the route-map approaches can include another flexibility of cooperating with stakeholders. Currently, however, the applications of DAPP are limited; thus, more applications that can explore its potential further and reduce its computational burden are strongly desired in the near future.

8.6 Closing Remarks

As seen in previous sections, RDM, ROA, or DAPP seem very promising when a DM problem is encountered in water resources planning under different climate change scenarios. In addition to these three novel and advanced DM methods, classical DM theories (e.g., CBA, CEA, MCA) are still good candidates if reasonable assumptions are made and/or when some proper modifications are made to capture the nonstationary characteristics and uncertainties of climate change. In addition to these methods, there are several other methods for DM problems under climate change that were not introduced in this chapter because only a few methods have been tested in the field of water resources. In some cases, using multiple alternatives can be more effective than choosing a single alternative, which is a core concept of "portfolio analysis" (WUCA 2010; Hunt and Watkiss 2013). Rooted in the financial market theory, portfolio analysis spreads risk over alternatives that are similar to the above novel DM methods that spread risk over either scenarios or time (Hunt and Watkiss 2013).

To choose the most appropriate DM method for a given DM problem, the strengths and weaknesses of candidate DM methods should be analyzed, which are well summarized in Watkiss and Hunt (2013). The choice of method also depends on the characteristics of the given problem and resources available to the decision maker. WUCA (2010) emphasized that decision makers should consider how to deal with probabilities, importance of quantitative results, willingness to invest time, money, external help, willingness to include stakeholders, and how to use outcomes. More recently, Watkiss and Hunt (2013) compared several DM methods with respect to the availability of the benefits information, the nature of climate information, the relevant time periods for adaptation, and the requirement of resources and experts. According to their analysis, MCA can work with qualitative benefit and climate information while CBA and ROA work with the monetary performance term and scenario probabilities; RDM and ROA are suited for long-term planning but are highly resource intensive.

As mentioned before, the novel DM methods are resource intensive mainly because climate change introduces additional sources of uncertainty. In other words, more experts, cost, time, and other types of efforts are generally required for CCA than for other conventional planning procedures. Therefore, these resource constraints can be a primary barrier to the success of advanced DM methods because reality usually suffers from a lack of resource security. In particular, continuous efforts on the public involvement procedure should be emphasized in the climate change era as stakeholders are usually reluctant to work with a complex and resource-consuming procedure like the above advanced DM methods. One should keep in mind that DM is not only an action to be implemented but also a whole cycle of processes.

Acknowledgment This research was supported by Basic Science Research Program through the National Research Foundation (NRF) funded by the Ministry of Science, ICT, and Future Planning (NRF-2013R1A1A2073677) and also supported by a grant from Advanced Water Management Research Program funded by the Ministry of Land, Infrastructure and Transport (14WMP-B082564-01) of Korea. Special thanks goes to Professor Joern Birkmann of University of Stuttgart and LG Yonam Foundation who supported the sabbatical leave of the first author when a major part of this chapter was written.

References

Adger WN (2006) Vulnerability. Glob Environ Chang 16(3):268–281

Adger WN et al (2005) Successful adaption to climate change across scales. Glob Environ Chang 15(2):77–86. doi:10.1016/j.gloenvcha.2004.12.005

Alliance W.U.C. (WUCA) (2010) Incorporating climate change uncertainties into water planning, decision support planning methods, San Francisco, CA, January

Birkmann J (2013) Measuring vulnerability to natural hazards: towards disaster resilient societies. In: Birkmann J (ed) Measuring vulnerability to promote disaster-resilient societies: conceptional frameworks and definitions. United Nations University, New York, pp 9–54

Bloom EW (2014) Changing midstream: providing decision support for adaptive strategies using robust decision making: applications in the Colorado River basin. Pardee Rand Graduate School, Rand Corporation, PhD Dissertation

Borison A, Hamm G (2008) Real options and urban water resource planning in Australia. Water Services Association of Australia (WSAA), April

Brown C (2013) Decision-scaling for robust planning and policy under climate uncertainty. Expert Perspectives Series Written for the World Resources Report, 2011

Brown C, Baroang KM (2011) Risk assessment, risk management, and communication. Methods for climate variability and change. Treatise Water Sci Elsevier 1:189–199

Brown C, Wilby RL (2012) An alternate approach to assessing climate risks. Eos 93(41):401–402. doi:10.1029/2012EO410001

Carlsson C, Fullér R (1996) Fuzzy multiple criteria decision making: recent developments. Fuzzy Sets Syst 78(2):139–153

Chung ES, Kim Y (2014) Development of fuzzy multi-criteria approach to prioritize locations of treated wastewater use considering climate change scenarios. J Environ Manag 146:505–516

Chung ES, Lee KS (2009) Prioritization of water management for sustainability using hydrologic simulation model and multicriteria decision making techniques. J Environ Manag 90(3): 1502–1511

Dessai S (2005) Robust adaptation decisions amid climate change uncertainties. PhD. University of East Anglia, Norwich
Dessai S, Hulme M (2004) Does climate adaptation policy need probabilities? Clim Policy 4(2):107–128
Dessai S, Hulme M (2007) Assessing the robustness of adaptation decisions to climate change uncertainties: a case study on water resources management in the East of England. Glob Environ Chang 17(1):59–72
Dessai S et al (2009) Climate prediction: a limit to adaptation. Adapting to climate change: thresholds, values, governance, pp 64–78
Dewar JA et al (1993) Assumption-based planning: a planning tool for very uncertain times. RAND CORP MR-114-A, Santa Monica
Dodgson JS et al (2009) Multi-criteria analysis: a manual. Department for Communities and Local Government, London
Ducey MJ, Larson BC (1999) A fuzzy set approach to the problem of sustainability. For Ecol Manag 115(1):29–40
Giles J (2002) Scientific uncertainty: when doubt is a sure thing. Nature 418(6897):476–478
Greco S et al (1999) The use of rough sets and fuzzy sets in MCDM. Multicriteria decision making. Springer US, Boston, pp 397–455
Grove D et al (2013) Adapting to a changing Colorado river: making future water deliveries more reliable through robust management strategies. AGU Fall Meeting Abstracts, vol 1, p 1407
Haasnoot M et al (2011) A method to develop sustainable water management strategies for an uncertain future. Sustain Develop 19(6):369–381. doi:10.1002/sd.438
Haasnoot M et al (2012) Exploring pathways for sustainable water management in river deltas in a changing environment. Clim Chang 115(3–4):795–819
Haasnoot M et al (2013) Dynamic adaptive policy pathways: a method for crafting robust decisions for a deeply uncertain world. Glob Environ Chang 23(2):485–498. doi:10.1016/j.gloenvcha.2012.12.006
Hall JW et al (2012) Robust climate policies under uncertainty: a comparison of robust decision making and info-gap methods. Risk Anal 32(10):1657–1672
Hallegatte S (2009) Strategies to adapt to an uncertain climate change. Glob Environ Chang 19(2):240–247. doi:10.1016/j.gloenvcha.2008.12.003
Hallegatte S et al (2012) Investment decision making under deep uncertainty-application to climate change. PRWP, 6193
Hunt A, Watkiss P (2013) Portfolio analysis: decision support methods for adaptation. MEDIATION Project, Briefing Note 5. Funded by the EC's 7FWP
Hurwicz L (1945) The theory of economic behavior. Am Econ Rev 35(5):909–925
Hyde KM et al (2005) A distance-based uncertainty analysis approach to multi-criteria decision analysis for water resource decision making. J Environ Manag 77(4):278–290
IPCC (2001) Climate change 2001: impacts, adaptation and vulnerability. Summary for Policy Makers, World Meteorological Organisation, Geneva
IPCC (2005) Guidance notes for lead authors of the IPCC fourth assessment report on addressing uncertainties. Cambridge University Press available via, Cambridge, UK, http://www.ipcc.ch/pdf/assessment-report/ar4/wg1/ar4uncertaintyguidancenote.pdf
Jeon SM (2013) A new approach of prioritizing the location for climate change adaptation strategy using fuzzy multi-criteria decision making. Seoul National University of Science and Technology, Master Thesis
Jeuland M, Whittington D (2013) Water resources planning under climate change: a "real options" application to investment planning in the Blue Nile (No. dp-13-05-efd)
Johnson TE, Weaver CP (2008) A framework for assessing climate change impacts on water and watershed systems. Environ Manag 43(1):118–134
Kang BS et al (2013) A sensitivity analysis approach of multi-attribute decision making technique to rank flood mitigation projects. KSCE J Civ Eng 17(6):1529–1539
Kelly PM, Adger WN (2000) Theory and practice in assessing vulnerability to climate change and facilitating adaptation. Clim Chang 47(4):325–352

Kim Y, Chung ES (2014) An index-based robust decision making framework for watershed management in a changing climate. Sci Total Environ 473:88–102
Kim Y, Chung ES (2015) Robust prioritization of climate change adaptation strategies using the VIKOR method with objective weights., JAWRA J Am Water Resour Assoc
Kim Y et al (2015) Iterative framework for robust reclaimed wastewater allocation in a changing environment using multi-criteria decision making. Water Resour Manag 29(2):295–311
Kunreuther H et al (2014) Integrated risk and uncertainty assessment of climate change response policies. Climate change 2014. Mitigation of climate change. Contribution of Working Group III to the Fifth Assessment Report of the Intergovernmental Panel on Climate Change, 2: 151–205
Kwadijk JCJ et al (2010) Using adaptation tipping points to prepare for climate change and sea level rise: a case study in the Netherlands. WIREs Clim Change 1(5):729–740
Lau KT et al (2006) The potential use of the Black-Scholes model in urban drainage risk management. Doctoral dissertation, MSc thesis, Department of Civil and Environmental Engineering, Imperial College of Science, Technology and Medicine
Lempert RJ, Grove DG (2010) Identifying and evaluating robust adaptive policy responses to climate change for water management agencies in the American west. Technol Forecast Soc Chang 77(6):960–974
Lempert RJ, Schlesinger ME (2000) Robust strategies for abating climate change. Clim Chang 45(3):387–401
Lempert RJ et al (2003) Shaping the next one hundred years: new methods for quantitative, long-term policy analysis. Tech. Rep RAND CORP MR-1626-RPC
Lempert RJ et al (2004) Characterizing climate-change uncertainties for decision-makers. Clim Chang 65(1):1–9
Lempert RJ et al (2006) A general, analytic method for generating robust strategies and narrative scenarios. Manag Sci 52(4):514–528
Lempert RJ et al (2013) Ensuring robust flood risk management in Ho Chi Minh City. PRWP, 6465
Lesnikovski AC et al (2015) How are we adapting to climate change? A global assessment. Mitig Adapt Strat Glob Chang 20(2):277–293. doi:10.1007/s11027-013-9491-x
Liu KF (2007) Evaluating environmental sustainability: an integration of multiple-criteria decision-making and fuzzy logic. Environ Manag 39(5):721–736
Loomes G, Sugden R (1982) Regret theory: an alternative theory of rational choice under uncertainty. Econ J :805–824
Loucks et al (2005) Water resources systems planning and management: an introduction to methods, models and applications. UNESCO, Paris
McMahon TA (2007) Review of Gould–Dincer reservoir storage–yield–reliability estimates. Adv Water Resour 30(9):1873–1882
McMahon TA (2011) A simple methodology for estimating mean and variability of annual runoff and reservoir yield under present and future climates. J Hydrometeorol 12(1):135–146
Metlay D (2000) From tin roof to torn wet blanket: predicting and observing groundwater movement at a proposed nuclear waste site. In: Sarewitz D (ed) Prediction: science, decision making, and the future of nature. Island Press, Covelo, CA, pp 199–208
Michailidis A, Mattas K (2007) Using real options theory to irrigation dam investment analysis: an application of binomial option pricing model. Water Resour Manag 21(10):1717–1733
Moss RH, Schneider SH (2000) Uncertainties in the IPCC TAR—recommendations to lead authors for more consistent assessment and reporting, In: Pachauri R, Taniguchi T, Tanaka K (eds) Guidance papers on the cross cutting issues of the third assessment report of the IPCC. World Meteorological Organization, Geneva, pp 33–51
Myers SC (1977) Determinants of corporate borrowing. J Financ Econ 5(2):147–175. doi:10.1016/0304-405X(77)90015-0
Neufville R (2003) Real options dealing with uncertainty in systems planning and design. Integrat Ass 4(1):26–34. doi:10.1076/iaij.4.1.26.16461
NPCC (2010) Climate change adaptation In: Rosenzweig C, Solecki W (eds) New York City: building a risk management response. Ann NY Acad Sci 1196:1–354

NRC (2010) Adapting to the impacts of climate change. Washington, DC. doi: 10.17226/12783
Orlove B (ed) (2009) Adapting to climate change: thresholds, values, governance. Cambridge University Press, Cambridge, pp 131–180
Pelling M (2011) Adaptation to climate change: from resilience to transformation. Routledge, London
Philles YA, Andriantiatsaholiniaina LA (2011) Sustainability: an ill-defined concept and its assessment using fuzzy logic. Ecol Econ 37(3):435–456
Porthin M et al (2013) Multi-criteria decision analysis in adaptation decision-making: a flood case study in Finland. Reg Environ Chang 13(6):1171–1180
Preston BL et al (2011) Climate adaptation planning in practice: an evaluation of adaptation plans from three developed nations. Mitig Adapt Strat Glob Chang 16(4):407438. doi:10.1007/s11027-010-9270-x
Ranger N et al (UKCIP) (2010) Adaptation in the UK: a decision making process. Environment Agency
Reeder T, Ranger N (2011) How do you adapt in an uncertain world?: lessons from the Thames Estuary 2100 project. WRI, Washington, DC
Ryu Y (2014) Application of real options valuation to water resources planning considering climate change uncertainty. Seoul National University, Master Thesis
Sanneh ES et al (2014) Prioritization of climate change adaptation approaches in the Gambia. Mitig Adapt Strat Glob Chang 19(8):1163–1178
Sarewitz D, Pielke RA Jr (2000) Breaking the global-warming gridlock. Atl Mon 286(1):55–64
Savage LJ (1954) The foundations of statistics, 2nd edn. Wiley, New York
Schneider SH (2001) What is 'dangerous' climate change? Nature 411(6833):17–19
Smit B et al (2000) An anatomy of adaptation to climate change and variability. Clim Chang 45(1):223–251
Song JY, Chung ES (2016) Robustness, uncertainty and sensitivity analyses of the TOPSIS method for quantitative climate change vulnerability: A case study of flood damage. Water Resour Manag. doi:10.1007/s11269-016-1451-2
Stainforth DA et al (2007) Confidence, uncertainty and decision-support relevance in climate predictions. Philos Trans R Soc A 365(1857):2145–2161
Steinschneider S, Brown C (2012) Dynamic reservoir management with real-option risk hedging as a robust adaptation to nonstationary climate. Water Resour Res. 48(5). doi: 10.1029/2011WR011540
Stevenson WJ, Ozgur C (2007) Introduction to management science with spreadsheets and student CD. McGraw-Hill Inc, New York
Tompkins EL et al (2010) Observed adaptation to climate change. UK evidence of transition to a well-adapting society. Glob Environ Chang 20(4):627–635. doi:10.1016/j.gloenvcha.2010.05.001
Turner SWD et al (2013) Linking climate projections to performance: a yield-based decision scaling assessment of a large urban water resources system. Water Resour Res 50(4): 3553–3567
Turskis Z, Zavadskas EK (2011) Multiple criteria decision making (MCDM) methods in economics: an overview. Technol Econ Develop Econ 2:397–427
UNFCCC (United Nations Framework Convention on Climate Change) (2002) Annotated guidelines for the preparation of national adaptation programs of action. United Nations, New York
Vadrevu KP et al (2010) Fire risk evaluation using multicriteria analysis – a case study. Environ Monit Assess 166(1–4):223–239
van der Pol TD et al (2014) Optimal dike investments under uncertainty and learning about increasing water levels. J Flood Risk Manag 7(4):308–318
Wald A (1949) Statistical decision functions. Ann Math Stat 20(2):165–205
Walker WE et al (2001) Adaptive policies, policy analysis, and policy-making. Eur J Oper Res 128(2):282–289. doi:10.1016/S0377-2217(00)00071-0
Walker WE et al (2013) Adapt or perish: a review of planning approaches for adaptation under deep uncertainty. Sustainability 5(3):955–979. doi:10.3390/su5030955

Watkiss P, Hunt A (2013) Method overview: decision support methods for adaptation, Briefing Note 1. Summary of Methods and Case Study Examples from the MEDIATION Project. Funded by the EC's 7FWP

Watkiss P et al (2013) Real options analysis: decision support methods for adaptation, MEDIATION Project, Briefing Note 4. Funded by the EC's 7FWP

Weaver CP et al (2013) Improving the contribution of climate model information to decision making: the value and demands of robust decision frameworks. Wiley Interdisciplinary Reviews. Clim Chang 4(1):39–60

Werners et al (2013) Adaptation turning points: decision support methods for adaptation, MEDIATION Project, Briefing Note 9. Funded by the EC's 7FWP

Wheeler SM (2008) State and municipal climate change plans: the first generation. J Am Plan Assoc 74(4):481–496. doi:10.1080/01944360802377973

Willows R, Connell R (2003) Climate adaptation: risk, uncertainty and decision-making. UKCIP Technical Report. UK Climate Impacts Programme

Woodward M et al (2014) Adaptive flood risk management under climate change uncertainty using real options and optimization. Risk Anal 34(1):75–92

Chapter 9
Adaptation to Climate Change: Green Development

Elpida Kolokytha

Abstract Mitigation and adaptation are both responses to climate change. This chapter provides the reader with information related to various policy measures for climate change, in water management. A theoretical part introduces issues concerning water resources management and sustainable water use. The chapter is mainly devoted to green development as a new development model for the twenty-first century that can provide adaptation to climate change and can guarantee economic prosperity, environmental protection of natural resources, and social equity. The main characteristics of green development are analyzed. Methods and applications of green development in water management are discussed as a solution to the impact of climate change on water. EU adaptation policies, in the context of green development, are discussed, and a case study from Greece demonstrates nonstructural measures and combined methodologies used to promote sustainable water management.

Keywords Adaptation policy • Green development • Water resources management • Climate change • EU adaptation

9.1 Introduction

Water is the most indispensable resource significantly affected by climate changes. Water, as an essential environmental resource, but also as a vital element of life that supports a variety of ecosystem services, plays an equally significant role in almost every economic activity and needs to be properly sustained. Natural climate variability and human-induced climate change have posed severe threats to natural resources. Extensive research has brought to light that the impacts of unrelenting climate change are both on the water cycle (T. Huntington 2006;

E. Kolokytha (✉)
School of Civil Engineering, Division of Hydraulics and Environmental Engineering, Aristotle University of Thessaloniki, 54124 Thessaloniki, Greece
e-mail: lpcol@civil.auth.gr

E. Kolokytha et al. (eds.), *Sustainable Water Resources Planning and Management Under Climate Change*, DOI 10.1007/978-981-10-2051-3_9

Trenberth 1999; Held and Soden 2000; Arnell et al. 2001) and on the water supply (Bates et al. 2008). There is a broad scientific consensus that immediate action must be taken. Other effects of climate change in water resources include air and water temperature increases, intensification of extreme events, sea level rise, and changes in coastal/ocean characteristics. The complexity of the different events creates nonlinear behavior in the entire system (Schneider and Lane 2006; Hansen et al. 2008) in ways that are unexpected and potentially irrevocable and need new knowledge and approaches to tackle.

Climatic drivers and, specifically, the uneven spatial and temporal distribution of precipitation and the extent of the rainfall (extreme events) are leading to a remarkable temporal and spatial variability in water resources worldwide. Periods of intense rainfall characterized by more runoff and less infiltration combined with increased evapotranspiration are expected to lead to groundwater depletion.

Temperature increases may have an effect on water quality and freshwater ecosystems in the form of changes in the aquatic species composition. Reduction in soil moisture, changes in land management, alterations in vegetation cover, all, are results of changes in temperature and precipitation which put the availability of water supplies in peril. The amount of freshwater availability is a function of multiple physical variables such as runoff, water quality, groundwater recharge, but also technical interventions such as water infrastructure. Changes in hydrology may not always have negative effects. Runoff, for example, may provoke erosion and severe quality problems as contaminants may be carried out by increased runoff, but at the same time, it may increase water resources availability. Water availability is likely to be further hampered by poor management, inefficient infrastructure, and overuse.

Other, non-climatic drivers, such as massive population growth of last century and also its tendency to rise near nine billion by 2050, have resulted in excessive water use. Technological innovations and consumption habits particularly in the developed countries will further increase water demand. The major increases, though, in water demand, will be in the emerging economies and developing countries (OECD 2012).

All the factors mentioned above are liable to affect key economic activities and development with a significant contribution to the quality of life, such as agricultural production and productivity, urban development, industry, tourism, and energy sector. The anticipated impact of these factors, on the food production and access to clean water together with the every so often occurrence of extremes events, could prompt uncontrolled population migrations, putting additional pressure on receiving countries. *"The number of refugees worldwide from environmental causes reached in 2014 its highest level since World War II,"* according to the official Global Risk website. In the Global Risk Report of 2015, issues such as water crises and extreme events related to water are ranked third regarding importance among global risks (Global risks 2015).

9.1.1 Water Resources Management and Climate Change

Sustainable management of water resources is currently challenging water managers throughout the world. Many of the issues related to water management today stem from the following factors:

- Changing priorities of water and environmental management goals over time
- The need for multidisciplinary inputs
- Uncertainties regarding future demand and supply
- Lack of adequate understanding of procedures that affect and are affected by the management of water and the environment
- The way institutions work
- Contemporary lifestyles and how water uses are perceived, especially in the "affluent West"

Engineering design and practice involves predicting and depicting future conditions with sufficient accuracy that the consequences of design choices can be evaluated. Hydrologists use frequency and regression analysis for the planning and operations of almost all water resources projects. The implicit assumption in all such analyses "is that the hydrologic record of the past is the best guide for the future" (Hirsch 2010, http://www.cwi.colostate.edu/NonstationarityWorkshop/SpeakerInfo/Hirsch_Abstract)

In statistical terms, the system is stationary. Climate change due to human-driven changes in the global atmosphere may not be subject to this rule (Kundzewicz et al. 2007; Milly et al. 2008). Whether stationarity exists or not is still open to debate (Lins and Cohn 2011; Koutsoyiannis 2011). If stationarity is "dead," new theories and tools need to develop to assist water management to take decisions.

A prominent example is the operation of current dams that had been designed under certain hydrological conditions that cannot efficiently respond to the spatial and temporal oscillations in water flows due to climate change. Increasing uncertainties reduce the predictability of the boundary conditions under which water management has to carry out.

It is not yet possible to estimate accurately climate uncertainties based on current modeling status, since their evolution is almost "a black box" for water planning and management needs (Stakhiv and Stewart 2010). Thus redesign of new infrastructure, customizing and adapting to new hydrological conditions, is an imperative need.

Integrated water resources management (IWRM) traditionally addresses challenges by incorporating spatial, temporal, economic, political, environmental, as well as administrative aspects of sustainability. It is a comprehensive methodology which, by the implementation of policies, laws, regulations, applied science and engineering, as well as "best management practices" in a consistent manner, promotes efficiency, effectiveness, and equity for all types of resources and all sectors.

Efficient management of risks related to climate change calls for an integrated approach linking technological, social, and economic development with the protection of natural ecosystems and with realistic projections of future climatic conditions. It is true, though, that significant progress had been made in the technological and engineering field in this sense. The major problem in the policy-making is how to establish the enabling environment, how to provide the means and resources, and how to integrate them into different sectors that economic prosperity, environmental integrity, and social coherence can be accomplished. The fundamental concern is not the exact amount of critical parameters such as rainfall and evaporation but rather the variability of these parameters from year to year or their seasonal modifications that should be incorporated in the decision-making process of water management. The real decision rules are determined by what society needs. A variety of parameters, other than detailed hydrologic information, are also crucial for decision-making, among them, economic priorities, cost–benefit criteria, legislations, personal views, etc. The great difficulty is encountered mainly with the existing institutional and regulatory framework and especially in the implementation of policy aspects, and not so much with hydrologic uncertainty. The problem is particularly intense in developing countries (Stakhiv 2010).

9.1.2 Mitigation and Adaptation Measures

Two types of response to climate change, mitigation and adaptation, play an equally important role in the transformation toward sustainability: on the one hand, economically effective measures for the reduction of GHG emissions (mitigation) (i.e., replacement of conventional fossil fuels such as coal with low-carbon power such as natural gas or even better renewable energy sources) and, on the other, planned as well as proactive planning and prediction of the impacts of climate change on water resources among which, i.e., using water conservation devices, promoting redistribution of crops with emphasis to set aside water-intensive ones, (adaptation) constitute the set of measures toward climate change.

The more successful the mitigation efforts are in cutting emissions, the less extensive the need for adaptation will be since mitigation reduces the root cause of the climate change problem. However, even if mitigation measures applied succeeded to limit and then reduce GHG emissions, the impact of climate change will challenge the planet for at least the next 50 years which provides the imperative need for adaptation measures in any case. "Limiting temperature increases to 2 °C requires as much as a 50 % reduction in global gas emissions by 2050" (OECD 2011c).

The objective of adaptation is to reduce the risk posed by the impact of climate change. Different strategies of adaptation exist. Climate change adaptation strategies are divided into planned adaptation (Füssel 2007) and autonomous adaptation (Wilk and Wittgren 2009). Planned adaptation explicitly takes climate change and variability into account, while autonomous adaptation (Malik et al. 2010) refers to measures that are not specifically climate related but are used to improve resilience to climate change as an additional effect. Also, critical is the issue of adaptation decisions seen as private (farmer, business, etc.) or public (government, joint agency). A limit to adaptation is a significant discussion that apart from technical and economic constraints entails issues of ethics, culture, and perception of risk and of course the diversity of goals of adaptation that complicates attempts to define limits. How people perceive issues of "risk," "need," and "habit" heavily depends on personal values that are different for various stakeholders in the society.

Depending on the primary purpose of the measures, the range of these measures can become quite extensive. Measures vary depending on the cause they need to confront (e.g., dealing with water scarcity, flood management, etc.) and also depending on different types of intervention that can be used with reference to legislative, administrative, financial, and other policies (Nõges et al. 2010).

While there is no single recipe for a successful climate adaptation policy, the OECD Green Growth Strategy (OECD 2011e) and earlier OECD work (OECD 2009a, 2011a; Duval 2008) summarize the following combination of key elements:

- National climate change strategies, as roadmaps for adaptation.
- Price-based instruments, for example, taxes on CO_2 emissions and subsidies for emission-reducing activities.
- Command and control instruments and regulations, for example, the Resource Conservation and Recovery Act to reduce environmental pressure, the "polluter pays principle," etc.
- Technology-based policies, including R&D and public research. Three types of technological instruments exist and should better work complementarily, hardware, software, and org-ware. Hardware refers to physical tools (reservoirs, rainwater harvesting equipment, etc.). Software refers to the processes, knowledge, and skills required in using the technology (like water recycling techniques), and organizational technologies, or org-ware, apply to the ownership and institutional arrangements about a technology (i.e., water pricing specifications) (Christiansen et al. 2011; UNFCCC 2014).
- Public awareness campaigns and information tools. Well-designed information-based instruments, such as energy/water efficiency labels on household appliances, combined with market-based and regulatory tools, constitute an effective approach (OECD 2007a, b, 2011d).

All the above set of tools, however, could not be successful if applied without taken into account how people and society perceive adaptation and to what extent they are ready to adapt to changes that are in line with broad cultural and ethical values. To gain consensus to implement adaptation measures, social acceptability and education are essential. A significant constraint on adaptation is funding, especially for underdeveloped countries that lack financial resources for adaptation. Also political will is important to choose the right adaptation measures fit into the overall context.

9.1.3 *Synergies Between Climate Change Adaptation and Mitigation (A + M)*

Indeed, there exist complex linkages between adaptation and mitigation. Some potential water management adaptation measures (e.g., desalination, pumping of deep groundwater, or water treatment) are very energy intensive, and their implementation would increase greenhouse gas emissions (Mata and Budhooram 2007). Afforestation can also contribute to climate change mitigation through carbon sequestration, which may affect, in a positive or negative way, biodiversity and ecosystem services. In general mitigation policies may reduce the impacts and consequently the need for adaptation to climate change. However, there are examples of mitigation measures (bioenergy) which may constrain adaptation options. Interactions between mitigation actions and adaptation to climate change need to be carefully examined and incorporated in any final adopted plan. Many authors have suggested that a more holistic approach to mitigation and adaptation would be more effective and efficient (Ayers and Huq 2008; Klein et al. 2007) and reduce tradeoffs between the two (Kane and Shogren 2000). Moser (2012) advocated for such a holistic approach stating that "the overlap of M + A demands a long-term, life-cycle, and systems perspective." Integration in a collaborative way and balance between mitigation and adaptation measures would provide a safe path toward the alleviation of climate impacts.

Planning processes that take place at higher levels that may include national laws, policies, and strategies, as well as financial means and measures can constitute an acceptable framework for "active cooperation" on mitigation and adaptation measures. Moreover institutional arrangements together with operational programs, projects, and initiatives throughout the country can enhance this effort (Lalisa et al. 2014).

9.2 Adaptation Strategies in the Context of Development

International evidence demonstrates how water scarcity (too little water), floods (too much water), and the lack of basic water and sanitation services and infrastructures coupled with poor water quality can hamper economic growth, lead to environmental degradation, and sharpen social disparities (OECD 2011a, http://www.un.org/waterforlifedecade/green_economy_2011/pdf/). The increase of the global temperature above 2 °C may provoke irreversible changes to the planet. Therefore, it is crucial to maintain the level of increase in global temperature below 2 °C. The global cost of adapting to climate change to a level lower than 2 °C from 2010 to 2050 has been estimated to be $75–100 billion each year (Margulis et al. 2010).

The concept of economic development also entails the notion of "water security." The twofold nature of water as both being a catastrophic and productive power is critical to understanding how investments in water security can support economic development. Reducing the negative economic costs associated with scarcity, flooding, and pollution, and taking advantage of the economic benefits of using water productively, can shove long-term economic development (Quick and Winpenny 2014). In this sense, natural disasters (floods, fires, etc.) should be imported into national capital accounts as a loss, while the protection and restoration of natural disasters, as well as investments in the natural environment, should be introduced as gains in governmental accounting.

According to Shipper and Lisa (2007), there are two approaches linking development and adaptation. In the first, adaptation contributes to development as it promotes the reduction of vulnerability and associated risks and takes climate change into consideration for development planning by improving the resilience of the system. In the second approach, development leads to better adaptation. Bearing in mind that vulnerability is related not only to climatic drivers but to other reasons as well, the most appropriate way is to integrate vulnerability reduction into general development policy.

There are cases where adaptation concerns conservation of status quo while others where the current situation is undesirable and hence adaptation measures are about progress. One can easily understand what is said above, considering the differences that occur between developed and developing countries. Significant disparities can be seen concerning infrastructure, capacity, economy, and sociopolitical implications in both cases.

In developed countries, high level of infrastructure combined with well-coordinated high-quality data and information bases can decrease vulnerability to natural disasters and reduce the adaptation measures, whereas in developing countries, inadequate or fragile infrastructure, low ethos of infrastructure maintenance, and lack of data and databases lead to the need for primary measures (most of them mitigation rather than adaptation measures) to deal with development and uncertainty of climate change impacts.

Speaking of capacity, limited administrative and scientific skills, and centralized systems, the absence of technological advances and lack of financial resources put a hindrance on developing countries in the implementation of adaptation measures. Moreover, high dependence on land and agricultural production and also external help and short term of planning perspective also intensify the already challenging environment in the developing countries.

Moreover, the different sociopolitical conditions, with high population growth, heavy environmental degradation, poorly informed public, centralized decision-making, and the imperative need for supervision in the underdeveloped countries, show the difficulties in adopting those measures.

The interdependency of water with other key domains (energy, food, and ecosystems) stipulates the "nexus" between them. It is important to stress that regional and sectoral conditions determine, at a significant level, the policy options.

9.2.1 Green Economy

Sustainable development is the alternative path between two fundamentally different models that had prevailed in the twentieth century: the "technocratic" one, according to which economic development is the major objective, while social and environmental aspects of development are of secondary importance having, consequently, a complementary role. Natural resources are the means for economic development, and science and technology can restore any apparent damage to the environment. Thus, the protection and restoration of natural systems, according to this model, depends solely on cost and economic feasibility, leading in many cases to the abandonment of the damaged systems, as extremely costly. This "supply-oriented management" practice has led to overexploitation and depletion of natural resources. On the other part, the "deep" ecological approach stated that economic development, through the overexploitation of natural resources, was the main responsible for the ecological disaster of the planet. The contribution of science and technology in human progress was considered as part of the environmental problem of destruction and ecological imbalance and not at all a condition for halting it. Basic in ecological criticism was the ongoing systematic pursuit of economic profit as the driving force of the economy.

In between those two models, sustainable development, as a novel guiding principle, had introduced the notion of "carrying capacity." Sustainable development is a balanced development, among the three pillars, namely, the economy, the environment, and the society, acting in an integrated way and not on a separate basis as up today. Sustainable development supports development that is compatible with the limitations imposed by nature. It introduces the carrying capacity of ecosystems as the "red line" for the management of the environment and stimulates the adjustment of economic development into nature's adaptive capacity. Sustainable development exemplifies the reorientation of economic activities toward resource conservation and protection. This "limited" "under condition" development, however, is incapable to effectively respond to current conditions of economic crisis, poverty, unemployment, and recession, which ask for "more development."

As it is set, in the twenty-first century, a twofold crisis is witnessed based on just one reason: an economic crisis and in the same time an environmental one, both caused by the dominant aggressive economic development model. The current foremost development model, based on the unbounded use of resources to meet demands, has directly resulted in overexploitation and overconsumption. Neither has it solved the problem of poverty (social sustainability) nor the problem of healthy ecosystem services (ecological sustainability). In fact, the current development model born in the era of globalization of the economy is based on the abolition of borders for the circulation of goods, capital, and labor, on the inequality of wealth contribution – in the hands of "the few" strong lobbies – and on the immense socioeconomic differences. This new context of economy and development did not answer to the problem of poverty alleviation (MDG goal) nor to that of preserving and enhancing the environment. On the contrary, it has caused severe deterioration and exhaustion of the global ecosystem, undermining hence not only the ability of future generations to meet their needs but posing an immediate risk even to today's survival of the inhabitants of our global system. In fact, the well-being of the 25 % of the world's population costs the depletion of the 70 % of global natural capital. And this means that if we decide to continue the same economic development policy, three planets like earth would be needed to sustain this development.

Undeniably, all multiple crises, namely, financial, economic, climate, energy, ecosystems, and demography of the twenty-first century, compel us to think of a radical transformation of the economy. The transition to sustainability calls for the shift to green economy–green development. The green economy appears as the follow-up to sustainable development. Clearly the principles are the same. The way to achieve them may differ.

9.2.1.1 Selected Definitions of Green Economy and Green Growth

European Environment Agency: "A green economy is one in which policies and innovations enable society to use resources efficiently, enhancing human well–being in an inclusive manner, while maintaining the natural systems that sustain us" (http://www.eea.europa.eu/themes/economy/about-green-economy, EEA 2012).

OECD: "Green growth means fostering economic growth and development while ensuring that natural assets continue to provide the resources and environmental services on which our well–being relies. To do this, it must catalyze investment and innovation that will underpin sustained growth and give rise to new economic opportunities" (http://www.oecd.org/greengrowth/48012345.pdf, 2011).

UNDP: "[...] new growth poles that can potentially contribute to economic recovery, good job creation, and reduced threats of food, water, energy, ecosystem and climate crises, which have disproportionate impacts on the poor" (http://content.undp.org/go/newsroom/2009/june/green-economy-a-transformation-toaddress-multiple-crises.en).

UNEP: "[...] a green economy is one that results in improved human well–being and social equity while significantly reducing environmental risks and ecological scarcities. In its simplest expression, a green economy can be thought of as one which is low-carbon, resource efficient and socially inclusive. Critical to attaining such an objective is to create the conditions for public and private investments to incorporate broader environmental and social criteria. A green economy is one whose growth in income and employment is driven by public and private investments that reduce carbon emissions and pollution, enhance energy and resource efficiency, and prevent the loss of biodiversity and ecosystem services" (www.unep.org/greeneconomy).

World Bank: "[...] green growth — that is, growth that is efficient in its use of natural resources, clean in that it minimizes pollution and environmental impacts, and resilient in that it accounts for natural hazards and the role of environmental management and natural capital in preventing physical disasters. And this growth needs to be inclusive" (http://siteresources.worldbank.org/EXTSDNET/Resources/Inclusive_Green_Growth_May_2012.pdf).

Green economy can, indeed, introduce a sustainable economic development model, where environmental aspects are no longer constraints, but rather become incentives for development and, therefore, should be incorporated into the development goals and objectives.

Whatever the definition of green development is, three objectives underline the green economy approach: improving *resource efficiency*, ensuring *ecosystem resilience*, and enhancing *social equity*.

The transition toward a "new development model" requires investment in the sustainability of ecosystem services upon which the world depends on and actions so that the environment can continue to be used for the benefit of current and future generations. The environment is the fundamental bottom line, as without a safe and stable environment we can have no economy or society. The sustainable use of natural resources and the environment has to be the core concept of economic growth, while the perception of endless resource exploitation should be replaced by resource efficiency. Crisis can turn into an opportunity, regarding fostering a new development model on the environment and with the quest of also investing in cultural and moral values for the pursuit of prosperity and people's well-being.

The new development model should rely on economic activities based on the comparative advantages of each country. This happens because the only way to balance economic, environmental, and social aspects of the development would be to base the economy upon activities that are fully compatible with the carrying capacity of the ecosystem. Hence, development should rely on products and activities fully compatible with the local conditions, such as the climate, the environment, the tradition, and the human resources. Instead of promoting, in the case of Greece, for example, the construction of water-consuming golf courses in Greek islands, which face severe water scarcity problems, Greece should invest in its Mediterranean diet and local agricultural products, such as olive oil, wine, grapes, and fruits, in its cultural heritage, biodiversity, and natural beauty. This is after all the principal of demand management, according to which demand should be adjusted to the limits of the carrying capacity of the systems. According to the premise of water sustainability, water demand should be in line with the real potential of water resources.

Moreover, the environmental objectives can no longer be a barrier but an incentive for development. The new development model needs to incorporate the reduction of consumption regarding demands, especially in the developed West, redirecting consumption toward more efficient paths and redistributing resources to the "poor" nations.

The twin challenges of resource efficiency and ecological resilience lie at the heart of achieving a green economy (SOER2015).

A transition to sustainability demands serious changes in the way people perceive the boundaries of the earth. A critical change is needed in "habits," "wants," and "needs" in the face of a diverse world (W. Adams, and S. Jeanrenaud, 2008).

The green economy is a total conversion of the current socioeconomic paradigm into a new development model, accepting environmental and social values equal to economic values according to UNEP (2011a). In economic terms, it involves the integration of social and environmental "externalities" into the market system, with all its consequences. The green economy will need to be interpreted and applied by national governments as a suite of policy measures selected and designed by national priorities and circumstances (UNDESA 2012).

Social issues such as sustainable consumption, equitable access to natural resources and ecosystem services, green jobs, and fair distribution of taxes are predominant to achieving a green economy.

The main characteristics of green economy are:

(a) Decoupling natural resource use and environmental impacts from economic growth

Decoupling is about "shifting from debt-financed consumption as the primary economic driver of our economies which proved to be unsustainable, to sustainability-oriented investments in innovation as the leading economic driver of our economies" (UNEP 2011b). This process of decoupling requires concentration both on resource efficiency innovations and environmental sustainability boundaries, expressed by the concept of "environmental flow" for securing the ecosystems' good status (EEA 2012).

Tools for achieving decoupling:

The application of environmental policy such as environmental certificates of use, Life Cycle Assessment (LCA), and Environmental Impact Assessment (EIA)
The introduction of "clean" technologies for the production of environmentally friendly products, for the reduction of the gas emissions
Public environmental awareness by establishing responsibility for users in case of environmental damage (polluter not only pays principle but also restores)

(b) Conservation through resource efficiency technologies that can enable users in different sectors to reduce water use and/or promote sustainable water use

Tools in this direction:

The positioning of public opinion through education and information campaigns
The encouragement of reuse and recycle

Also thanks to the use of improved and more efficient technologies, lower inputs of material and energy result to lower output of pollutants and higher environmental protection. Examples of such technology are drip irrigation systems, bioclimatic buildings, electric cars, etc.

A major drawback, though, is the "rebound effect." Efficiency gains in resource use may lead to more use of other resources (indirect rebound) or more use of the same resource (direct rebound). Efficiency improvements might bring about outputs cheaper. Some illustrative examples of "rebound effect" are the use of energy-saving devices (for heating) which can be left open several hours because they consume small amounts of energy. In the case of water, the spread of drip irrigation and the consequent saving in water/unit may end up to the expansion of irrigated land, putting in danger the groundwater reserves.

(c) *Creation of employment* (green jobs) through projects and environmentally friendly activities. Green jobs also refer to implementing and evaluating sustainable business practices, to promote job opportunities for environmental services and also to provide education on how to turn "green" in business. Sectors, where green jobs are possible, are sustainable agriculture, ecotourism, water treatment and management, environmental law, and "clean" energy. Specifically, in the water sector, green jobs are those that save water and energy while alleviating pressure on water infrastructure. Some examples include plumbers, water auditors, and low-flush water equipment installation.
(d) *Interference of the state* to secure economic development, environmental protection, and social equity. Economic instruments, such as taxes, subsidies, and tax exemptions, are used together with legislative instruments, namely, environmental quality standards (min allowable limits), and financial penalties.

In addition to these policies are the combination of capacity, the dissemination of good policy practice, and general education and awareness to make sure that green measures are well designed, implemented, enforced, and understood, without causing unintended impacts or being prevented by practical or political challenges (UNEP 2011).

Opposing green growth, there is a growing interest in the idea of "degrowth" (décroissance). "Degrowth" is an ethical concept of how the world needs to change. "A society of infinite growth is impossible in a finite world" Latouche, S. (2004). The degrowth concept relates the social and ecological questions of distribution, with the reduction of consumption and production to secure social rights for everyone.

The degrowth movement refers mainly to the reduction of overconsumption of energy and materials and advocates that economic degrowth is a necessary step in the transition to a more sustainable society. It connects micro-practices with macroeconomic concepts and a solidarity-based economy of "commons." For degrowth economy, the fundamental criterion is of sharing resources and not their individual possession. The idea of a just degrowth economy invests in everlasting ethical values of equity, solidarity, and freedom beyond the dominance of the existing ruling economic system. The current lifestyle can provide bigger houses but smaller families, more conveniences but less time to enjoy, more degrees but less understanding, more knowledge but less judgment, and significant progress in new medications but less "health."

9.2.2 Sustainable Use of Water in a Green Economy

The transition to a green economy requires new thinking about water and its management. Water needs to be seen as an integral part of the ecosystem, and as such, solutions should revolve around maintenance of ecosystem services, reallocation among sectors, true cost price, and public engagement in all stages of decision-making. Using water efficiently, making it available to all at a reasonable cost, and ensuring environmental sustainability of ecosystems consist the "puzzle" of sustainable water management.

In fact, sustainable water use requires the management of the water balance, also known as water inventory, in a river basin. Water use should rely solely on the actual consumption of available renewable water resources to serve water needs for various purposes. Sustainable water management should integrate all sectors to guarantee that all relevant water uses are satisfied within the sustainability limits of the river basin. The application of the fundamental principles of sustainable water management guarantees the sustainability of the resource. IWRM can act as a tool for implementing sustainable water use by incorporating the various aspects of all uses in an integrated and not fragmented way, by implementing demand management and accurate pricing policies, and by enabling the active participation of all interested users (stakeholders). Most significant differences in IWRM goals between the developed and the developing countries are summarized in Table 9.1.

It should be put emphasis on these differences because they are critical for the stimulation of the green economy since there exist major priority differences between countries and even between river basins. The role of water in the green economy is implicitly argued in UNEP, GWP, WWC, UN-Water, and other international organizations and forums.

There are certain measures applied in the water sector to favor both ecological and social sustainability. The most common types of measures are (UN-Water 2011):

- Economic instruments such as green taxes, green accounts, water markets, and pollution rights
- Innovative technology such as drip irrigation in agriculture, waste reuse in water treatment, and conservation technologies for urban use and tourism such as low-water-pressure devices
- Financing of water infrastructure in the form of reform of charges for water-related services and public–private partnerships

Table 9.1 Differences in IWRM goals between the developed and the developing countries

Developed countries	Developing countries
Coherent planning	Poverty reduction
Resource use efficiency/bulk private sector investments	Access to clean water/water supply and sanitation basic infrastructure
Flood control/irrigation	Flood control/irrigation/drainage
Watershed protection and management	Waterborne diseases
Hazard risk reduction plans	Rural development plans
Participatory planning	Water user associations for operation and management
Advanced technologies	"Proper" technologies

- Improved water resources planning with the emphasis on water utilization and protection, policy integration and institutional and legal frameworks, and water governance across all economic sectors

There is also ample criticism on whether green economy truly enhances sustainable water use and management. Those opposing the idea of this development model argue that too much emphasis is given on the role of the private sector and too little on the responsibilities of the state to supervise the sustainable use of natural resources (Die 2012).

Three primary drivers are behind water investments: expanding the water supply, increasing water supply efficiency, and improving water quality. All entail innovative companies specializing in water technologies that could represent some of the world's biggest capital growth opportunities. "The global market for environmental services and products" *is expected to triple by 1.37 trillion $/year to 2.47 trillion $ until 2020.* Primary sectors involved are energy, water, transport, and wastewater (Allianz Global Investors 2013). Today the water industry is considered as the third most profitable business after oil and electricity (Global Water Intelligence 2013).

Another risk reported is that of imposed prices for water, which may threaten food security, obstruct access to water supply and sanitation, and increase poverty (Die 2012). Privatization of water for municipal use is one indicative example that has led to unbearable increases in the water prices (200 % up in Bolivia).

The critical interactions that concern the water-energy-food nexus should be carefully analyzed, conflict of objectives must be discussed, and priorities should be set. Many of the measures proposed are in question concerning their effectiveness. In arid regions, for example, freshwater is being produced by the desalination of seawater using fossil energy sources. The replacement of conventional sources by renewable ones, such as solar radiation, abundantly found in many dry countries, could convert this negative example into a positive one if renewable solar energy is used (Die 2011).

As water is central both to the quest for sustainability and to economy, it will also be central to innovative solutions for the "greening" of the economy. Organic farming and ecotourism are examples of economic activities that harmonize perfectly with the spirit of a "green" water policy. The application of any adaptation measures, however, will not be achieved with supply management options such as large structural projects, (dams and river diversions), but rather with the right combination of a series of nonstructural measures and policies that ensure the integration of environmental considerations into economic goals.

Sectors like agriculture, water supply/treatment infrastructure, tourism, energy, and industry are interrelated to water efficiency practices. The list of policy responses and the methodologies used are not exhaustive, but the objective is to get an idea of possible methods and applications in the water sector about green development and green growth.

Among the requirements for a successful policy reform are the acquisition of relevant data and meticulous study and analysis for evaluating policies. Methods may be quantitative (cost–benefit analysis, integrated assessment modeling, and multi-criteria analysis, DPSIR, LFA) or qualitative (surveys and participatory approaches) (IPCC2014).

New metrics is available, such as ecological, carbon, and water footprint (WF), which is easy to communicate and hand over the urgency of the global problem, raising public awareness. In the case of the water problem, the water footprint can successfully calculate the direct and indirect water use. However, no single best method but rather a combination of structural and nonstructural methods can provide a comprehensive decision policy. One should carefully consider the overall context, the boundary conditions, and different complexities to avoid poor policy decisions.

9.2.3 Applications

9.2.3.1 Organic Farming as a Form of "Green" Agriculture

Organic farming is the process of producing food naturally. This method excludes the use of synthetic chemical fertilizers and modified organisms to facilitate the growing of crops, being an environmentally friendly approach.

The accreditation of a product that comes from "organic farming" is provided on a certain label. When it comes to selecting food products, certification is needed, to prevent the incorrect usage of the term "organic." Certification procedures help to ensure that the relevant standards regarding production and processing have been met.

Advantages of this method are that the food produced in this manner is considered to:

- Be of higher quality.
- Have higher nutritional value as opposed to producing food using modern, industrial practices.
- Contain no chemicals, artificial fertilizers, or pesticides.
- Contain no genetically engineered or altered substances or organisms.
- Minimize soil erosion.

Organic farming uses less water than conventional farming and replenishes the soil with vital nutrients. Also, the reliance on natural pesticides allows an organic farm to use wasps arriving at certain seasons during the year to eliminate other pests.

A farm in Jordan performed a study, finding that the water in the underground aquifer beneath the farm could irrigate 400 dunums of land, but through the use of water saving techniques, the farm uses the same level of water for irrigating 2,500 dunums (Luck, T. 2010). Concurrent measures for supporting organic farming in Jordan from the Jordan River Foundation were:

The establishment of a legislative framework for organic farming, the examination of already existing organic farming sites, the increasing awareness and knowledge of organic farming through workshops, and the supporting organic farming in both the public sector and NGOs (JRF 2007)

Despite the growing world demand, organic farming has not taken off in Jordan and has remained a niche product mainly due to a lack of appropriate policy incentives and institutional support, training on innovative agricultural practices and awareness. More information can be found in UNEP (2011), "Towards a Green Economy in Jordan."

9.2.3.2 Trading of Water Rights

In Australia, in the Murray–Darling Basin, which is under severe scarcity conditions, an expanding market for the trading of water use rights has enabled water to be allocated efficiently among users, generating significant economic gains (http://www.un.org/waterforlifedecade/green_economy_2011/pdf/resume_day_1.pdf). The Australian government introduced the National Water Initiative, under which water trading allows scarce water resources to be transferred to their most productive uses. Over a decade, a progressive water reform has provided the framework for the country's market-based approach, illustrating the lengthy process of setting up the infrastructure necessary for a market. Two critical success factors were the decoupling of water rights from land rights and making water rights proportional shares of available resources rather than fixed volumes. Environmental sustainability is ensured through the purchasing of water rights for the environment (UN 2011).

9.2.3.3 Karnataka Watershed (Sujala) Project, India

This example is but one of the many programs financed by the World Bank, on watershed management and poverty alleviation in rain-fed areas of India. A system approach, with a focus on soil and water conservation and sustainable resource use, combined with participatory planning and participation to improve local livelihoods, gender equity, and community capacity, was performed in India. Monitoring and evaluation were an essential facet of the program. The Indian Space Research Organization (Antrix) conducted the work. A set of different technological tools, namely, remote-sensing data with on-the-ground monitoring techniques combined with a household survey with baseline and control group, participatory observations, regional studies, and other case studies, were used. Indicators to measure quantitative and qualitative issues, before, during, and at the end of the project, were studied. The work also included a complete database, with reliable data and timely information useful to monitor the progress of the project (financial and physical) all levels. In the end, a broad range of data together with a set of reports were given to program managers and beneficiaries (World Bank 2013).

9.3 Adaptation Policies

Adaptation needs to be structured across sectors, in multilevel and interregional activities bringing together stakeholders with different levels of expertise, interests, and values (Grothmann 2011; Lebel et al. 2010). Adaptation policies can be classified (Table 9.2) according to various types of services.

In defining adaptation policies, it is extremely useful to identify priority measures. In doing that, it is important to explore which adaptation measures are of the highest priority and which complementary actions should follow. Successful adaptation should take into account not only the impacts of climate change but also how other non-climatic (socioeconomic, political, etc.) stresses affect the system and take measures for both cases.

Local conditions and local stakeholders are the most relevant to choose the analytical tools and approaches of critical economic, environmental, and social drivers. The interactions between the different sectors, the scale, and speed of change required as well as the actions that need to be taken over time, along with uncertainties, should shape the process of adaptation. These measures will confront high vulnerability and uncertainty. Problems are encountered because most of the time the government decides the adaptation measures, but local stakeholders need to implement them.

The results of the vulnerability assessment are the basis of the measures chosen. Of great importance is the involvement of stakeholders because any ranking of adaptation measures will involve those that affect and are being affected by those actions. Active engagement of stakeholders will provide accurate feedback of needs and the level of acceptance.

Table 9.2 Adaptation policies according to various types of adaptation

Adaptation			
Purpose	Planned (refers to a deliberate action to respond to change)	Autonomous (by sectors, by regions)	
Time component	Proactive (before the change to minimize the effects)	Reactive (after the change to adjust to new conditions)	
Space component	Local	National	International
Motive/interest	Public (government and all relevant public agencies)	Private (individuals, business, corporations)	
Type of measures (responses)	Structural (infrastructure)	Legal (laws, regulations)	Financial (taxes, subsidies)
		Institutional (establishment of relevant bodies to guide response to climate change)	Technological
Temporal scope	Short term (direct effect)	Long term (future impacts)	

To achieve short-term "win–win" strategy and also to support long-term changes, the application of a combination of policy instruments is crucial. Short-term changes will increase the strength of existing water systems, while long-term changes will deal with plans of high uncertainty. Measures may include price signals for water conservation, regulations, and standards that sustain changes in various practices and information and education programs to build public awareness.

9.3.1 Stakeholder Process

Decisions in water resources management are uncertain, and rational arguments arise within interested parties because the members of each party understand a problem from different viewpoints, even when they have similar interests in achieving a goal. To gain consensus, the participation of stakeholders in the decision-making is central especially in the implementation stage (Gardner at al. 2009).

"The term "stakeholder" in climate change studies refers to policy makers, scientists, administrators, communities, and managers in the economic sectors most at risk" (Engaging Stakeholders in the Adaptation Process. http://www4.unfccc.int/nap/Country%20Documents/General*)*. In this context, stakeholders can be brought together from both public and private sectors to develop a shared understanding of the issues and to create adaptations.

Benefits of Participation

- Better informed decisions: Stakeholders can express and exchange different views and find common ground in discussions that may lead to better-informed decisions.
- Transparency and wide representation in a democratic way in decisions: By involving people who are affected by the decision, a broader agreement can be sought, which will potentially increase support for implementation.
- Cooperative action, the capacity to work together and to create mutual trust.

Stakeholder analysis improves the understanding of the economic, social, and political impact of adaptation measures on interested groups, namely, authorities from local and regional administrations and decision power links. The engagement of various stakeholders requires a uniform awareness of climate change impacts for attaining consensus for implementation of the chosen adaptation policy.

True participatory processes across sectors and scales are meaningful when stakeholders at different levels of action can express their opinions, share their knowledge, and elaborate in real decision-making. Top-down strategies are no more the solution to the problem. Adaptation strategies will only work if they fit local conditions. Stakeholder processes include questionnaires, small group discussions, roundtables, application of CVMs, etc.

Basic keywords for stakeholder and participatory processes:

Stakeholders – Who are they? Why are they essential for consultation? What is their status of influence in decision-making?
Priority issues – Of high risk, high vulnerability, high sensitivity to climate changes
Techniques – Which techniques and methods will be most effective in communicating with the different stakeholders?
Documentation – How will the results of the process be used?

9.3.2 Vulnerability Assessment

Several aspects related to water management compile the vulnerability issue. Environmental degradation, lack of ecosystem resilience, economic inequalities, insufficient funding, lack of civil resistance, inadequate social protection, lack of preparedness, and not enough training are among the most common reasons for the high vulnerability of water resources management in the era of climate change. Underdeveloped regions are traditionally more vulnerable to climate change since their level of adaptation is relatively low.

The vulnerability of a given system is related to its physical exposure to climate change effects and its adaptive capacity to these conditions (Allen 2005). Systems that have a lower potential impact from changes in climate and climate variability, and those that have a higher adaptive capacity, are considered less vulnerable to climate change and vice versa.

A common methodology to assess vulnerability is by using DPSIR framework. The main drivers lead to pressure parameters that determine the state of the system, the potential impact of changes, and, of course, the response in the sense of adaptation measures. Relevant indicators measure the sensitivity of the system and the exposure to rank the system regarding high or low vulnerability to complex risks.

9.3.3 Review of Institutional and Regulatory Framework

There are almost 804 laws and policies according to Global Legislative 2015 from 99 countries that represent the 93 % of GHG emissions.

"A framework law is defined as a law, or regulation with equivalent status, which serves as a comprehensive, unifying basis for climate change policy, addressing multiple aspects and issues of climate change mitigation or adaptation (or both) in a holistic, overarching manner" (Global Legislative 2015, http://www.lse.ac.uk/GranthamInstitute/legislation/the-global-climate-legislation).

To effectively implement adaptation policies, we need to have strong institutions (national ministries, regional management entities, local authorities, etc.) which can adjust to changes needed to reinforce their adaptive capacity. Institutions are responsible for a variety of actions that are related to climate change. They allocate water; control infrastructure, such as dams and networks; and apply laws and policies (for droughts, floods), so they should be resilient and adaptive. Institutions can orient behaviors and constrain policy-making since they also influence political decisions. Of course at the same time, they can promote policy implementation of adaptation measures in many ways.

A good idea would be to establish new independent sectoral bodies dedicated to work on adaptation to climate change. These bodies may consult institutions how to apply necessary adjustments in all levels efficiently.

Regulatory adaptation measures to cope with climate change mainly consist of reviewing existing legislation and regulations to address climate impacts. For example, to deal with water scarcity problems, abstraction limits or restrictions on water uses are imposed. Also updating regulations on land use and standards for urban planning may address flood risk.

9.4 EU Adaptation Policies

9.4.1 EU Policies to Confront Climate Change

The European Union has a long history of environmental policies and has produced a considerable number of legal and guidance documents, forming a robust integrated water resources management system. Although integration is a central notion in EU, "EU policy mitigation and adaptation policies are likely to remain separate endeavors" (Responses-Frans Berkhout et al. (2013)).

Specifically, The European Union's mitigation strategy, known as "20-20-20 Energy and Climate Package," adopted in January 2008, is an example of a comprehensive and legally binding climate strategy with three different objectives:

1. Reach a reduction of GHG emissions by at least 20 % compared to 1990 by 2020, with a commitment to increase it to 30 % with a satisfactory international agreement.
2. 20% of energy is coming from renewable sources by 2020, supplemented by 10% of renewable transport fuel.
3. Reduce the European Union's energy consumption by 20% compared to the baseline in 2020.

On 23 October 2014, the European Council (2014) endorsed a binding common target for the 28 member states of the European Union of "at least 40% domestic reduction in greenhouse gas emissions by 2030 compared to 1990."

By defining a clear, long-term goal to limit the increase of global temperature to 2°C or better to 1.5°C by the decrease of greenhouse gas emissions and zero overall global emissions in the second half of this century, representatives of 196 countries, in Paris, (COP21), have set a path to decouple the prosperity and development from fossil fuel use. Maybe the EU needs to redefine its mitigation goals, accordingly.

Sectoral policies were enhanced with complementary adaptation measures. Sectors that are mainly affected by climate change and discussed below are water, agriculture, and biodiversity.

A set of key policy documents, related to water, namely, the WFD, Floods Directive, EU Adaptation, White Paper, and adaptation to climate change in water management, followed by relevant regulatory and economic policy instruments, such as Nitrates Directive, CAP, Urban Waste Water Directive, Natura 2000, Integrated Coastal Zone Management, etc., work complementarily as a roadmap to adaptation. R&D programs (FP7, FP8, etc., Joint Research Center) and financial mechanisms for funding (LIFE, EU Cohesion Funds, etc.) also support this effort (OEDC 2013). In detail:

- The EU's Biodiversity Strategy (COM (2010) 244 final) aims at pausing the loss of biodiversity and the degradation of ecosystem services by 2020 and restoring them.
- The White Paper "Adapting to climate change: Towards a European framework for action" (COM (2009) 147 final) sets out the EU framework for adaptation to climate change, including objectives and actions.
- The Soil Thematic Strategy (COM(2006) 231) and the proposed Soil Framework Directive (COM(2006) 232 final) have as the main objective the protection and sustainable use of soil resources.

For successful adaptation, cross-sectoral coordination needs to be further developed.

9.4.2 The EU Water Framework Directive

The Water Framework Directive (WFD) is a legal act (Directive 2000/60/EC) based on the principles of IWRM. The overall system provided by the WFD is based on the central concept of integration. Integration is a fundamental notion for all water resources, environmental and ecological objectives, water uses, functions and values, interdisciplinary analyses and expertise, and different decision-making levels within a common policy framework. The ultimate goal is "the good ecological status" for water bodies. Physical rather than administrative boundaries in the sense of management at river basin level represent a significant innovation (Kolokytha E. 2011).

The EU WFD through the 3Ps signifies an important step toward the sustainable use of water resources in Europe:

- Planning and integrated management
- Pricing and actual cost recovery
- Participation and improved decision-making

It is a principal legislative instrument that indirectly promotes climate adaptation policies through its step-by-step procedure of the different planning periods of implementation. The analysis of identified environmental pressures (Art. 5) in conjunction with the climate impact sensitivity of the programs of measures incorporated in the river basin management plans is the first step. The program of measures refers to a set of operations, including legal, control, and administrative initiatives, contained in the river basin management plans, contributing to the implementation of the WFD. Each measure may include various actions or initiatives that can be used to mitigate the effect of pressures on water caused by the different sectors.

The set of measures should be effective, sustainable, and cost-efficient to respond to changing conditions. In the second planning cycle, climate change impacts should be directly taken into account. A variety of complementary directives, addressing sectoral differences (CAP for the agricultural sector, energy directive in energy, etc.) and also specific impacts of climate change (Flood & Drought Directive, Biodiversity Directive), are present to cover the issue of climate change.

Opportunities given to the WFD river basin management planning for developing climate change adaptation policies have provided the reasoning for a guidance document, under the Common Implementation Strategy (CIS) (EC 2009). This guidance document sets out a framework to reduce EU's vulnerability to the impact of climate change.

9.4.3 The Common Agricultural Policy (CAP)

Agriculture in Europe accounts for around 33 % of total water use and is the largest source of nutrient pollution in water" (EEA 2012).

The vital role water plays in agricultural activities and the need for water protection render the necessity to look for synergies in modern agriculture and water policies. CAP, an important financial instrument, has been through a history of 50 years of reforms, trying to incorporate all global challenges, to sustain agriculture in Europe (Fig. 9.1). "Green" agriculture requires physical capital assets, investments in research, and capacity building to enhance soil fertility to achieve sustainable water use, crop and livestock diversification, and appropriate farm level automatization.

9.4.3.1 Typology of Terms in CAP

GAEC standards: The obligation to maintain land in good agricultural and environmental condition refers to a range of standards related to soil protection, maintenance of soil organic matter and structure, avoiding the deterioration of habitats, and water management.

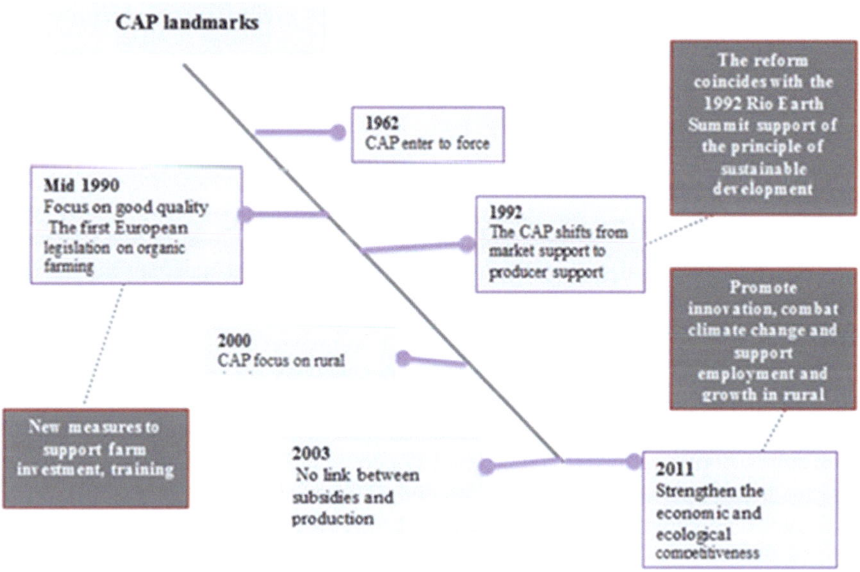

Fig. 9.1 CAP landmarks

Cross–compliance: A mechanism that ties direct payments to farmers and some rural development payments to compliance with a series of rules relating to the environment, food safety, etc., to maintain GAEC.

Direct payments: Payments granted directly to farmers under an income support scheme.

The need to address the challenges of climate change has been recognized, particularly in its rural development policy (the European Agricultural Fund for Rural Development – EAFRD). Currently, two CAP instruments (cross–compliance and the European Agricultural Fund for Rural Development) are used to encouraging good farming practices that are compatible with the protection of the environment and in line with related environmental legislation (EU Nitrate Directive http://eur-lex.europa.eu/legal-content/EN/TXT/?uri=CELEX:31991L0676).

In 2013, in fact, the CAP had introduced the main three objectives of viable food production, sustainable use of natural resources, and climate action. Territorial development challenges presented by climate change are signaled as an important priority for the CAP, in keeping with the priorities of the EU 2020 strategy (http://ec.europa.eu/agriculture/cap-overview/2012_en.pdf).

In the CAP, aid is provided at rural development measures promoting environmentally sustainable farming practices, like agri-environment schemes. Also a reduction in support payments is promoted for those farmers who don't respect environmental laws (http://ec.europa.eu/agriculture/envir/index_en.htm).

The most relevant EU policies to tackle extremes events (floods and droughts) are:

9.4.4 The EU Floods Directive (Directive 2007/60/EC)

The Floods Directive (2007/60/EC) is included in a larger "Flood Action Program" for the assessment and management of flood risks aimed at reducing the adverse consequences for human health, the environment, and cultural and economic activity associated with floods in Europe.

The Flood Directive's goal is to reduce and manage the risk from floods. The directive requires member states to perform a preliminary assessment by 2011 to detect the river/coastal basins which may have the risk of probable flooding. For such zones, flood risk maps should be developed by 2013, while by 2015, flood risk management plans should be made focused on prevention and preparedness. The directive is in line with the Water Framework Directive.

9.4.5 The EU Water Scarcity and Drought Strategy (COM (2007) 414)

The EU Water Scarcity and Drought Strategy (COM (2007) 414) aims at the identification of the extent of water scarcity and droughts in Europe and functions as a roadmap for the assessment of existing and potential selected measures for tackling water scarcity and droughts.

To confront the risk from extreme events, also CAP, through specific cross-compliance regulations (tillage practices to hold moisture in the soil), is indirectly linked.

9.4.6 The Biodiversity Strategy (COM (2011) 244)

In 2011, the EU adopted an ambitious strategy setting out measures and investments (six targets and twenty actions) to pause the loss of biodiversity and ecosystem services in the EU by 2020.

The synthesis of the review of the strategy for water scarcity and droughts, the analysis of the implementation of the Water Framework Directive, and the review of the vulnerability of environmental resources produced "The Blueprint to Safeguard Europe's Water Resources" (COM(2012)673 which is consistent with Europe 2020.

9.5 From Theory to Practice

9.5.1 A Case Study from Greece

An application of nonstructural adaptation measures to cope with climate change impacts in agriculture is performed. The water footprint is a new metrics that can easily communicate issues of water scarcity and water pollution to a broad audience (raising awareness) to stress these water problems. The new EU Common Agricultural Policy (2014–2020) is an important adaptation policy for the "greening" of agriculture. These two adaptation instruments are tested in Mygdonia river basin, Central Macedonia, Greece. The water balance of the Mygdonia basin is constantly negative, indicating that water management in the area is being unsustainably performed. The WF assessment can alter the change of behavior of the main stakeholders, the farmers, and urges the agricultural community to take appropriate measures to cope with water scarcity and water pollution problems in the area. Also by taking into consideration the provisions of the new CAP, the formulation of two possible scenarios provides viable solutions for the sustainable agriculture in the area.

9.5.1.1 The Area Under Study

The Mygdonia basin covers a total area of 2.026 km^2 and comprises two subbasins, namely, Koronia and Volvi lakes. The economic development of Mygdonia basin depends primarily on agricultural activity while cattle breeding is also well developed. The water-consuming crops, the excessive use of fertilizers, and the wasteful irrigation systems have led to the almost vanishing of Lake Koronia and significant widespread water pollution of Lake Volvi. The highly negative ($-20x10^6$ m^3/y) annual water balance of Mygdonia basin (Greek Ministry of Development 2008) reveals the unsustainable water management of the basin.

http://kyrcha.info/2013/04/18/tutorials-calculating-the-fractal-dimension-of-the-greek-coastline-1-25

The agricultural area covers a total of 406.000 acres scattered between 14 municipalities. The cultivation of wheat, sunflower, and maize is among the most encountered crops, occupying the area (Fig. 9.2). The calculation of the total water footprint (WF) of all crops in Mygdonia basin is presented, by analyzing its components (blue, green, gray). Data of farmland area and crop production were provided by the Agricultural Department of Central Macedonia, Greece, for the year 2011.

Blue Water Footprint refers to the volume of excess water from surface and groundwater resources being used during production,which is the irrigated water in the case of crops. Green water footprint expresses the amount of rainwater used during production. Gray water footprint reveals the amount of freshwater

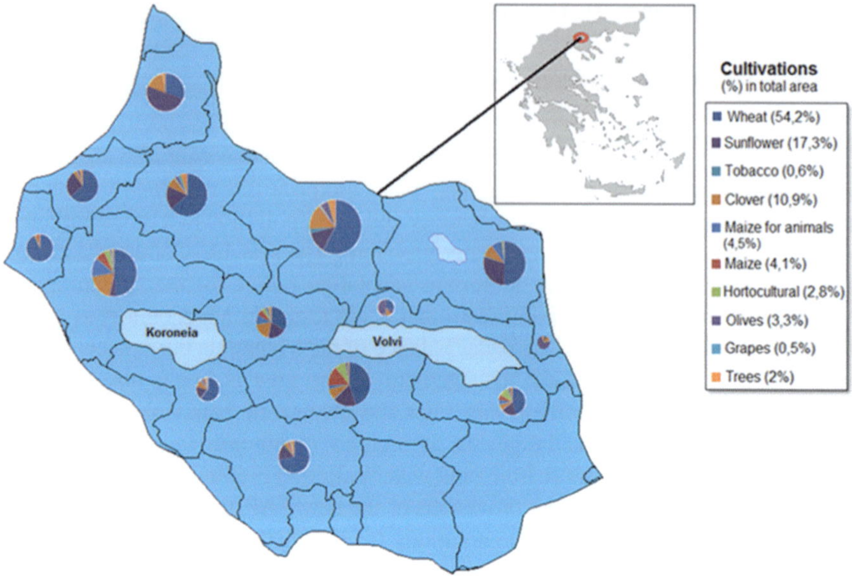

Fig. 9.2 Crop distribution in Mygdonia Basin

Table 9.3 Typology of methodology for the calculation of WF of crops

Water footprint		
$WF = WF\text{green} + WF\text{blue} + WF\text{gray}$ (m3/ton)		
WFgreen	WFblue	WFgray
$WF\text{green} = CWUg/Y$	$WF\text{blue} = CWUb/Y$	$WF\text{gray} = \left[\frac{a*AR}{C\max - C\text{nat}}\right]/Y$
where	where	
$CWUg = \Sigma Ug$	$CWUb = \Sigma Ub$	
$Ug = \min(ETc,\ Peff)$	$Ub = \max(0, ETc - Peff)$	
$ETc = Kc * ETo$		
$ETo = p\,(0.46 * T\text{mean} + 8)$		

that is required to dissolve a load of pollutants based on existing water quality standards (Mekonnen & Hoestra 2011, http://waterfootprint.org/media/downloads/Mekonnen-Hoekstra-2012-WaterFootprintF).

Total water footprint (WF) in m^3/tons of the major crops along with its components (green, blue, and gray water footprint) was calculated based on the methodology applied by Mekonnen et al. (2011). Critical parameters are crop evapotranspiration and yield, required for the estimation of the green and blue water footprint. Those parameters have been calculated following the method and assumptions provided by Allen et al. (1998) and also applied by FAO.

To calculate the total WF, one needs to calculate each one of the three WF components, expressed by the equations in Table 9.3 separately.

Where

CWU_g: green water use (m^3/acre)
CWU_b: blue water used by the crop (m^3/acre)
U_g: monthly green water demand (m^3/acre)
U_b: monthly blue water demand (m^3/acre)
Y: crop yield (ton/acre)
P_{eff}: effective rainfall (mm/month)
ET_o: reference crop evapotranspiration (mm/day) as an average for a period of 1 month
ET_c: potential evapotranspiration of the cultivation (mm/day)
T mean: daily mean temperature (°C)
p: mean daily percentage of annual daytime hours
K_c: crop coefficient that depends on climate, type of crop, and stage of development
C_{max}: maximum acceptable concentration of nitrogen (kg/m^3)
C_{nat}: the natural concentration of nitrogen in the receiving water body (kg/m^3)
a: nitrogen that leaches or runs off
AR: nitrogen application rate (kg/acre)

The total use of green water (CWU_g) for every crop is the sum of the monthly green water demand (U_g), yearly. Its value depends on the crop evapotranspiration and the total available amount of soil moisture.

Total blue water use (CWU_b) is calculated only for the irrigation period, and it is the sum of all monthly blue water used by the crop during that season. Blue water consumption of a crop refers to the amount of irrigated water that is used to fully fulfill crop water requirements.

The potential crop evapotranspiration (ET_c) is calculated based on the Blaney–Criddle method. The influence of the climate on crop water needs is given by the reference crop evapotranspiration (ET_o). Effective rainfall (P_{eff}) is calculated using the USDA-SCS method that takes into account the mean rainfall of the area, crop evapotranspiration, and the depth of moisturized soil (USDA 1993).

Crop coefficient values for basic crops are given by the Food and Agriculture Organization (FAO) (http://www.fao.org/docrep/x0490e/x0490e0b.htm#crop%20coefficients).

T mean and p values are taken from Lagada weather station (http://penteli.meteo.gr/stations/lagadas/).

The percentage of the contaminant, which penetrates the water system, varies, according to the bibliography, between 3% and 10%, mainly depending on water permeability of the soil. The reference area is taken equal to 5%, due to medium water permeability of the soil. According to 2/2600/2001 law in Greece, the C_{max} NO_3 is considered 50 mg/l for surface recipients. Due to lack of data, C_{nat} was considered 0 (Kolokytha E. 2014).

9.6 Results and Discussion

9.6.1 WF Analysis of All Cultivations

For each crop, the WF is calculated following the methodology explained in Tables 9.4a and 9.4b. The results for clover cultivation are presented in Fig. 9.3.

The most water-intensive crops (those with large blue WF) are sunflower, clover, and maize that cover around 37% of the total land (Fig. 9.4). Tobacco, although having large blue WF, occupies a tiny area of cultivation and hence has no special influence on the results. Concerning pollution, maize cultivation has the major impact on the environment due to the bulk use of fertilizers.

Table 9.4a WF calculation per crop cultivation

Month	Kc	T_{mean} (°C)	P	ET_o	Et_c (mm/day)	Etc (mm/month)	Peff (mm/month)	Ug (mm/month)	Ub (mm/month)
April	0,78	11,8	0,3	4,05	3,16	94,86	17,52	17,52	77,34
May	0,93	17	0,32	5,08	4,73	141,85	40,99	40,99	100,86
June	1,02	22,5	0,34	6,26	6,38	191,44	8,86	8,86	182,59
July	1,01	26,6	0,33	6,69	6,76	202,72	1,00	1	201,72
August	0,95	24,9	0,31	6,04	5,74	172,26	16,02	16,02	156,24
September	0,84	22,6	0,28	5,17	4,34	130,16	52,03	52,03	78,13
October	0,63	12,9	0,25	3,50	2,21	66,22	30,64	30,64	35,58

Irrigation period: April to October

Table 9.4b Components of WF calculation per crop cultivation

Y (ton/στρ.)
1,38
WFgreen (m^3/ton)
121,06
WFblue (m^3/ton)
603,23
WFgrey (m^3/ton)
196,56
WF (m^3/ton)
920,85

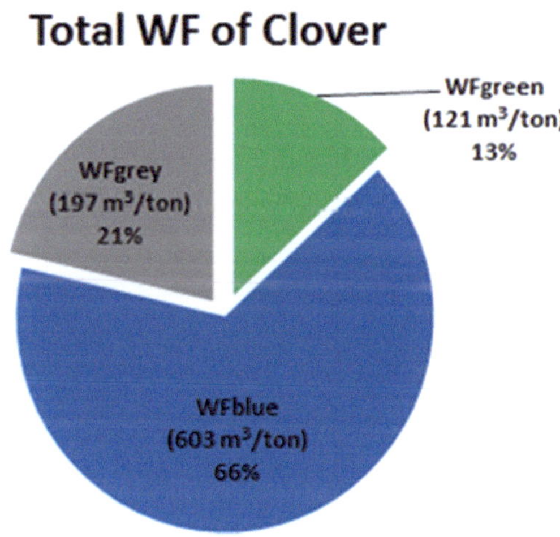

Fig. 9.3 Total WF of Clover

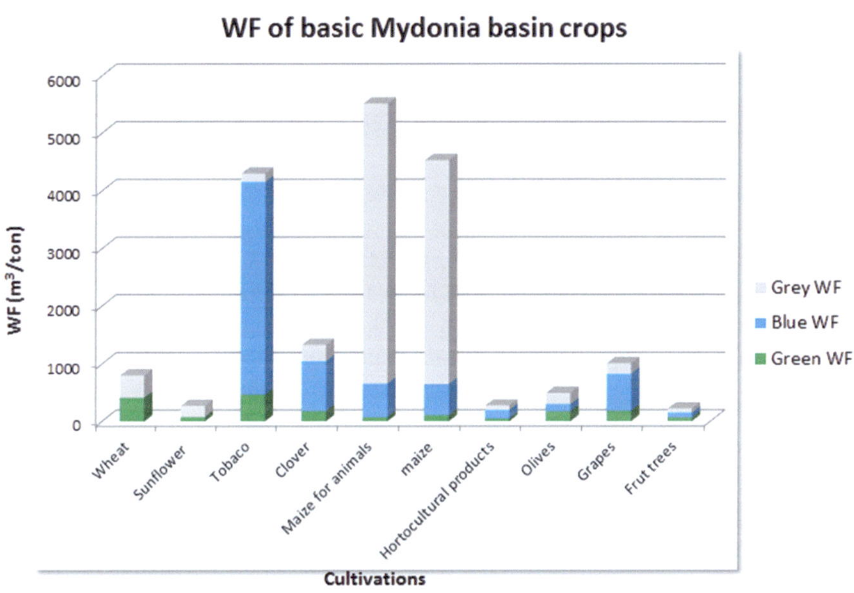

Fig. 9.4 Total WF, by component, of major crops in the area under study

9.6.2 *Water Demand for Agriculture*

To be able to assess data related to water demand, the water footprint was used as an indicator of water consumption, and the total water demand of the region was found. Crop performance is equal to Y = produced tons of crop/acres of the crop.

Table 9.5 Green, blue and grey water in Mygdonia basin

Municipality	Green Water (m^3)	Blue Water (m^3)	Grey Water(m^3)	Total (m^3)
Assiros	508.101,49	329.614,21	543.479,69	1.381.195,39
Evagelismos	281.999,56	642.563,81	367.930,04	1.292.493,41
Sxolari	2.354.962,15	5.526.408,44	37.098.223,38	44.979.593,97
St. Vasilios	1.908.209,71	1.928.375,68	5.569.187,89	9.405.773,28
Vasiloudi	589.736,24	554.258,17	2.432.714,91	3.576.709,33
Gerakarou	815.778,02	2.769.293,89	10.968.069,66	14.553.141,56
Lagadikia	813.206,12	2.279.922,31	12.766.019,52	15.859.147,96
Analipsi	461.118,29	1.294.182,01	6.088.943,40	7.844.243,70
Iraklio	682.081,71	3.138.931,77	16.209.832,44	20.030.845,91
Kavalari	2.826.445,04	12.384.752,45	51.174.697,36	66.385.894,86
Kolhiko	1.330.382,85	4.821.711,02	31.885.870,96	38.037.964,84
Lagada	507.777,22	1.640.524,38	4.457.100,49	6.605.402,09
Lagina	588.124,23	988.521,38	3.123.448,45	4.700.094,06
Perivolaki	706.443,52	3.188.584,28	10.964.434,68	14.859.462,48
Chrisavgi	280.304,19	959.770,36	6.942.674,34	8.182.748,89
Liti	623.428,50	857.956,52	2.526.481,22	4.007.866,24
Drimos	903.363,11	547.510,62	2.386.046,23	3.836.919,96
Melissohori	1.417.236,60	97.476,33	1.717.999,21	3.232.712,14
Total	**17.598.698,55**	**43.950.357,63**	**207.223.153,86**	**268.772.210,05**

In the baseline scenario, for the total of the arable acres and in order to achieve the performance measured in the year 2011, 48.000.000 + 128.000.000 = 176.000.000 cubic meters of water is required by all crops to satisfy this demand. Of these, 48 million is satisfied through precipitation and the remaining amount of 128 million is covered through the exploitation of surface and groundwater resources (irrigation water). In fact, the need for irrigation water per year is approximately 128 million cubic meters (almost 30% of total water), whereas of rainfall, we get 48 million cubic meters (11%) (Table 9.5). More than half the water used is polluted and corresponds to 252 million cubic meters of diluted water to dissolve pollution (59%). Knowing that, based on studies on climate change in the Mediterranean region, a further 20% reduction in rainfall is expected which means that water-intensive crops will need even more water to grow up, making the situation worse. The results in Table 9.3 form the baseline scenario.

Concerning the blue water, clover and maize are responsible for over 80% of the total blue water demand in the area, according to the % of crops cultivated. On the other hand, maize is the most polluting cultivation and the main cause for pollution of Lake Volvi, since its cultivation is mainly concentrated around Lake Volvi.

From the above analysis, it is clear that the systematic cultivation of water-intensive crops and the bulk use of fertilizers are responsible for the qualitative and quantitative degradation of the area.

There is a wide spectrum of provisions provided by the EU CAP 2014–2020 for the "greening" of agriculture in EU. The most suitable ones to be applied in Mygdonia basin are the following:

- Direct financial aid will be offered from the EU to farmers who cultivate in an environmentally friendly way by changing crops and reducing fertilizers.
- Also, the financial aid will focus on organic farming and creation of fallow areas. Crops that are preferred are nonirrigated crops and tree farming.

Taking into account those measures of the new CAP, two possible scenarios toward green growth were tested.

In the first scenario, water-intensive crops like clover, sunflower, tobacco, and maize are being replaced by grassland, while some of them are put into fallow. Concurrently, a reduction of 20% in fertilizers will be applied in the remaining crops as it is proposed by the new CAP provisions.

The second scenario has the following characteristics: All water-intensive crops will be replaced by 30% of grassland, 30% of trees, and 40% of rain-fed crops for animal food. Also, a reduction of 30% in fertilizer will be applied.

There is a significant increase in green water demand and a similar decrease in blue water demand due to the use of rain-fed crops. Also gray water demand is significantly reduced due to the reduction of the applied fertilizers. Protection of the environment and more sustainable water management and area development are the outcomes of this solution.

9.7 Conclusions-Discussion

WF analysis ends up with conclusions that are easy to be understood by the broad audience and hence providing the means to formulate a response strategy. By studying the WF, we can quantify total and actual water consumption and evaluate the sustainability of the water consumption spatially and at a particular time. The water footprint, or at least its blue and green component, measures resource use and not environmental impact.

The main advantage of this metric, as shown by the results, is that it quantifies pollution. Hence, it is easy to communicate to farmers appropriate changes, needed to reorient agriculture to a more sustainable and "green" direction and provide a true potential to use water in a more efficient way.

The results of the analysis of the WF of the major crops in Mygdonia basin have shown that there is an imperative need for crop redistribution. Increased water needs in combination with the extensive use of fertilizers are responsible for both the quantitative problem in Koronia basin and the qualitative one in Volvi basin. Organic farming is a viable solution to reduce the gray water footprint since it excludes or drastically limits the use of manufactured fertilizers, pesticides, and other chemicals that largely increase water pollution. It is important though to mention that maize

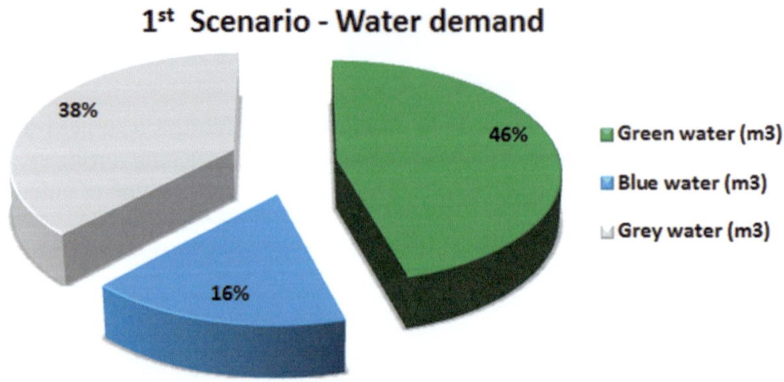

Green water (m^3)	Blue water (m^3)	Grey water (m^3)	Total
32.428.918,25	11.750.453,43	26.827.420,79	71.006.792,47

Fig. 9.5 Scenario 1

and maize for animals cannot be set aside and should remain in the area since it is the basic food for the livestock.

In scenario 1 (Fig. 9.5), the focus is on drastically reducing the cultivated area, through the conversion of much of the land into pastures and fallow lands. With the goal of achieving the same crop performance to that of 2011, only from the remaining crops, 44,000,000 cubic meters of water are required (32,000,000 + 12,000,000) in total. Green water consumption was not reduced as much as the blue, and this is because crops that were set aside were grounded substantially on irrigation water.

In the second scenario (Fig. 9.6), changes mainly concerned the redistribution of crops. In this case, the reduction in the arable area was not that great, but the emphasis was given on the replacement of crops, adding trees as well as dried leguminous crops suitable for animal feed. New water requirements were larger and specifically (47,000,000 + 21,000,000) reached 68,000,000 cubic meters. In this case, it was observed that green water consumption remained at the same level as in the baseline scenario, and this is because the new crops that were chosen utilize water from precipitation to achieve their given crop performance. By applying the two abovementioned scenarios, the abolition of sunflower and clover leads to significant water conservation of approximately 115.000.000 m^3 of irrigated water.

To achieve water conservation and provide "green" solutions in agriculture in the direction of "green development," a two-step procedure may be adopted.

The more the crop performance increases, the more the WF decreases functioning as a tool for maximization of crop yield (ton/acre), which in turn may lead to excessive water use. Therefore, the best way is to achieve the maximization of crop efficiency (ton/m^3) instead of crop yield. The improvement of the efficiency of irrigation schemes and the control of water leakages will increase crop efficiency. Those technical interventions can contribute up to 10% water saving.

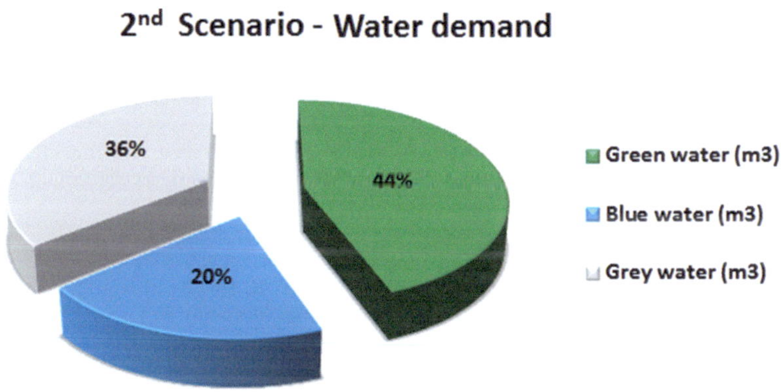

Green water (m^3)	Blue water (m^3)	Grey water (m^3)	Total
46.738.912,05	21.108.121,03	37.607.516,07	105.454.549,15

Fig. 9.6 Scenario 2

It is expected that climate change will decrease the current levels of precipitation in the area, and blue and green water will change dramatically. In the case of severe water deficits, radical changes regarding the selection of crops as well as the total decrease of cultivated land will be needed.

Of course, in times of economic crisis, such changes will, most probably, encounter the opposition of the farmers who are reluctant to new reforms that may differentiate or even risk their current income. In this case, we need the active involvement of all stakeholders using accurate and detailed information, as well as a suitable economic policy in the form of providing economic incentives to end up to consensus.

In parallel, more contemporary irrigation methods, regulations for borehole control, and restrictions on the fertilizer usage should be provided to improve the negative water balance in the basin.

It seems that the applied economic model should be changed. EU Common Agricultural Policy could be used to sustain the rural development in the area. By subsidizing activities such as agro-tourism that promote traditional local products that are compatible with the local climate, CAP could work as an effective viable tool toward the sustainable development of agriculture in the area, sustaining, hence, the farmers' income.

Acknowledgment Special thanks to I. Tzelatidis, civil engineer, who supported the case study section with material from his master thesis (2013), "The concept of the Water Footprint as an assessment tool for the management of water resources. Application to agricultural crops in Mygdonia river basin," Dept. of Civil Engineering, AUTH, Greece.

References

Adams WM, Jeanrenaud SJ (2008) Transition to sustainability: towards a humane and diverse world. IUCN, Gland, 108 pp. ISBN 978-2-8317-1072-3

Allen RG, Pereira LS, Raes D, Smith M (1998) Crop evapotranspiration, Guidelines for computing crop water requirements, Irrigation and Drainage Paper 56, Food and Agric. Organization of the United Nations, Rome, 300 pp

Allianz Global Investors (2013), Water: key 21st century growth opportunity, white paper series, www.allianzgi.com

Arnell NW, Liu C, Compagnucci R, da Cunha L, Hanaki K, Howe C, Mailu G, Shiklomanov I, Stakhiv E (2001) Hydrology and water resources. In: McCarthy JJ, Canziani OF, Leary NA, Dokken DJ, White KS (eds) IPCC climate change 2001: impacts, adaptation & vulnerability, The third assessment report of Working Group II of the Intergovernmental Panel on Climate Change (IPCC), 1000. Cambridge University Press, Cambridge, UK, pp 133–191 (1000 pages)

Ayers JM, Huq S (2008) The value of linking mitigation and adaptation: a case study of Bangladesh. Environ Manage. doi: 10.1007/s00267-008-9223-2. Springer Science+ Business Media, LLC 2008

Bates BC, Kundzewicz ZW, Wu S, Palutikof JP (eds) (2008) Climate change and water. Technical paper of the Intergovernmental Panel on Climate Change. IPCC Secretariat, Geneva, 210 pp

Berkhout Frans et al (2013) European responses to climate change: deep emissions reductions and mainstreaming of mitigation and adaptation. REsponsEs project policy Brief. Institute for Environmental studies (IvM), v University Amsterdam, Amsterdam

Christiansen L, Olhoff A, Trærup S (eds) (2011) Technologies for adaptation: perspectives and practical experiences, UNEP Risø Centre, Roskilde

Die (2012) Sustainable water management through green economy? Briefing paper 5/2012, German Development Institute

Die (2011) Water-energy-food – do we need a nexus perspective? The Bonn Nexus conference by Dr. Ines Dombrowsky, German Development Institute/Deutsches Institut für Entwicklungspolitik (DIE)

Duval, R. (2008) A taxonomy of instruments to reduce greenhouse gas emissions and their interactions. OECD Economics Department Working Papers, No. 636, OECD Publishing. doi: 10.1787/236846121450

European Commission, Directive 2007/60/EC of the European Parliament and of the Council of 23 October 2007 on the assessment and management of flood risks. Off J Eur Commun, L 288, 6.11.2007, p. 27

EurActive, EU News and Policy Database (2013) CAP 2014–2020: a long road to reform, http://www.euractiv.com/cap/cap-reform-2014-2020-linksdossier-508393

European Council (2014) European Council (23 and 24 October 2014): conclusions on 2030 climate and energy policy framework, Brussels, Belgium. Available at: http://www.consilium.europa.eu/uedocs/cms_data/docs/pressdata/en/ec/145356.pdf

European Environment Agency (2014) Annual European Union greenhouse gas inventory 1990–2012 and inventory report 2014 (Submission to the UNFCCC Secretariat) – Technical Report 09/2014. Brussels, Belgium. Available at: http://www.eea.europa.eu/publications/european-union-greenhouse-gasinventory-2014

European Environment Agency, EEA (2012) Report No 1/2012, Towards efficient use of water resources in Europe

European Union (EU) (2000) EU water framework directive. Directive 2000/60/EC of the European Parliament and of the Council of 23 October 2000 establishing a framework for Community action in the field of water policy. Off J 327, 22/12/2000, P. 001–0073

European Commission (2009) River basin management in a changing climate, common implementation strategy for the water framework directive, (CIS) Guidance Document No. 24, ISBN 978-92-79-14298-7, 2009

European Commission, Directive 2007/60/EC of the European Parliament and of the Council of 23 October 2007 on the assessment and management of flood risks. Off J Eur Commun, L 288, 6.11.2007, p. 27
European Commission, Communication to the European Parliament and the Council – Addressing the challenge of water scarcity and droughts in the European Union, COM/2007/04141 final, 2007
European Commission, Communication to the European Parliament and the Council, Our life insurance, our natural capital: an EU biodiversity strategy to 2020. Our life insurance, our natural capital: an EU biodiversity strategy to 2020, COM/2011/244
Füssel Hans-Martin EEA (2007) Adaptation planning for climate change: concepts, assessment approaches and key lessons. Sustainab Sci. doi:10.1007/s11625-007-0032-y
Global Risks Report (2015) 10th edition, World Economic Forum, Switzerland, 2015. http://www3.weforum.org/docs/WEF_Global_Risks_2015_Report15.pdf
Global Water Intelligence (2013) Vol. 1, www.globalwaterintel.com
Gardner J, Dowd A-M, Mason C, Ashworth P (2009) A framework for stakeholder engagement on climate adaptation. CSIRO Climate Adaptation Flagship Working paper No. 3. http://www.csiro.au/resources/CAF-working-papers.html
Greek Ministry of Development (2008) 2nd Water Master Plan of National Park of Lake Koronia and Volvi
Grey D, Sadoff C (2007) Sink or swim? Water security for growth and development. Water Policy 9:545–571
Grothmann T (2011) Governance recommendations for adaptation in European urban regions: results from five case studies and a European expert survey. In: Otto Zimmermann K (ed) Resilient cities – cities and adaptation to climate change – proceedings of the Global Forum 2010. Springer, Hamburg, pp 167–175
Hansen J, Sato M, Kharecha P, Beerling D, Berner R, Masson-Delmotte V, Pagani M, Raymo M, Royer DL, Zachos JC (2008) Target atmospheric CO2: where should humanity aim? Open Atmos Sci J 2:217–231. doi:10.2174/1874282300802010217
Held IM, Soden BJ (2000) Water vapor feedback and global warming. Annu Rev Energy Environ 25:441–475
Hirsch RM (2010) A perspective on non-stationarity in water management". Workshop on Non-Stationarity, Hydrologic Frequency Analysis and Water Management, Boulder
Huntington TG (2006) Evidence for intensification of the global water cycle: review and synthesis. J Hydrol 319:83–95
Intergovernmental Panel on Climate Change (IPCC) (2014) Climate change 2014: mitigation of climate change. Contribution of working group III to the fifth assessment report of the Intergovernmental Panel on Climate Change. In: Edenhofer O, Pichs-Madruga R, Sokona Y, Farahani E, Kadner S, Seyboth K, Adler A, Baum I, Brunner S, Eickemeier P, Kriemann B, Savolainen J, Schlömer S, von Stechow C, Zwickel T, Minx JC (eds). Cambridge University Press, Cambridge, UK/New York. Available at: http://www.ipcc.ch/report/ar5/wg3/
Jordan River Foundation (JRF) (2007) Organic farming initiative. Retrieved from www.jordanriver.jo/program_capacity_organic.asp
Kane S, Shogren JF (2000) Linking adaptation and mitigation in climate change policy. Clim Change 45(1):75–102, http://dx.doi.org/10.1023/a:1005688900676
Klein RJT, Huq S, Denton F, Downing TE, Richels RG, Robinson JB, Toth FL, Parry ML, Canziani OF, Palutikof JP, Van der Linden PJ, Hanson CE (eds) (2007) Inter-relationships between mitigation and adaptation. Climate change 2007: impacts, adaptation and vulnerability. Contribution of Working Group II to the Fourth Assessment Report of the Intergovernmental Panel on Climate Change, Cambridge University Press, Cambridge, UK, pp 745–777
Kolokytha E (2011) The European Union water framework directive, a driving force for shared water resources management. In: Ganoulis J, Aurelli A, Fried J (eds) Transboundary water resources management, a multidisciplinary approach, Wiley-VCH, pp 52–60. ISBN 978-3-527-33014-0

Kolokytha E (2014) Agricultural development in Lake Koronia. The role of the water footprint of major crops in combating climate change. 3rd IAHR Europe Congress, e-proceedings, Porto -Portugal

Koutsoyiannis D (2011) Hurst-Kolmogorov dynamics and uncertainty. J Am Water Resour Assoc 47(3):481–495

Kundzewicz ZW, Mata LJ, Arnell N, Döll P, Kabat P, Jiménez B, Miller K, Oki T, Şen Z, Shiklomanov I (2007) Freshwater resources and their management. Climate change 2007: impacts, adaptation and vulnerability. In: Parry ML, Canziani OF, Palutikof JP, van der Linden PJ, Hanson CE (eds) Contribution of Working Group II to the Fourth Assessment Report of the Intergovernmental Panel on Climate Change. Cambridge University Press, Cambridge, UK, pp 173–210. http://www.ipcc.ch/pdf/assessment-report/ar4/wg2/ar4-wg2-chapter3.pdf

Lalisa A, Duguma A, Susan W, Wambugua, Peter A, Minang A, Meine van Noordwijk (2014) A systematic analysis of enabling conditions for synergy between climate change mitigation and adaptation measures in developing countries. Environ Sci Policy 42:138–148

Latouche (2004) Degrowth economics: why less should be so much more. Le Monde Diplomatique, November 2004 (http://mondediplo.com/2004/11/14latouche)

Lebel L, Grothmann T, Siebenhüner B (2010) The role of social learning in adaptiveness: insights from water management. Int Environ Agreements Polit Law Econ 10(4):333–353

Lins HF, Cohn TA (2011) Stationarity: wanted dead or alive? J Am Water Resour Assoc 47(3):475–480. doi: 10.1111/j.1752-1688.2011.00542.x

Luck T (2010) Local vineyard showcases virtues of organic farming. The Jordan Times Jordan River Foundation (2007) Organic Farming Initiative. Retrieved from www.jordanriver.jo/program_capacity_organic.asp. Taken from Towards a Green Economy in Jordan. A scoping study, UN-Ministry of Environment 2011

Malik A, Qin X, Smith S (2010) Autonomous adaptation to climate change: a literature review, Institute for International Economic Policy Working Paper Series, Elliott School of International Affairs. The George Washington University, Washington, DC

Margulis S, Narain U, Pandey K, Cretegny L, Bucher A, Schneider R, Hughes G, Essam T, Mearns R (2010) The costs of developing countries adapting to climate change. New methods and estimates. The Global Report of the Economics of Adaptation to Climate Change Study, Consultation Draft. World Bank, Washington, DC. See http://beta.worldbank.org/climatechange/content/economics-adaptation-climate-change-study-homepage

Mata LJ, Budhooram J (2007) Complementarity between mitigation and adaptation: the water sector. Mitig Adapt Strateg Glob Chang 12(5):799–807. doi:10.1007/s11027-007-9100-y

Mekonnen MM, Hoekstra AY (2011) The green, blue and grey water footprint of crops and derived crop products. Hydrology Earth System Science, Twente Water Centre, University of Twente, Enschede

Milly PCD, Betancourt J, Falkenmark M, Hirsch RM, Kundzewicz ZW, Lettenmaier DP, Stouffer RJ (2008) Stationarity is dead: whither water management? Science 319:573–574. doi:10.1126/science.1151915

Moser S (2012) Adaptation, mitigation, and their disharmonious discontents: an essay. Clim Change 111(2):165–175, http://dx.doi.org/10.1007/s10584-012-0398-4

Nachmany M, Fankhauser S, Davidová J, Kingsmill N, Landesman T, Roppongi H, Schleifer P, Setzer J, Sharman A, Singleton CS, Sundaresan J, Townshend T (2015) The 2015 global climate legislation study. A review of climate change legislation in 99 countries: summary for policy-makers, Grantham Research Institute on Climate Change and the Environment was established-LSE, Globe, IPU. Grantham Research Institute: http://www.lse.ac.uk/grantham/, GLOBE:http://globelegislators.org/, IPU: http://www.ipu.org/

Nõges T, Nõges P, Cardoso Ana C (2010) Review of published climate change adaptation and mitigation measures related with water. Eur Commiss Joint Res Centre Inst Environ Sustainability. doi:10.2788/18203

OECD (2007a) OECD principles for private sector participation in infrastructure. OECD Publishing. doi: 10.1787/9789264034105-en

OECD (2007b) Instrument mixes for environmental policy. OECD Publishing. doi: 10.1787/9789264018419-en

OECD (2009a) The economics of climate change mitigation: policies and options for global action beyond 2012, OECD Publishing. doi: 10.1787/9789264073616-en

OECD (2011a) Delivering on green growth. In OECD, towards green growth. OECD Publishing. doi: 10.1787/9789264111318-7-en

OECD (2011b) Virginie Marchal, Rob Dellink (ENV) Detlef van Vuuren (PBL) Christa Clapp, Jean Château, Eliza Lanzi, Bertrand Magné (ENV), Jasper van Vliet "The OECD Environmental Outlook to 2050" OECD Environment Directorate (ENV) and the PBL Netherlands Environmental Assessment Agency (PBL)

OECD (2011c) Virginie Marchal, Rob Dellink (ENV) Detlef van Vuuren (PBL) Christa Clapp, Jean Château, Eliza Lanzi, Bertrand Magné (ENV), Jasper van Vliet "The OECD Environmental Outlook to 2050" OECD Environment Directorate (ENV) and the PBL Netherlands Environmental Assessment Agency (PBL). Ch. 3

OECD (2011d) Greening household behavior: the role of public policy. OECD Publishing. doi: 10.1787/9789264096875-en

OECD (2011e) Towards green growth, OECD green growth studies. OECD Publishing. doi: 10.1787/9789264111318-en

OECD (2012) Environmental outlook to 2050: the consequences of inaction.. ISBN 978-92-64-122161

Quick T, Winpenny J (2014) Topic guide: water security and economic development. Evidence on Demand, UK, http://www.evidenceondemand.info/topic-guide-water-security-and-economic-development

Ray PA, Brown CM (2015) Confronting climate uncertainty in water resources planning and project design. The Decision Tree Framework, World Bank Group, Washington, DC

Schneider SH, Lane J (2006) An overview of 'Dangerous' climate change. In: Schellnhuber HJ (ed) Avoiding dangerous climate change. Cambridge University Press, Cambridge, UK, Available for download from: http://www.defra.gov.uk/environment/climatechange/research/dangerouscc/index.htm

Shipper E, Lisa F (2007) Climate change and adaptation: exploring the linkages. Tyndall Center for Climate Change Research Working Paper 107

SOER (2015) EEA, the European environment—state and outlook 2015/Synthesis

Stakhiv E, Stewart B (2010) Needs for climate information in support of decision-making in the water sector. Procedia Environ Sci 1:102–119

Trenberth KE (1999) Conceptual framework for changes of extremes of the hydrological cycle with climate change. Climatic change 42, 327–339. Allen Consulting (2005) Climate change risk and vulnerability. Australian Greenhouse Office, Department of the Environment and Heritage, Canberra, Australia

Tzelatidis I (2013) The concept of the Water Footprint as an assessment tool for the management of water resources. Application to agricultural crops in Mygdonia river basin. Master thesis, Department of Civil Engineering, AUTH

UNDESA (2012) A guidebook to the Green Economy issue 2: exploring green economy principles

UNEP (2011a) Towards a green economy: pathways to sustainable development and poverty eradication. www.unep.org/greeneconomy, ISBN: 978-92-807-3143-9

UNEP (2011b) Enabling conditions, supporting the transition to global green economy. United Nations Environment Program

UNFCCC (2014) Report of the workshop on technologies for adaptation. Langer Eugen, Bonn, Germany, 4 March 2014. Available from: http://unfccc.int/ttclear/pages/

UN-Water (2011) A water toolbox or best practice guide of actions: a contribution from UN-Water Conference on "Water in the Green Economy in Practice: Towards Rio + 20

USDA, US Department of Agriculture (1993) Part 623. National engineering handbook. Irrigation water requirements. Chapter 2. 2–1. (210-vi-NEH, September 1993). United States. Department of Agriculture Soil Conservation Service. Part 623 National. Engineering Handbook. (ftp://ftp.wcc.nrcs.usda.gov/wntsc/waterMgt/irrigation/NEH15/ch2.pdf)

Wilk J, Wittgren HB (eds) (2009) Adapting water management to climate change. Swedish Water House Policy Brief Nr. 7. SIWI, 2009

World Bank (2013) The Karnataka Watershed (Sujala) Project. Independent Evaluation Group. The World Bank. Available at: http://ieg.worldbankgroup.org/webpage/ Karnataka-watershed-Sujala-project (taken from Lessons from country experience, Green Growth in Practice – www.ggbp.org

Chapter 10
Adaptation to Climate Change: Institutional Analysis

Ana Cristina Souza da Silva, Carlos de Oliveira Galvão, Márcia Maria Rios Ribeiro, and Tafnes da Silva Andrade

Abstract There are many challenges for water management under climate change. Institutional capacity for enabling adaptation is one of those challenges, which have to consider uncertainties, participatory processes, and monitoring. Elinor Ostrom, a Nobel winner, produced many relevant contributions to understanding institutional governance. Her work pointed to requirements of adaptive governance, institutional design principles for local common pool resources systems, and social-ecological framework analysis. Recently, Ostrom's institutional principles have been extended for the governance of adaptation to climate change in the water sector. Adaptation in water sector is a continuous process of learning. Drought management in the present and past is also a way of learning considering experiences on institutions dealing with this challenge. This chapter illustrates how Ostrom's principles, in the context of a drought management experience in Brazil, might provide a continuous way for assessing if institutions are capable to play their roles in the process of adaptation to climate variability and change.

Keywords Climate variability • Drought • Water management • Water supply • Reservoir • Ostrom

10.1 Introductory Concepts

There are many challenges for water management under climate change. Institutional capacity is one of those challenges, which have to consider uncertainties, participatory processes, and monitoring. Elinor Ostrom, a Nobel Prize winner, produced many relevant contributions to understanding institutional governance. Her work pointed to requirements of adaptive governance, institutional design

A.C.S. da Silva • C. de Oliveira Galvão (✉) • M.M.R. Ribeiro
Federal University of Campina Grande, UFCG, 58429-900 Campina Grande, PB, Brazil
e-mail: ccristina24@yahoo.com.br; galvao@dec.ufcg.edu.br; marcia.ribeiro@ufcg.edu.br

T. da Silva Andrade
Environmental Agency of the State of Pernambuco, Pernambuco, Brazil
e-mail: tafnesandrade@yahoo.com.br

E. Kolokytha et al. (eds.), *Sustainable Water Resources Planning and Management Under Climate Change*, DOI 10.1007/978-981-10-2051-3_10

principles for local common pool resources systems, and social-ecological framework analysis. Huntjens and collaborators recently extended Ostrom's institutional principles for the governance of adaptation to climate change in the water sector. This chapter is largely based on their works and is organized in three parts: the first briefly introduces some basic concepts on institutions, adaptation, adaptive water management, and governance; the second revises very shortly Ostrom's framework and design principles for institutional analysis; the final part shows the application to a water management case. The first two parts do not intend to present in detail or extensively the Ostrom's theoretical framework, since there is a rich literature to which the reader can consult. Our main objective is to add an application of those principles to a drought-related water management problem. Through this example, we intend to help the readers to use this extraordinary framework in other applications in the water sector.

The following paragraphs present some basic concepts, very concisely, to enable the reader to easily understand the remaining topics.

10.1.1 Institutions

Institutions can be defined "as formal and informal rules that are understood and used by a community" (Hess and Ostrom 2005). "*Rules* are linguistic statements similar to norms but rules carry an additional, assigned sanction if forbidden actions are taken and observed by a monitor. For rules to exist, any particular situation must be linked to a rule-making situation and some kind of monitoring and sanctioning must exist. Rules may be crafted in any of a wide diversity of collective-choice or constitutional-choice arenas in local, regional, national, or international domains" (Ostrom 2013).

10.1.2 Governance

According to Graham et al. (2003), "governance is a process whereby societies or organizations make their important decisions, determine whom they involve in the process and how they render account." "For a governance regime to deal with the current and anticipated impacts of climate change it first needs to have a policy or strategy in place. From this perspective, the output of a governance system is not only defined by its physical interventions, but also by means of its management interventions" (Huntjens et al. 2012).

10.1.3 Adaptation

> Adaptation is a process of deliberate change in anticipation of or in reaction to external stimuli and stress. Adaptation involves change. Adaptation is, therefore, standard practice in the human world as individuals, communities, and societies adjust their activities,

life courses, and locations to take advantage of new opportunities. But adaptation is often imposed on societies and localities because of external undesirable change. Efforts to respond to these changes frequently entail reducing vulnerability and enhancing the capacity to adapt, in effect, to enhance the resilience of people and places, localities, and ways of life (Nelson et al. 2007).

10.1.4 Adaptive Water Management

Adaptive Water Management adds value to the Integrated Water Resources Management approach. The central contribution of Adaptive Water Management (AWM) within the context of Integrated Water Resources Management (IWRM) is that it provides added value through explicitly embracing uncertainty. AWM acknowledges the complexity of the systems to be managed and the limits in predicting and controlling them. This implies an integrated management approach which adopt a systemic perspective rather than dealing with individual problems in isolation (Van der Keur et al. 2010).

10.2 Ostrom's Contributions for Institutional Analysis

Elinor Ostrom was the 2009's Nobel Prize in Economic Sciences. The Nobel Prize acknowledges her contribution as (Elinor Ostrom – Facts 2015): she "challenged the conventional wisdom by demonstrating how local property can be successfully managed by local commons without any regulation by central authorities or privatization. ... She gathered information through field studies and then analyzed this material. In her book 'Governing the Commons', from 1990, she demonstrated how common property can be successfully managed by user associations and that economic analysis can shed light on most forms of social organization. Her research had great impact amongst political scientists and economists."

10.2.1 Institutional Design Principles for Local Common Pool Resources Systems

According to Ostrom (1990), "the term 'common-pool resource' (CPR) refers to a natural or man-made resource system that is sufficiently large as to make it costly (but not impossible) to exclude potential beneficiaries from obtaining benefits from its use. To understand the processes of organizing and governing CPRs, it is essential to distinguish between the resource system and the flow of resource units produced by the system, while still recognizing the dependence of the one on the other." She identified a set of sustainable and robust institutional principles for good governance of CPR by individuals and communities (Ostrom 1990; Ostrom 2005), while Huntjens et al. (2012) extended Ostrom's principles for the governance of adaptation to climate change for the water sector:

10.2.1.1 Principle 1 – Clearly Defined Boundaries

Ostrom (1990) defined this principle as "Individuals who have rights to withdraw resource units from CPR must be clearly defined, as must boundaries of the CPR itself." The principle was extended by Huntjens et al. (2012) as "Completeness of water-user stakeholders in the adaptation process and clarity about who has rights to use water resources in the case of droughts."

10.2.1.2 Principle 2 – Congruence Between Appropriation and Provision Rules and Local Conditions

Ostrom (1990) defined this principle as "Appropriation rules restricting time, place, technology, and/or quantity of resource units are related to local conditions and to provision rules requiring labor, material, and/or money." Ostrom (2005) also presented the principle as "Proportional equivalence between benefits and costs: rules specifying the amount of resource products that a user is allocated are related to local conditions and to rules requiring labor, materials, and/or money inputs."

> The second design principle is that the rules-in-use allocate benefits proportional to inputs that are required. If a group of users is going to harvest from a resource over the long run, they must devise rules related to how much, when, and how different products are to be harvested. They also need to assess the costs of operating a system on users. When the rules related to the distribution of benefits are made broadly consistent with the distribution of costs, participants are more willing to pitch in to keep a resource well-maintained and sustainable . . . (Ostrom 2005).

Huntjens et al. (2012) specified the principle for adapting to climate change in large river basins as "Equal and fair (re-)distribution of risks, benefits and costs, requiring engagement with, and strong representation of, groups likely to be highly affected or especially vulnerable."

10.2.1.3 Principle 3 – Collective-Choice Arrangements

Ostrom (1990) defined this principle as "Most individuals affected by the operational rules can participate in modifying the operational rules." Huntjens et al. (2012) extended this principle as follows: "To enhance the participation of those involved in making key decisions about the system, in particular on how to adapt."

10.2.1.4 Principle 4 – Monitoring

According to Ostrom (1990), "Monitors, who actively audit CPR conditions and appropriator behavior, are accountable to the appropriators or are appropriators." Huntjens et al. (2012) defined this principle as "Monitoring and evaluation of the process: providing a basis for reflexive social learning and supporting accountability."

"Appropriator can be used to refer to anyone who appropriates resource units from some type of resource system. In many instances appropriators use or consume the resource units they withdraw. Appropriators also use resource units as inputs into production processes" (Ostrom 1990).

10.2.1.5 Principle 5 – Graduated Sanctions

Ostrom (1990) considers that "appropriators who violate operational rules are likely to be assessed graduated sanctions (depending on the seriousness and context of the offense) by other appropriators, by officials accountable to these appropriators, or by both."

10.2.1.6 Principle 6 – Conflict Resolution Mechanism

This principle is presented by Ostrom (1990) as follows: "Appropriators and their officials have rapid access to low-cost arenas to resolve conflicts among appropriators or between appropriators and officials." Huntjens et al. (2012) added: "Including timing and careful sequencing, transparency, trust-building, and sharing of (or clarifying) responsibilities."

10.2.1.7 Principle 7 – Minimal Recognition of Rights to Organize

Ostrom (1990) described this principle as "the rights of appropriators to devise their own institutions are not challenged by external governmental authorities."

10.2.1.8 Principle 8 – Nested Enterprises

For resources that are parts of larger systems, Ostrom (1990) established another principle: "Nested enterprises: appropriation, provision, monitoring, enforcement, conflict resolution, and governance activities are organized in multiple layers of nested enterprises." Huntjens et al. (2012) presented this principle as "Nested enterprises/polycentric governance in a multi-level context, as functional units to overcome the weakness of relying on either just large-scale or only small-scale units to govern complex resources systems."

The principles "Robust and flexible process" and "Policy learning" were added by Huntjens et al. (2012) to the above eight Ostrom principles, for the complex systems adaptation, such as large basins under uncertain conditions, such as climate change:

10.2.1.9 Principle 9 – Robust and Flexible Process

Huntjens et al. (2012) defined this principle as "institutions and policy processes that continue to work satisfactorily when confronted with social and physical challenges but which at the same time are capable of changing." Based on other studies, they indicate five characteristics that might build robust and flexible processes: organizational redundancy, flexibility to include new initiatives, confidence-building, integration of intersectoral policies or "integrating adaptation" ("climate adaptation does challenge integration"), and sequential and time dilemmas.

10.2.1.10 Principle 10 – Policy Learning

According to Huntjens et al. (2012), policy learning is related to "policy and institutional adjustments based on commitment to dealing with uncertainties, deliberating alternatives and reframing problems and solutions."

10.2.2 Adaptive Governance

Dietz et al. (2003) proposed some requirements of adaptive governance: *provide necessary information*; *deal with conflicts*; *induce compliance with rules*; *provide physical, technical, and institutional structure*; *and encourage adaptation and change*. They associated those requirements with Ostrom's general principles for robust governance of environmental resources:

"*Providing information*. Environmental governance depends on good, trustworthy information about stocks, flows, and processes within the resource systems being governed, as well as about the human-environment interactions affecting those systems. Effective governance requires not only factual information about the state of the environment and human actions but also information about uncertainty and values."

"*Dealing with conflict*. Sharp differences in power and in values across interested parties make conflict inherent in environmental choices."

"*Inducing rule compliance*. Effective governance requires that the rules of resource use are generally followed, with reasonable standards for tolerating modest violations."

"*Providing infrastructure*. The importance of physical and technological infrastructure is often ignored. Infrastructure, including technology, determines the degree to which a commons can be exploited (e.g., water works and fishing technology), the extent to which waste can be reduced in resource use, and the degree to which resource conditions and the behavior of humans users can be effectively monitored."

"*Be prepared for change*. Institutions must be designed to allow for adaptation because some current understanding is likely to be wrong, the required scale of organization can shift, and biophysical and social systems change."

Considering that Huntjens et al. (2012) adapted Ostrom's principles to deal with the adaptation to climate change, those principles might attend to the adaptive governance requirements. The composition of robust strategies that consider institutional adaptation might consider possible relations between those principles and requirements.

10.2.3 Social-ecological Analysis Framework

According to Ostrom (2009), "all humanly used resources are embedded in complex, social-ecological systems (SESs), ... and a common, classificatory framework is needed to facilitate multidisciplinary efforts toward a better understanding of complex SESs." Anderies et al. (2004) use the term SES to refer to "the subset of social systems in which some of the interdependent relationships among humans are mediated through interactions with biophysical and non-human biological units."

Anderies et al. (2004) raise some interesting questions to base their proposal of *a framework to analyze the robustness of SESs from an institutional perspective*: "How do institutional arrangements affect the robustness of SESs? Why do some systems survive in highly varying environments over time and others collapse? Which attributes of the institutions are more likely to lead to the creation of robust SESs? How do these attributes depend on the underlying ecological system?"

Later, Ostrom recommends the following guidelines when applying the framework:

- The choice of relevant second or deeper levels of variables for analysis (from the large set of variables at multiple levels) depends on the particular questions under study, the type of SES, and the spatial and temporal scales of analysis (Ostrom 2009).
- At a broad level, one can begin to organize an analysis of how attributes of (i) a resource system (e.g., lake), (ii) the resource units generated by that system (e.g., water), (iii) the users of that system, and (iv) the governance system jointly affect and are indirectly affected by interactions and resulting outcomes achieved at a particular time and place. Using such a framework also enables one to organize how these attributes may affect and be affected by the larger socioeconomic, political, and ecological settings in which they are embedded, as well as smaller ones (Ostrom 2007).

10.3 Ostrom's Institutional Analysis of the Brazilian Water Policy from a Drought Management Perspective

Adaptation in water sector is a continuous process of learning. Drought management in the present and past is also a way of learning considering experiences on institutions dealing with this challenge. The following sections of this chapter are intended to illustrate how Ostrom's principles, in the context of a drought management experience in Brazil, might provide a continuous way for assessing if and when institutions are capable to play their roles in the process of adaptation to climate variability and change.

10.3.1 Droughts

Drought is a recurrent feature of climate that occurs in all climatic zones, from wet to dry, and varies from one region to another, more likely observed in areas of high precipitation variability. It differs from aridity to be a temporary situation, which may last from months to years. In arid areas the aridity is permanent.

Drought is a complex phenomenon difficult to quantify, whose conditions develop slowly and gradually and do not usually involve structural damage, being its onset and duration difficult to be determined. In addition, its impacts spread across different sectors of the economy and affect large areas and populations. Anthropogenic activities, such as inappropriate land use and water management, can lead to worsening drought impacts. Changes in land use and land drainage have altered the hydrologic regimes in terms of water quality and water balances, which may have repercussions on the vulnerability to drought.

IPCC (2014) defines drought as an extreme climatic event characterized by below-normal precipitation over a time period relative to its local normal condition, long enough to cause a serious hydrological imbalance. But drought definition is not a closed issue. Although there is consensus that the drought is the result of lack of precipitation, which produces water deficit for a specific activity or group, drought phenomenon was not a precise definition. It can be analyzed from different perspectives related to what is appropriate to the activity, time, and place under consideration. Droughts can be classified into meteorological, agricultural, hydrological, and socioeconomic.

Meteorological drought is a period of months to years with below-normal precipitation, in comparison to some average amount of rainfall, leading to other types of droughts. Periods of meteorological drought are identified in relation to actual precipitation from average amounts on monthly, seasonal, or annual time scales. It must be defined regionally by the regional climatic variation conditions.

The agricultural sector is the first to be affected by drought because of its dependence on stored soil moisture. Agricultural drought impacts are associated with soil water deficits resulting in loss of yield and have characteristics of meteorological and hydrological drought.

Hydrological drought results from long periods of low precipitation that impacts streamflow, reservoir, lake, and groundwater levels, as well as recharge rates, producing significant societal impacts. Hydrological drought impacts lag the occurrence of meteorological drought because it takes some time for precipitation deficits to accumulate in the hydrological system components. Land use also can influence hydrological drought and may change both water infiltration in the soil and runoff patterns.

Socioeconomic drought is associated with supply and demand of some water-dependent economic good such as water supply, fishing, and hydropower, among others. It occurs when water supply is unable to meet economic or social demands due to weather-related factors.

The non-implementation of risk management strategies, with the consequent concentration only on crisis management strategies, and the growing demand for water by society also contribute to exacerbated negative socioeconomic impacts of droughts, making it even more difficult to meet the demands during all drought events, especially in areas with high climate variability and high population density.

Risk management strategies for coping with droughts are somewhat equivalent to adaptation strategies in coping with climate change and, thus, a challenge to the institutions dealing with it. The identification of deficiencies and failures, capacities, and strengths of such institutions is very important to design and improve them, to better manage the processes of dealing with climate variability and change.

10.3.2 The Brazilian Semiarid Region

The large semiarid northeastern region of Brazil (Fig. 10.1), covering about 1 million km^2, is characterized by frequent drought periods, which have caused different impacts on its water resources availability and socioeconomic systems. Annual rainfall presents a high interannual variability and amounts ranging from 400 to 800 mm. Average temperature is 27 °C, and evaporation rates are higher than 2000 mm/year. It is one of the most vulnerable regions to climate change in Brazil, with possible decrease of rainfall and increase in temperature and, consequently, evapotranspiration.

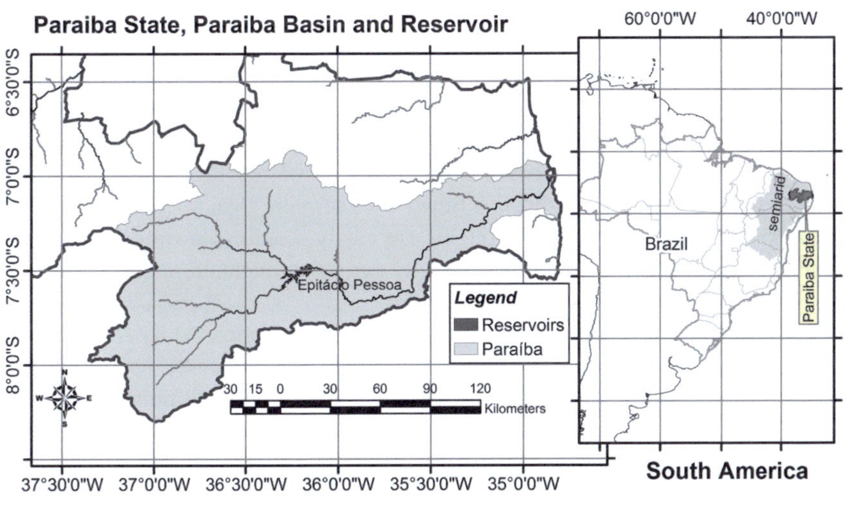

Fig. 10.1 Brazilian semiarid, Paraiba River Basin and Epitácio Pessoa Reservoir

Historically, populations and communities have dealt with climate variability through storage, in small reservoirs and shallow aquifers, of water from rainfall and intermittent streams and rivers during the 3-month rainy season for general use during the dry season. Agriculture was mostly rain-fed and also used some small-scale irrigation for the intra-season dry spells.

With population and cities' growth (currently, in 2014, 23.5 million inhabitants), water supply, either for human consumption or livestock and agriculture, became a huge social and economic problem. One of the most important measures to cope with this new context was the construction of large surface reservoirs for water storage from wet years to dry years, with the creation of the *National Department Against Droughts* (DNOCS), in 1909. Those reservoirs allowed the settlement in large- and medium-sized cities during the twentieth century and also the establishment of several irrigation districts for food production.

10.3.3 Epitácio Pessoa Reservoir

The Epitácio Pessoa Reservoir is located in the Paraíba River Basin, State of Paraíba, located in the semiarid region of Brazil (Fig. 10.1). Its drainage basin is very impacted by human actions such as deforestation (Galvao et al. 2001). This reservoir was built by the Brazilian Federal Government during the period of 1951–1957 and represents the main source for supplying a population of about 500 000 inhabitants, including Campina Grande city, an important educational and economical center for the region. The water reservoir users are a water supply company and irrigators. Only the water supply company is an authorized user by the National Water Agency, due to limitation of its regularized discharge.

This reservoir has experienced extreme water shortages, which demanded adaptation measures, such as suspending the allocation of water for irrigation and greatly reducing urban supply. These measures led to different impacts on water users, which required new efforts on social adaptation (Grande et al. 2014). One of the drought periods took place in 1997–2003. More recently (2012–2016), the region suffered another severe drought period, the biggest water crisis that ever happened around the reservoir. The high climatic variability of the region can be better observed in Fig. 10.2, which shows a time series of more than 40 years of rainfall and reservoir storages. Both big droughts are easily identifiable from this record. The reservoir has also lost part of its capacity due to sedimentation. From 2004 the new capacity of 411 million m^3 was estimated.

The reservoir's regularized discharge is 1.85 $m^3.s^{-1}$, estimated by the National Water Agency. The Paraíba State Water Resources Plan estimates its regularized discharge as 1.23 $m^3.s^{-1}$. This controversy is one of the institutional problems related to the reservoir management. In 2012, at the beginning of the drought period, estimates of reservoir's water withdrawal were approximately 2.39 $m^3.s^{-1}$, where 1.44 $m^3.s^{-1}$ was used for urban water supply and 0.95 $m^3.s^{-1}$ for irrigation. These numbers clearly demonstrate that, despite the occurrence of one of the major

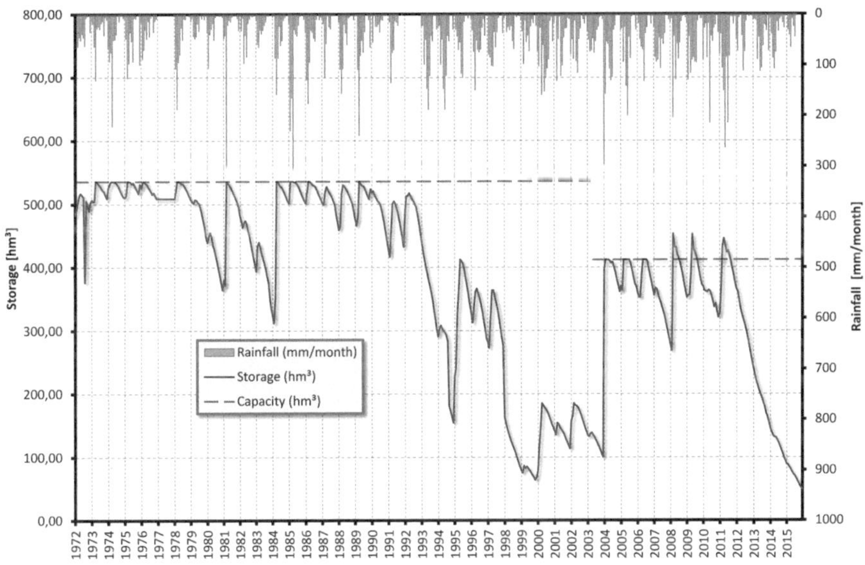

Fig. 10.2 Historical time series of water storage at Epitácio Pessoa Reservoir

meteorological and hydrological droughts ever, overexploitation of the reservoir played the principal role in consuming its water availability.

10.3.4 The Brazilian Water Policy and Management System

In Brazil, the Federal Law 9433 (1997) established the National Water Resources Policy and created the National Water Resources Management System (SINGREH). This Law addresses as main principles: water as a public and economic good, water planning at the basin level, decentralized and participatory management, priority for human water demands in drought periods, and multiple usages for water must be allowed.

Five management instruments were defined by the Law: water resources plans (at national, state, and river basin levels, to underpin and guide the implementation of the National Water Resources Policy and the management of water resources, which define management actions, programs, projects); a system for classification of water bodies according to their water quality and preponderant use (WBC); concession of water rights (CWR); bulk water charges, with the revenue to be invested in the basin (BWC); and the Water Resources Information System (WRIS).

The SINGREH is composed of a set of water management organizations including, at national and state levels, deliberation and regulation organisms (national and state's Water Resources Councils), policy implementation and operation organisms (the National Water Agency – ANA – and state's water agencies, which supervise, monitor, and evaluate the actions and activities resulting in the fulfillment of the federal and state's legislation related to water resources), participatory and decentralized management organisms (river basin committees), and technical organisms (river basin agencies). The river basin committees and national and state's water councils are composed of elected members representing water users; federal, state public, and municipal agencies; and civil society organizations, which have activities in the basin. The Law 9433 establishes that water conflicts must be, in first instance, discussed and solved at the basin committee.

The water resources plans must be revised and updated frequently (usually at each five years). They are approved by the basin committees (basin plans), by the state's water council (basin and state's plans), and by the national water council (basin and national plans). The committees and councils are also responsible to monitor, supervise, and evaluate the implementation of the plans.

10.3.5 Institutional Analysis

This chapter focuses on the institutional analysis of the process of adaptation to climate change. It considers that the high climatic variability of the Paraíba River Basin, with drastic changes in rainfall and runoff amounts between clusters of dry and rainy years, can be a proxy of an altered climate. Droughts happening during the clusters of dry years are also proxies of possible societal impacts of climate change. The performance of the water policy and management system in coping with the high climate variability and its consequences can be taken as their capacity to lead the necessary process of adaptation to a future climate.

As seen in the previous sections, the institutional water management system was not able to deal with the 2012–2016's drought in the basin, even after the previous experience of the 1997–2003's drought. The following institutional analysis can show which problems were responsible for the system's failure.

Following the Ostrom's framework for the analysis of SES, the case is studied starting with the identification of the SES for the Paraíba River Basin. After that, the analysis assesses the adherence of the institutions to the Ostrom's institutional design principles in two analytical aspects: the contents of the water policy and management system and their performance during the past drought events (Fig. 10.3).

10.3.5.1 The Social-Ecological System Identification

The socio-ecological system is identified and characterized according to its (Fig. 10.4) social, economic, and political settings; resource system; resource units; users; and governance. Also, the climate variability of the studied period is characterized and the impact of this variability on water availability. All the SES's components interact, as shown in the figure.

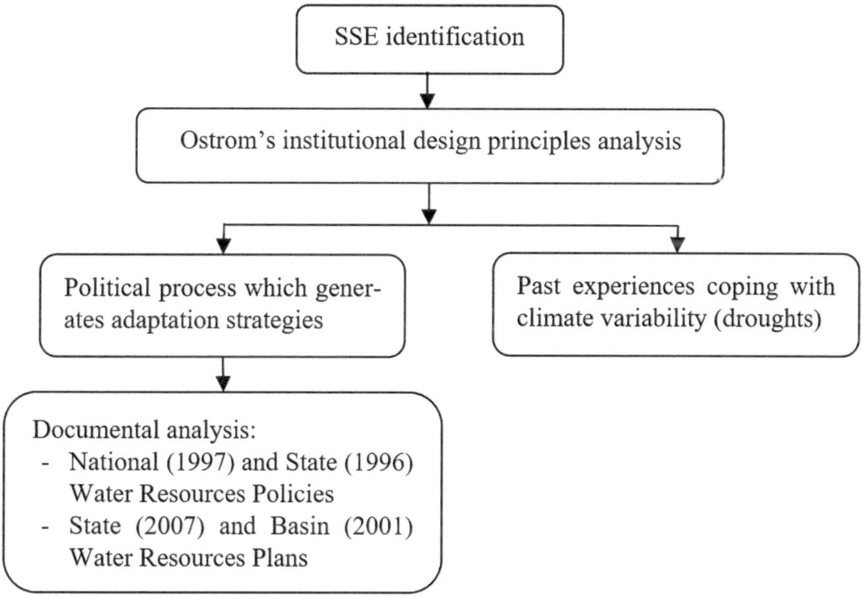

Fig. 10.3 Analytical process for the institutional analysis

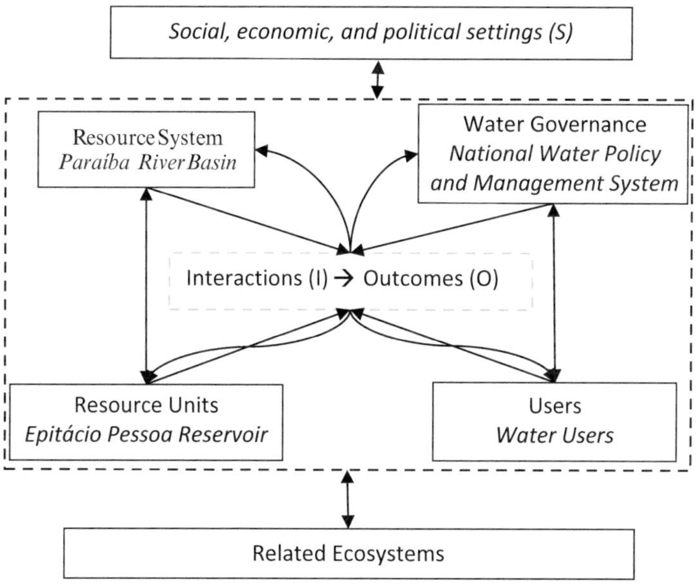

Fig. 10.4 SES configuration for the Paraiba River Basin and Epitácio Pessoa Reservoir (Based on Ostrom 2007)

Water Governance

The national water policy and management system defines the water governance of the resource system and its scope. The water policy defines the management relations between the water users, river basin, water reservoir, and the water management and policy itself.

Resource System

In the Brazilian case, the river basin is the territorial unit for the Water Resources National Policy implementation and the National Water Resources Management System activities. The Paraíba River Basin is, then, the SES's resource system.

Resource Units

The Epitácio Pessoa Reservoir is the SES's resource unit, since it is the hydraulic structure responsible for making water available to the users.

Water Users

The SES users are a water supply company and irrigators. Only the water supply company is an authorized user by the National Water Agency.

Related Ecosystems

Basin's climate variability, terrestrial ecosystems, land cover and use (vegetation and agriculture), and aquatic (reservoir lake) ecosystems highly influence reservoir water availability and quality.

Social, Economic, and Political Settings

Social, economic, and political settings might influence water governance and challenge its rules, in some cases. It influences also water users and uses. In this case the reservoir supplies a city that has increased its population, industries, irrigation, and also water demands.

10.3.5.2 Ostrom's Institutional Design Principles Analysis

Once the SES has been characterized, the analysis of the consistency between Ostrom's institutional design principles and the Brazilian Water Policy and Management system can be performed (Fig. 10.5).

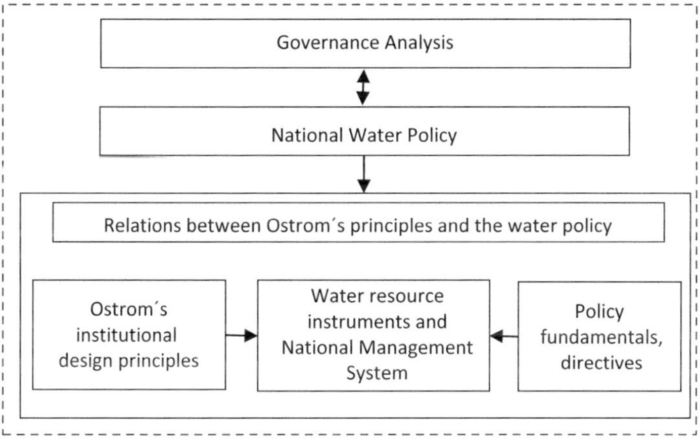

Fig. 10.5 A summary of Ostrom's principle analysis considering the Brazilian National Water Policy

Table 10.1 presents the results of the analysis, where each institutional principle is presented with the aspects of the case that are consistent, or not, with the principle.

Table 10.2 shows the main inconsistencies found in the analysis and the principles they refer to. The fragility of the collective-choice arrangements, such as the basin committees and water resources councils, is the most influential factor that challenges governance of adaptation in the case, followed by the lack of enforcement of the water management instruments. The table also shows these factors as nested reasons to the failure of the governance. The biased behavior of water agencies toward centralization can be nested into fragility of the collective-choice arrangements. Problems with monitoring and sanctions can be associated to the lack of enforcement of the management instruments. These types of clustering can help in identifying the most important problems to be addressed toward improving governance of the management system.

The following main conclusions can be drawn from the study:

The analysis of the water policy and management system shows that they are almost completely consistent with the institutional principles proposed by Ostrom and by Huntjens, as a democratic, participatory, and decentralized governance of the water resources. There is only one desirable mechanism not yet considered in the policy and management system: financial compensation to users when they lose, partially or fully, their rights to withdraw water, preventing an equal and fair (re-)distribution of risks, benefits, and costs (Principle 2).

On the other hand, the analysis of the 20-year period (1997–2016) including two drought and one rainy periods shows that the water governance was actually centralized, preventing social learning toward adaptation to climate variability and change. The main causes are identified as lack of enforcement of the water management instruments, deficient monitoring of the management process, fragile

Table 10.1 Results of the institutional analysis

Institutional principle	Consistency	Inconsistency
1. Clearly defined boundaries	Plans, CWR, and WBC: clearly define the CPR capacity (total availability in quantity and quality), users, limits to withdraw, rules, and criteria for water use in case of droughts; this information is of public access	There is lack of enforcement of plans, CWR, and WBC, and BWC is not yet implemented. Withdraw limits were not followed, and unauthorized users did withdraw water from the CPR
	BWC: clearly define charges for withdraw and changes in case of droughts	There is conflict about the reservoir's regularized discharge value between the National Water Agency and the State Water Plan
	IS: collects, processes, and makes available primary data on the CPR	Fragility of committee and councils
2. Congruence between appropriation and provision rules and local conditions	Plans, CWR, and WBC: define the appropriation and provision rules for the basin and CPR generally (plans, WBC) and for each user (CWR). CWR for one user is only approved by the Water Agency if the appropriation and provision rules are justified in relation to the local conditions	There is lack of enforcement of plans, CWR, and WBC, and BWC is not yet implemented
	BWC: defines charges considering congruence between user characteristics and local conditions	Fragility of committee and councils
	Committee and councils: designed to allow engagement with, and strong representation of, groups likely to be highly affected or especially vulnerable	There are no compensation mechanisms for users during exceptional situations, such as droughts
	In exceptional situations, such as droughts, there are adjustment mechanisms to adapt the implementation of the instruments	Equal and fair (re-) distribution of risks, benefits, and costs in case of droughts was not effective (see discussion on water justice in the basin by Grande et al. 2014)
3. Collective-choice arrangements	Committee and councils are the collective-choice arrangements. This principle, and related arrangements, is in the core of the Brazilian Water Policy (decentralization and participation)	Fragility of committee and councils Agencies try to centralize the decision and implementation processes
	Plans are defined and approved by the committee and councils	

(continued)

Table 10.1 (continued)

Institutional principle	Consistency	Inconsistency
4. Monitoring	IS and plans: provide appropriate information for monitoring	Monitoring has not been effectively used to enforce appropriate water use and user behavior or to audit the institutional management system itself
	Committee and councils: must monitor and audit CPR conditions, appropriator behavior, and the execution of the plans; they are accountable to the appropriators; their compositions include appropriators	Fragility of committee and councils
5. Graduated sanctions	Plans, CWR, WBC, and BWC: define graduate sanctions for appropriators in case of rules violation	Sanctions have not been effectively applied
	Brazilian law: defines graduate sanctions for public officials, representatives, and authorities	
6. Conflict resolution mechanism	Committees are the first-instance bodies for conflict resolution, followed by the councils	Fragility of committee and councils
7. Minimal recognition of rights to organize	This principle is in the core of the Brazilian Water Policy (decentralization and participation)	Fragility of committee and councils
	Users and appropriators are free to constitute associations and other collective arrangements and can be part of the committee and councils. CWR and BWC, for example, can be applied to these collective arrangements	Agencies try to centralize the decision and implementation processes
	Plans are defined and approved by the committee and councils	
8. Nested enterprises	Basin Committee and State and National Councils, as well as Basin, State, and National Plans, are designed as multiple layers and nested enterprises	Plans are not effectively implemented
		Fragility of committee and councils
		Agencies try to centralize the decision and implementation processes

(continued)

Table 10.1 (continued)

Institutional principle	Consistency	Inconsistency
9. Robust and flexible process	Users, appropriators, and representatives from government and society form the committee and councils. Committees are also represented in the councils	The 2012–2016's drought demonstrated that this principle is absolutely not consistent
	Policy and plans can be modified by the collective-choice arrangements	Sequential and time dilemmas are not discussed properly. Confidence-building process is weak
	Policy and plans recognize the necessity of integration of intersectoral policies	Fragility of committee and councils
10. Policy learning	Frequent reassessment and review of plans and management instruments, as well as committee and council's composition	The 2012–2016's drought demonstrated that this Principle is absolutely not consistent, particularly considering the occurrence of the 1997–2003's drought
		Fragility of committee and councils

Table 10.2 Main factors challenging governance at Paraíba River Basin

Factor	Institutional principle
Fragility of committee and councils	1, 2, 3, 4, 6, 7, 8, 9, 10
Agencies try to centralize the decision and implementation processes	3, 7, 8
There is lack of enforcement of plans, CWR, and WBC, and BWC is not yet implemented	1, 2, 8, 10
Monitoring has not been effectively used to enforce appropriate water use and user behavior or to audit the institutional management system itself	1, 2, 4
Sanctions have not been effectively applied	5
Equal and fair (re-) distribution of risks, benefits, and costs in case of droughts was not effective	2

committees and councils jointly with a centralization attitude by the agencies, and sanctions not applied to users or to managers.

Actually, Ribeiro et al. (2012), analyzing the participatory process in the Paraíba River Basin Committee, show that, recently, motivation and decision-making capacity were reduced. The authors recognize that "difficulties are increased by the lack of government institutions willing to share their power and/or able to promote public debates, and by the society's traditional lack of involvement in decision making. As a consequence, the very existence of spaces for discussion and participatory decision-making, like the basin committees, does not guarantee the success of the implemented water management model."

10.4 Conclusions and Prospects

This chapter presented Ostrom framework for institutional analysis, applied to a common problem in water resources management: drought management and water supply. The social-ecological system subject of our analysis is located in a region of high climatic and hydrological variability, which can serve as an illustration on what can be found in the future under climate change. The institutional problems present in our case are similar to those present in the efforts for adaptation to climate change. We can close this chapter with a recommendation from Ostrom (2010):

> To successfully address climate change in the long run, the day-to-day activities of individuals, families, firms, communities, and governments at multiple levels—particularly those in the more developed world—will need to change substantially. Encouraging simultaneous actions at multiple scales is an important strategy to address this problem.

Acknowledgments The authors acknowledge funding from the Brazilian agencies CNPq, CAPES, and FINEP; R&D Projects BRAMAR/FINEP; INCT Climate Change; and Brazilian Research Network on Global Climate Change (Rede Clima).

References

Anderies JM, Janssen MA, Ostrom E (2004) A framework to analyze the robustness of social-ecological systems from an institutional perspective. Ecol Soc 9(1):18

Dietz T, Ostrom E, Stern PC (2003) The struggle to govern the commons. Science 302:1907–1912

Elinor Ostrom – Facts (2015) *Nobelprize.org*. Nobel Media AB 2014. Web. 11 Nov 2015. http://www.nobelprize.org/nobel_prizes/economic-sciences/laureates/2009/ostrom-facts.html

Galvao CO, Rego JC, Ribeiro MMR, do Patrocinio Tomaz Albuquerque J (2001) Sustainability characterization and modelling of water supply management practices. IAHS Publ 268:81–88

Graham J, Amos B, Plumptre T (2003) Principles for good governance in the 21st century. Policy Brief 15:1–6

Grande M, Galvão C, Miranda L, Rufino I (2014) Environmental equity as a criterion for water management. IAHS Publ 364:519–525

Hess C, Ostrom E (2005) A framework for analyzing the knowledge commons. In: Understanding knowledge as a commons, theory to practice, ed. Library and Librarians' Publication

Huntjens P, Lebel L, Pahl-Wostl C, Camkin J, Schulze R (2012) Institutional design propositions for the governance of adaptation to climate change in the water sector. Glob Environ Chang 22:67–81

IPCC (2014) Climate change 2014: synthesis report. Contribution of Working Groups I, II and III to the Fifth Assessment Report of the Intergovernmental Panel on Climate Change [Core Writing Team, Pachauri RK, Meyer LA (eds)]. IPCC, Geneva, 151 pp

Nelson DR, Adger WN, Brown K (2007) Adaptation to environmental change: contributions of a resilience framework. Annu Rev Environ Resour 32:395–419

Ostrom E (1990) Governing the commons: the evolution of institutions for collective action. The Political Economy of Institutions and Decisions, ed. Cambridge University Press, New York

Ostrom E (2005) Understanding institutional diversity. Princeton University Press, Princeton

Ostrom E (2007) A diagnostic approach for going beyond panaceas. Proc Natl Acad Sci 104(39):15181–15187

Ostrom E (2009) A general framework for analyzing sustainability of social-ecological systems. Science 325:419–422
Ostrom E (2010) A multi-scale approach to coping with climate change and other collective action problems. 1(2):27–36. http://thesolutionsjournal.com/print/565. Solutions for a sustainable and desirable future. Accessed 26 Dec 2015
Ostrom E (2013) Do institutions for collective action evolve? Springer Science Business Media, New York
Ribeiro MAMF, Vieira ZMCL, Ribeiro MMR (2012) Participatory and decentralized water resources management: challenges and perspectives for the North Paraíba River Basin Committee – Brazil. Water Sci Technol 66(9):2007–2013
Van Der Keur P, Jeffrey P, Boyce D, Pahl-Wostl C, Hall A (2010) Adaptive water management in terms of development and application within IWRM. In: The adaptive resource management handbook. Earthscan, London

Part IV
The Way Ahead

Chapter 11
Climate Change Impacts and Water Resource Management and Planning

Elpida Kolokytha, Carlos de Oliveira Galvão, and Ramesh S.V. Teegavarapu

Abstract Sustainable water resource planning for climate change is a process of assessing risks related to climate change, evaluating and selecting strategies that are based on current hydrological knowledge, using climate models, monitoring existing conditions, and providing guidelines for adaptation and the optimal use of available water resources for the benefit of the society. This chapter provides a road map to show how climate projections can be incorporated into water management decisions. In particular, it explores which projections to use and how to relate them to planning assumptions on water supplies, water demands, and operating restraints, and it also provides useful recommendations on how to improve water resource planning and management and on how to proceed with effective water policy actions considering climate variability and change.

Keywords Water resource management and planning • Climate change • Sustainable water policy

11.1 Introduction

The Earth is experiencing a rapid climate warming which will affect natural and human systems. In the scientific debate on the recent and projected climate at the global scale (IPCC 2014), the unanimous agreement of the Paris climate conference COP21 puts an end to the attempt to challenge the anthropogenic influences on

E. Kolokytha (✉)
School of Civil Engineering, Division of Hydraulics and Environmental Engineering, Aristotle University of Thessaloniki, 54124 Thessaloniki, Greece
e-mail: lpcol@civil.auth.gr

C. de Oliveira Galvão
Federal University of Campina Grande, UFCG, 58429-900 Campina Grande, PB, Brazil
e-mail: galvao@dec.ufcg.edu.br

R.S.V. Teegavarapu
Department of Civil, Environmental and Geomatics Engineering, Florida Atlantic University, Boca Raton, FL, USA
e-mail: rteegava@fau.edu

E. Kolokytha et al. (eds.), *Sustainable Water Resources Planning and Management Under Climate Change*, DOI 10.1007/978-981-10-2051-3_11

climate change and the leniency of economic growth as a key factor in global warming. Undeniably, all multiple crises, namely, financial, economic, climate, energy, ecosystems, and demography of the twenty-first century, compel us to think of a radical transformation of the economy. The transition toward a "new development model" requires investment in the sustainability of ecosystem services upon which the world depends on. The sustainable use of natural resources and the environment has to be the core concept of economic growth, while "resource efficiency" should replace the perception of endless resources exploitation.

The entire planet, as expressed by the 196 leaders of the world amid the global recession and economic crisis, agreed to adopt a comprehensive action plan to limit the rise in global average temperature to a maximum of 2 °C or even up to 1.5 °C by mainly reducing the global greenhouse gas emissions (UNFCCC 2015). The reduction of the global greenhouse gas emissions, while indeed a solution to the major problem of climate change, does not seem to give way to solving the other problems related to the disruption of the ecological balance and the depletion of natural resources, among which water resources are critical. They agreed, among others, to "enhance linkages and create synergy between, inter alia, mitigation, adaptation, finance, technology transfer and capacity building" (UNFCCC 2015).

Over the past decades, a consensus had been formed on integrated water resource management (IWRM) as an appropriate approach to address pressures on water resources. Significant pressures, such as expanding populations and urbanization, the mounting expansion of economic activities, environmental impacts of human intervention in nature, climatic change, and galloping rise of water demand, all articulate the premises of a desirable, sustainable development by promoting comprehensive approaches to structural and nonstructural responses to persistent water resource problems.

Globally, climate change is likely to increase water demand while witnessing a decrease in water supplies. The high variance of the spatial and temporal distribution of precipitation and the extent of the rainfall (extreme events) is leading to a significant temporal and spatial variability in water resources affecting severely water availability worldwide. This shifting balance would challenge water resource management to concurrently meet the needs of society but also maintain ecosystem integrity. Changes in temperature will impact both water quality and quantity and a significant number of related activities. Climate change in general, though, may have a positive as well as an adverse effect on water resources. The predicted changes in availability, quality, and accessibility of water will have vital consequences for human populations, by influencing important sectors of the economy such as agriculture and food security, health, energy, industry, navigation, urban use, and biodiversity, and will also increase conflict over water resources.

This book is a contribution of distinguished scientists from almost all continents that aims to explain issues related to water resource planning and to explore new ways of decision-making and management to adapt to the impacts of evolving climate change phenomenon (Fig. 11.1). It covers a range of issues from hydrologic modeling to management and policy responses. It brings updated theory, methodologies, and tools and provides case studies to understand how to adapt to climate change and turn theory into practice.

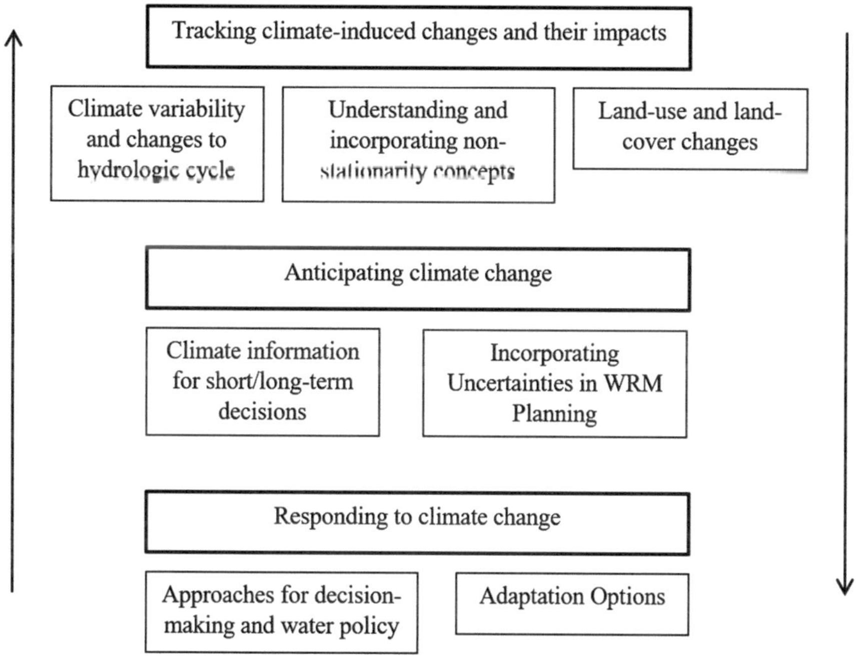

Fig. 11.1 Issues in water resource planning and management related to climate change and two-way feedbacks among them

11.2 Tracking Climate Change Impacts

Scientists need to explore and answer the question of "what kinds of changes are happening and how well do we know climate-related changes to include them in our planning and management analyses?" "How deep is our knowledge concerning extreme events (floods/droughts) and their changes in occurrence and magnitude?"

11.2.1 Climate Variability and Changes to Hydrologic Cycle

Climate variability refers to deviations of climate statistics over a period of time. Internal climate variability refers to what is frequently called "unforced" climate changes or "modes" of variability, and these modes represent climate anomalies that originate in a particular region and have significant impacts elsewhere around the globe (Cronin 2009). Teleconnections are defined as linkages between climate anomalies at some distances from each other. R. Teegavarapu in Chapter 1 has discussed these teleconnections manifested by coupled oceanic-atmospheric oscillations that influence hydroclimatic variables in different regions of the world. Even though influence of climate variability on precipitation extremes and characteristics is the main focus in Chap. 1, emphasis should also be on evaluating links of these oscillations to variations in temperature, sea level, and streamflows (both low and high) in different regions of the world. Oscillations ranging from inter-year to decadal and multi-decadal time scales have enormous influence on hydroclimatic variable extremes and characteristics (Teegavarapu 2013; Teegavarapu et al. 2013). Often one or more oscillations influences are noted in one single region (Goly and Teegavarapu 2014). While general influences of these oscillations on different hydroclimatic variables are known, clear understanding of nonuniform spatial and temporal influences and how regional hydroclimatology limits the extents of these influences is critical for water resource management and hazard mitigation due to extremes.

Any study evaluating climate variability influences on different hydroclimatic variables should attempt to assess:

- General characteristics of the variables
- Trends and variations in variables at different temporal levels
- Interannual and intra-annual variations
- Changing extremes over time
- Changes in temporal occurrences of these extremes
- Changes and trends in temporally aggregated variable values
- Spatial variation of trends
- Variations in the summary statistics of observations in two or more temporal windows coinciding with the temporal windows of different phases of oscillations
- Indices that are specific to a particular hydroclimatic variable

Evaluation of the influences of inter-year and decadal and multi-decadal oscillations on precipitation extremes and characteristics reported in Chap. 1 highlights the main issues surrounding hydrologic design and water resource management, and they include:

1. Natural climate variability along with climate change will have implications on hydrologic design due to changing precipitation extremes.
2. Existing hydrologic and hydraulic design procedures need to be evaluated and revised to consider extremes and also antecedent conditions that may lead to catastrophic failures of existing infrastructure.
3. Nonuniform spatial and temporal variations in availability of water, occurrences of precipitation extremes (temporal shifts), and changes to characteristics will impact the frequency of flood and drought conditions.
4. Knowledge of climate variability is critical for short- and long-term forecasts of hydroclimatic variables thereby aiding sustainable water resource management.
5. Understanding climate variability and change influences will also help in developing climate change-sensitive and climate change-adaptive hydrologic design, rehabilitation (ad hoc and long term) of existing infrastructure, and long-term assessment of risks associated with occurrences of hydroclimatic extremes.

11.2.2 Understanding and Incorporating Non-stationarity Concepts

Climate change is believed to possibly increase the frequency and magnitude of extreme events (IPCC 2012). Global warming may lead to an intensification of the hydrologic cycle owing to the greater moisture-holding capacity of the atmosphere (Meehl et al. 2007). This increases the likelihood of more frequent high precipitation events that influence the risk of pluvial flooding. On the other hand, increasing the variability of precipitation and increased evapotranspiration because of rising temperatures may lead to more frequent droughts. Such changes imply that the past is not a good predictor for the future, as it used to be, leading to the well-discussed "stationarity is dead" (Milly et al. 2008). Indeed, as growing population gets exposed to the risk of extreme events, hydrologists face the challenge of understanding the spatiotemporal behavior of extremes, especially under rapidly changing conditions as expressed by A. Mondal and P. Mujumdar in Chap. 2. A set of new techniques is discussed in Chap. 2 which incorporates non-stationarity concepts.

Traditionally, hydrologic designs rely on tail quantiles of extremes, such as the "100-year flood." As explained by A. Mondal and P. Mujumdar, non-stationarity influences not only the magnitudes of tail quantiles but also the uncertainties associated with their estimation. Given such uncertainties and the assumptions involved, another critical question arises "whether non-stationary models are indeed more suitable for characterizing transient extremes as compared to the traditional concepts based on stationarity, and whether that leads to a more robust assessment of the risk of a hazardous event." A comprehensive analysis which brings into focus the issues and uncertainties in modeling nonstationary extremes vis-à-vis traditional approaches based on stationarity lacks more or less in the hydrologic literature.

Typically, large temporal scale variations (such as trend) in hydrologic data are treated as deterministic, whereas small temporal scale variations are modeled as random components though they are clearly not measurement errors and do not come from repeatable experiments (Koutsoyiannis 2006). Understanding the cause of a trend can allow modeling of future development plans. Careful consideration should be given to the distinction between trends and persistence. Matalas (1990) commented on the problem of distinguishing trend from persistence by saying that "... a trend in the short run may be part of an oscillation in the long run." Datasets of only a few decades can easily point to strong trends that are simply an artifact of some quasiperiodic variation (Hirsch 2011). The use of very long hydrologic records (datasets) is a "safe path" to these problems since those records contain multiple realizations of many oscillations. Continuity of records is another critical issue, which compels the maintenance of reliable observational networks.

11.2.3 Land Use and Land Cover Changes

Of course, such changes have different drivers, not only climate but also land use/land cover changes, dam operations/removal, groundwater depletion, or other human interventions which significantly affect decision-making (Sivapalan and Samuel 2009). The scientific hydrology community is paying more and more attention to several aspects, related to both physical and societal factors, of the changing dynamics of the water cycle in relation to rapidly changing human systems (Montanari et al. 2013). Changes and their explanations and technical applications in the water sector are crucial, and R. Ranzi, P. Caronna, and M. Tomirotti in Chap. 3 aim at contributing to this streamline, underlining the fact that anthropogenic effects, such as land use changes, are drivers to be included in both analyses of trends and development of adaptation and mitigation measures to climate change. They investigated possible reasons for decreasing annual runoff in river basins of the Central Southern Alps over the past 150 years, including precipitation, temperature, and land use changes. The results show that runoff changes might be caused mainly by increased agricultural water demands and land use changes. The afforestation that occurred in the past decades has increased evapotranspiration and interception losses, enhanced also by the higher temperatures.

11.3 Anticipating Climate Change

Recognizing the already occurring impacts of climate and land use on the hydrological cycle and water systems is the first step toward planning and decision-making for an improved water resource management. However, since changes will continue to happen in a long-term horizon, climatologists and hydrologists have, in the last decades, developed models to anticipate their future impacts, which continue to deserve research and development efforts for their improvement.

11.3.1 Climate Information Needed for Short-/Long-Term Decisions

Incorporating climate information means translating climate projection information into key planning assumptions (e.g., supplies, demands, and constraints in the context of resource management or infrastructure safety assessments) (Brekke 2011).

Projections of precipitation change are less reliable than projections of temperature change, and projections of conditions in water basin level are less reliable than for larger areas. Model projections and statistical analysis provide high level of confidence in statements concerning less important aspects of hydrology (the central tendency) and high level of uncertainty in the most important aspects such as extreme events (Hirsch 2011).

The most comprehensive projections of future global climate conditions are provided by the global circulation models (GCMs). There is a problem of "scale" though since outputs from GCMs are typically available at spatial scales of tens to hundreds of kilometers. Global circulation models (GCMs) are simulation-based mathematical models that assess climate change which are key components for studying the hydrological implications on water resources at the river basin level. Selection of suitable GCMs among the available is necessary for reliable and practical implementation at the basin level. K. S. Raju, D. N. Kumar, and N. Babu I in Chap. 4 explain the necessity for evaluating GCMs based on relevant performance metrics for possible ranking and perform the evaluation of 45 GCMs.

11.3.2 Incorporating Uncertainties in WRM Planning

Water resource planners always have dealt with uncertainties in relation to climate variability, many times referred to as "predicted uncertainty," fairly well. But now new sources of uncertainty related to rapid and potentially "irreversible" changes in state due to the climate need new information, advanced tools, and new techniques. A deep understanding of the processes concerning climate as a driver of both supply and demand needs to be developed in order to limit the range of uncertainty and to be able to take the appropriate decisions in the context of climate change which many times can even be characterized by "unknown" results. As Kundzewicz (2011) said, *we know increasingly well that we do not know enough.*

Climate projections provide a broad range of possible climate futures. The degree of uncertainty in climate projections and the weakness to determine their predictive capabilities are so high that trying to estimate or accepting a possibility that one future is more likely to occur than another would probably result in poor decision-making. The same may happen if long-term decisions are based on short-term information. The links between climate and hydrology as well as the impacts of ecosystem changes on the quantity and quality of water need to be further analyzed.

11.4 Responding to Climate Change

Two types of responses to climate change, mitigation and adaptation, play an equally important role in the transformation toward sustainability. The more successful mitigation efforts are in cutting emissions, the less extensive the need for adaptation will be since mitigation reduces the root cause of the climate change problem. However, even if mitigation measures applied succeeded to limit and then reduce GHG emissions, the impact of climate change will challenge the planet for at least the next 50 years which provides the imperative need for adaptation measures in any case. The role of adaptation, whether reactive or anticipatory, spontaneous or planned, is crucial for assessments of potential impacts of climate change.

11.4.1 Approaches for Decision-Making and Water Policy

According to Stakhiv and Stewart (2010), there are five major types of interventions to respond to climate change:

The first one concerns new investments or capacity expansion of reservoirs, irrigation systems, water supply, and wastewater treatment plans. The second refers to the operation, monitoring, and regulation of existing systems to accommodate new uses or conditions. The third one deals with the maintenance and major rehabilitation of existing systems such as dams, irrigation systems, and wetlands. One major category signifies nonstructural measures to alter water demands such as demand management practices, e.g., regulations, legislation, pricing, information campaigns, education, and public awareness processes for existing systems. Finally, get familiar with new efficient "green" technologies such as drip irrigation, biotechnology.

Although the impact of climate change on hydrological systems is considered to vary from region to region, changes in hydrological systems associated with changing climate may threaten the reliability of existing water resource management system developed so as to adapt to the current hydrological system. From this point of view, the sophistication of reservoir management is also considered to be important to enhance existing water resource management systems. Although the ability of reservoir systems to control water-related disasters can be improved with many measures including construction of new reservoirs and enlargement of an existing reservoir's capacity to store or release water, sophistication of real-time operation of existing reservoirs would be an attractive option. D. Nohara and T. Chori in Chap. 5 depict and provide a method to design and assess integrated reservoir operation for both flood control and water utilization with consideration of imperfect real-time hydrological predictions to enhance the capability of an existing multipurpose reservoir.

Flood control is one of the most important issues of reservoir operation. S. Oishi in Chap. 6 contributes to this important issue by demonstrating techniques and theories for managing water resources using advanced weather forecasting. It deals with the possibility of adopting adaptive countermeasures to manage the water resources by making the most of the existing structure.

11.4.2 Adaptation Options

The transition to sustainability calls for the shift to a new development model that of green economy-green development. The green economy appears as the follow-up to sustainable development. E. Kolokytha in Chap. 9 explains the "new development model" and its characteristics and provides a set of structural and nonstructural measures for adaptation to climate change impacts on water resources.

Adapting to the changing climate is an immediate challenge that requires choosing a successful strategy. Y.O. Kim and E.S. Chung in Chap. 8 review classical decision-making methods and discuss their limitations when applied to climate change adaptation planning. However, the various advanced decision-making methods are resource intensive; thus, a continuous administrative effort and institutional as well as technical supports are required for their success.

Identifying the best adaptation response is difficult. Risk management techniques help to overcome these problems. Risk management is presented as a decision-making framework that assists in the selection of optimal strategies (according to various criteria) using a systems approach that has been well defined and accepted in public decision-making. In the context of adapting to climate change, the risk management process offers a framework for identifying, assessing, and prioritizing climate-related risks and developing appropriate adaptation responses. S. Simonovic in Chap. 7 presents risk management as a practical approach to climate change adaptation as it relates to the risk of flooding. The developed methodology can be effectively used to assess the risk associated with various options and, therefore, provides support for climate change adaptation through risk management. The results from the application of the methodology lead to better policy and informed decision-making.

There are a series of limitations on the development of adaptation actions due to the lack of appropriate data, poor decision-support mechanisms, and limited institutional capacity. The issues of institutional capacity and governance are discussed by A. Silva, C. Galvao, M. Ribeiro, and T. Andrade in Chap. 10 which demonstrates the required conditions for the institutions to be capable of playing the right role in the process of adaptation to climate variability and change.

11.5 Looking Ahead

As we have analyzed and tried to give light to the different shortcomings, gaps, and difficulties that arise from the impact of climate change on water resources, a critical question should be answered: What can be done to improve planning and address climate change challenges?

11.5.1 Understanding and Incorporating Non-stationarity in Climate-Change Sensitive Hydrologic Designs

Understanding influences of climate variability and change on hydroclimatic variables is essential for future operations of hydrosystems and sustainable water resource management. Climate change and variability are expected to bring shifts to hydroclimatic extremes both at global and regional scales. Agencies dealing with management of water and environmental systems and developing adaptation policies in view of changing climate should be prepared to:

1. Address short- and long-term trends in hydroclimatic variables especially precipitation and streamflow and other essential climatic variables.
2. Evaluate the occurrences, variability, and sudden changes of extremes (considering frequency and magnitude) in space and time and develop sustainable and climate change-sensitive hydrologic designs.
3. Assess the influences of climate variability on hydroclimatic variables at different spatial and temporal scales considering the extent of regional climatology influences.
4. Understand changes in trends and attribute or separate them based on natural variability or anthropogenic influences.
5. Develop adaptation and rehabilitation plans and procedures for water resource infrastructure to handle expected future changes and episodic events (e.g., recurring floods and droughts)

11.5.2 Improving Climate Change Projections and Knowledge Gaps

In the case of water, so far, scientists have relied on global circulation models that provide information on general climate patterns whose projections are downscaled to fit local contexts. However, as suitable as this method is for setting the broader context, it does not incorporate the local vulnerabilities to climate change necessary for accurate decision and policy choices since climate change is a global phenomenon that has mainly regional and local spatiotemporal impacts. Efforts should revolve around methods to forecast climate change at the scale of the river basin, which is the autonomous hydrological unit that implicates all water inputs and respective uses. Multi-scale modeling approaches can be considered, where redundancies in the representation of the processes, in different models, should be avoided and feedbacks of influences can be adequately modeled.

In order to quantify uncertainty and risk, predict change, and create climate scenarios which incorporate the different social, economic, environmental, and engineering aspects of water management in climate models, wide-ranging and accurate long-term measurements should be provided. Monitoring need to be enhanced to confront data scarcity as well as new data sources (satellite observations, remote sensing technologies) should be used.

11.5.3 Opportunities to Improve WRM and Planning

In defining adaptation policies, it is extremely useful to identify priority and robust measures. In doing that, it is important to explore which adaptation measures are of the highest priority and which complementary actions should follow. Successful adaptation should take into account not only the impacts of climate change but also how other non-climatic (socioeconomic, political, etc.) stresses impact the system and take measures for both cases. Local conditions and local stakeholders are the most relevant to choose the analytical tools and approaches of critical economic, environmental, and social drivers. The interactions between the different sectors, the scale, and the speed of change required as well as the actions that need to be taken over time, along with uncertainties, should shape the process of adaptation.

11.5.4 Cross-Sector Assessment and Action

The sustainability and the appropriateness of water management decisions are significantly affected by the implication of nonstationary climate projections. Underestimation and overestimation of critical water parameters bound our ability to plan for the future and put in peril decisions interrelated with other sectors (energy, navigation, water supply) where water is central.

Relevant to that is how poorly governmental policy makers understand that many of the primary economic activities and a great number of relevant sectors (energy, agriculture, biodiversity) rely on adequate water supplies. Political will and considerable financial resources are needed to deal with some basic preparatory measures, future strategies, and plans.

Sustainable water management is not a "static" target but rather a "pathway" of decisions which need to incorporate all relevant factors, from economic to environmental, social, technical, and political, while this procedure may reverse or challenge previous decisions taken under other circumstances.

11.5.5 Assessing Performance of AWRM

All the set of tools mentioned earlier, however, could not be successful if applied without taking into account how people and society perceive adaptation and to what extent they, and their institutions, are ready to adapt to changes that are in line with broad cultural and ethical values. Adaptive Water Resources Management (AWRM) is an already established concept, and several approaches have been devised to implement it. However, there is still a lack of experience in applying methods, tools, criteria, and indicators to timely evaluate its performance in adapting to change while keeping an adequate quality of life and preserving societies' values. Those instruments should be integrated to the water resource plans and management systems reassessments.

11.5.6 *Potential Match of Theory and Practice: Research Results Should Turn to Action*

A great array of achievements has been accomplished in the field of research on climate impacts into water systems. There are a considerable number of research projects financed by leading agencies related to water and water management which do not adequately communicate their results to decision-makers. Successful adaptation of water resources to climate change calls for investment in human capital. In the era of climate change, water professionals (planners, designers, operators, researchers) need to function as "knowledge brokers" to understand the different ways of thinking and the different contexts in which information can be used. Scientists, who have the knowledge, should communicate with those who use knowledge, such as those working in the water agencies, policymakers, other stakeholders, and the general public. Proper communication should ultimately foster effective public participation toward adaptive water management.

References

Brekke LD (2011) Bureau of reclamation, technical service center, Lead Author, "Addressing Climate Change in Long-Term Water Resources Planning and Management: User Needs for Improving Tools and Information" (Also available online at www.usbr.gov/climate/userneeds)

Contributing Authors, Bureau of Reclamation: Chuck Hennig and Curt Brown, Research and Development Office, David Raff, Policy and Administration, Rod Wittler, Mid-Pacific Region

Contributing Authors, U.S. Army Corps of Engineers: Kathleen White and J. Rolf Olsen, Institute for Water Resources, Edwin Townsley, South Pacific Division, David Williams, Tulsa District, Fauwaz Hanbali, Hydrologic Engineering Center

Cronin TM (2009) Paleoclimates: understanding climate change past and present. Columbia University Press, New York

Goly A, Teegavarapu RSV (2014) Individual and coupled influences of AMO and ENSO on regional precipitation characteristics and extremes. Water Resour Res 2014:50

Hirsch RM (2011) A perspective on non-stationarity and water management. J Am Water Resour Assoc (JAWRA) 47(3):436–446. doi:10.1111/j.1752-1688.2011.00539.x

Intergovernmental Panel on Climate Change (IPCC) (2012) Summary for policymakers. In: Field CB, Barros V, Stocker TF, Qin D, Dokken DJ, Ebi KL, Midgley PM (eds) Managing the risks of extreme events and disasters to advance climate change adaptation. A special report of working groups I and II of the Intergovernmental Panel on Climate Change. Cambridge University Press, Cambridge, UK/New York, pp 1–19

Intergovernmental Panel on Climate Change (IPCC) (2014) Climate change 2014: mitigation of climate change. In: Edenhofer O, Pichs-Madruga R, Sokona Y, Farahani E, Kadner S, Seyboth K, Adler A, Baum I, Brunner S, Eickemeier P, Kriemann B, Savolainen J, Schlömer S, von Stechow C, Zwickel T, Minx JC (eds) Contribution of working group III to the fifth assessment report of the Intergovernmental Panel on Climate Change. Cambridge University Press, Cambridge, UK/New York. Available at: http://www.ipcc.ch/report/ar5/wg3/

Koutsoyiannis D (2006) Nonstationarity versus scaling in hydrology. J Hydrol 324(1):239–254

Kundzewicz ZW (2011) Comparative assessment: fact or fiction, workshop: including long term climate change in hydrologic design. World Bank, Washington, DC, 21c Nov 2011

Matalas NC (1990) What statistics can tell us. In Waggoner PE (ed) Climate change and U.S. Water Resources. AAAS Panel on Climatic Change. Wiley, New York, pp 139–149

Meehl GA, Stocker TF, Collins WD, Friedlingstein P, Gaye AT, Gregory JM, . . . Raper SC (2007) Global climate projections. In: Solomon S, Qin D, Manning M, Chen Z, Marquis M, Averyt KB, MHL (eds) Climate change 2007: the physical science basis. Contribution of Working Group I to the Fourth Assessment Report of the Intergovernmental Panel on Climate Change. Cambridge University Press, Cambridge, UK/New York

Milly PCD, Betancourt J, Fallkenmark M, Hirsch RM, Kundzewicz ZW, Lettenmaier DP, Stouffer RJ (2008) Stationarity is dead: whither water management? Science 319:573–574. doi:10.1126/science.1151915

Montanari A, Young G, Savenije HHG, Hughes D, Wagener T, Ren LL, Koutsoyiannis D, Cudennec C, Toth E, Grimaldi S, Blöschl G, Sivapalan M, Beven K, Gupta H, Hipsey M, Schaefli B, Arheimer B, Boegh E, Schymanski SJ, Di Baldassarre G, Yu B, Hubert P, Huang Y, Schumann A, Post D, Srinivasan V, Harman C, Thompson S, Rogger M, Viglione A, McMillan H, Characklis G, Pang Z, Belyaev V (2013) "Panta Rhei—Everything Flows": change in hydrology and society—The IAHS Scientific Decade 2013–2022. Hydrol Sci J 58(6):1256–1275

Sivapalan M, Samuel JM (2009) Transcending limitations of stationarity and the return period: process-based approach to flood estimation and risk assessment. Hydrol Process 23:1671–1675

Stakhiv E, Stewart B (2010) Needs for climate information in support of decision-making in the water sector. Procedia Environ Sci 1:102–119

Teegavarapu RSV (2013) Floods in a changing climate: extreme precipitation. Cambridge University Press-UNESCO (United Nations Educational Scientific and Cultural Organization), January, 2013, 285 p

Teegavarapu RSV, Goly A, Obeysekera J (2013) Influences of Atlantic multi-decadal oscillation on regional precipitation extremes. J Hydrol 495(2013):74–93

UNFCCC-United Nations Framework Convention on Climate Change (2015) Adoption of the Paris Agreement. http://unfccc.int/resource/docs/2015/cop21/eng/l09r01.pdf, 32 pp

Index

E. Kolokytha et al. (eds.), *Sustainable Water Resources Planning and Management Under Climate Change*, DOI 10.1007/978-981-10-2051-3

MIX
Papier aus verantwortungsvollen Quellen
Paper from responsible sources
FSC® C105338

If you have any concerns about our products,
you can contact us on
ProductSafety@springernature.com

In case Publisher is established outside the EU,
the EU authorized representative is:
Springer Nature Customer Service Center GmbH
Europaplatz 3, 69115 Heidelberg, Germany

Printed by Libri Plureos GmbH
in Hamburg, Germany